Probability and Statistics

Volume II

Didier Dacunha-Castelle
Marie Duflo

Probability and Statistics

Volume II

Translated by David McHale

Springer-Verlag
New York Berlin Heidelberg Tokyo

Didier Dacunha-Castelle
Université de Paris-Sud
Equipe de Recherche Associée
au C.N.R.S. 532
Statistique Appliqué Mathématique
91405 Orsay Cedex
France

Marie Duflo
Université de Paris-Nord
93430 Villetaneuse
France

David McHale (*Translator*)
Linslade, Leighton Buzzard
Bedfordshire LU7 7XW
United Kingdom

With 6 Illustrations

AMS Classification 60-01

Library of Congress Cataloging-in-Publication Data
Dacunha-Castelle, Didier.
Probability and statistics.
Translation of: Probabilités et statistiques.
Includes bibliographies and index.
1. Probabilities. 2. Mathematical statistics.
I. Duflo, Marie. I. Title.
QA273.D23 1986 519.2 85-25094

Printed and bound by R.R. Donnelley and Sons, Harrisonburg, Virginia.
Printed in the United States of America.

9 8 7 6 5 4 3 2 1

ISBN 0-387-96213-1 Springer-Verlag New York Berlin Heidelberg Tokyo
ISBN 3-540-96213-1 Springer-Verlag Berlin Heidelberg New York Tokyo

Note: *In this second volume of* Probability and Statistics, *reference is sometimes made to Volume* I. *These references appear in the form* [Vol. I. X.Y.Z.] *and refer to Chapter* X, *paragraph* Y, *sub-paragraph* Z. *Similarly, for the references to Volume* II, *denoted only* [X.Y.Z.].

INTRODUCTION

How can we predict the future without asking an astrologer?

When a phenomenon is not evolving, experiments can be repeated and observations therefore accumulated; this is what we have done in Volume I. However history does not repeat itself. Prediction of the future can only be based on the evolution observed in the past. Yet certain phenomena are stable enough so that observation in a sufficient interval of time gives usable information on the future or the mechanism of evolution.

Technically, the keys to asymptotic statistics are the following: laws of large numbers, central limit theorems, and likelihood calculations. We have sought the shortest route to these theorems by neglecting to present the most general models. The future statistician will use the foundations of the statistics of processes and should satisfy himself about the unity of the methods employed. At the same time, we have adhered as closely as possible to present day ideas of the theory of processes. For those who wish to follow the study of probabilities to postgraduate level, it is not a waste of time to begin with the least difficult technical situations.

This book for final year mathematics courses is not the end of the matter. It acts as a springboard either for dealing concretely with the problems of the statistics of processes, or to study in depth the more subtle aspects of probabilities.

Finally, let us note that a more classical probability course can easily be organized around Chapter 2 which is central, on Chapter 4 on Markov chains and Chapters 5 to 8 for the important parts which do not call on statistical concepts.

CONTENTS

SUMMARY OF VOLUME I

Censuses

Census of two qualitative characteristics
Census of quantitative characteristics
First definitions of discrete probabilities
Pairs of random variables and correspondence analysis

Heads or Tails. Quality Control

Repetition of n independent experiments
A Bernoulli sample
Estimation
Tests, confidence intervals for a Bernoulli sample, and quality control
Observations of indeterminate duration

Probabilistic Vocabulary of Measure Theory. Inventory of The Most Useful Tools

Probabilistic models
Integration
The distribution of a measurable function
Convergence in distribution

Independence: Statistics Based on the Observation of a Sample

A sequence of n-observations – Product measure spaces
Independence
Distribution of the sum of independent random vectors
A sample from a distribution and estimation of this distribution
Non-parametric tests

Gaussian Samples, Regression, and Analysis of Variance

Gaussian samples
Gaussian random vectors
Central limit theorem on $\mathbb{R}^k$
The χ^2 test
Regression

Conditional Expectation, Markov Chains, Information

Approximation in the least squares sense by functions of an observation
Conditional expectation – extensions
Markov chains
Information carried by one distribution on another

Dominated Statistical Models and Estimation

Dominated statistical models
Dissimilarity in a dominated model
Likelihood

Statistical Decisions

Decisions
Bayesian statistics
Optimality properties of some likelihood ratio tests
Invariance

Chapter 0
INTRODUCTION TO RANDOM PROCESSES

Objectives

As indicated by its number this short chapter is not very appealing. It contains general definitions on processes and some measure theory tools. We recommend an initial superficial reading. The definitions and terms will appear more natural after the study of some processes, in the following chapters.

0.1. Random Evolution Through Time

In order to study a random phenomenon which does not change, the natural idea which we have exploited in Volume I was to repeat identical and independent observations. If a study is to be made of the changes in unemployment, prices, or the output from a machine which is wearing out, we can again carry out a sequence of observations (each year, or each day, ...). However for this sort of problem the hypothesis of independence usually does not make any sense. We are led to consider the observations $(X_1, ..., X_n)$ as extracts from an infinite random sequence (or process in discrete time) (X_n): n, the index of the time of observation, may be taken to be in $\mathbb{N}$, 0 being the start of the phenomenon, or in $\mathbb{Z}$ if it is considered not to have a beginning. We can also think of the evolution of a system when the time t varies continuously; t

varies in $\mathbb{R}_+$ if there is a beginning, and in $\mathbb{R}$ otherwise.

Using these observations, various sorts of problems can be posed:

Prediction of the future values of the sequence;

Identification of the type of evolution under consideration, then **estimation** or **tests** bearing on the parameters of the model;

Filtering or prediction using partial observations of the process.

A totally chaotic evolution does not allow a suitable answer to any of the preceding problems. We are led, for sequences of observations, to various types of hypotheses:

Stationarity where the choice of the time origin does not change anything. This is an hypothesis of stability through time which allows the future to be predicted (Chapter 1).

For a real sequence, **tendency to increase or to decrease** (Chapter 2).

Chain evolution, where the memory extends only to the last observation (Chapter 4).

For regular enough models, when the duration of the observations increases, we obtain asymptotic theorems related to the processes or to the statistical methods employed. The statistical point of view is covered in Chapter 3. On the other hand, the question of the choice of the length or type of experiment is important.

As the observations proceed, we shall be able to decide, in the light of the previous results, whether to continue the experiment or not, or to modify it. This will be the object of Chapter 5.

In Chapters 6 to 8 tools will be given which allow the study of evolutions in continuous time.

0.2. Basic Measure Theory

0.2.1. Processes

We study a random phenomenon through time. The time space T is most often $\mathbb{N}$ or $\mathbb{Z}$ in Chapters 1 to 5 (sequences of observation), and $\mathbb{R}_+$ in Chapters 6 to 8 (continuous evolution in time). If the phenomenon takes its values in a measurable space $(E,\mathcal{E})$ an observation is a function $t \longmapsto x_t$ from T into E.

Definition 0.2.1. We are given: a space T, the **time space**, a measurable space $(E,\mathcal{E})$, the **state space**.

(a) A **random function** $(X_t)_{t\in T}$ defined on a measurable space $(\Omega,\mathcal{A})$, taking values in $(E,\mathcal{E})$ is given by a family of measurable functions X_t from $(\Omega,\mathcal{A})$ into $(E,\mathcal{E})$.

(b) A **stochastic process** (or **process**) $(\Omega, \mathcal{A}, P, (X_t)_{t\in T})$ is given by a random function $(X_t)_{t\in T}$ and a probability P on a measurable space $(\Omega, \mathcal{A})$.

Trajectories. Let $(X_t)_{t\in T}$ be a family of functions from Ω into E. The observation of a realization of the phenomenon is the observation, for an ω, of the function $t \longmapsto X_t(\omega)$, i.e., of the element $X(\omega) = (X_t(\omega))_{t\in T}$ of E^T.

To say that $(X_t)_{t\in T}$ is a random function defined on $(\Omega,\mathcal{A})$ taking values in $(E,\mathcal{E})$ is equivalent to one of the following equivalent statements (cf. Volume I [4.1] to verify their equivalence):

- for every t and every $\Gamma \in \mathcal{E}$, $(X_t \in \Gamma) = \{\omega;\ X_t(\omega) \in \Gamma\}$ is in $\mathcal{A}$;
- for every $(t_1, \ldots, t_n) \in T^n$ and every rectangle $\Gamma_1 \times \ldots \times \Gamma_n \in \mathcal{E}^n$, $\{(X_{t_1}, \ldots, X_{t_n}) \in \Gamma_1 \times \ldots \times \Gamma_n\}$ is in $\mathcal{A}$;
- for every $(t_1, \ldots, t_n) \in T^n$ and every Γ in the product σ-algebra $\mathcal{E} \otimes \ldots \otimes \mathcal{E} = \mathcal{E}^{\otimes n}$ of E^n, $\{(X_{t_1}, \ldots, X_{t_n}) \in \Gamma\}$ is in $\mathcal{A}$;
- for every **rectangle** of E^T, i.e., for every product $\Pi_{t\in T}\Gamma_t$ of elements of $\mathcal{E}$ all identical to E except for a finite number of them,

$$X^{-1}\left[\prod_{t\in T}\Gamma_t\right] = \{\omega;\ X_t(\omega) \in \Gamma_t \text{ for all } t\} \text{ is in } \mathcal{A};$$

– X is measurable from $(\Omega,\mathcal{A})$ into $(E,\mathcal{E})^T$ with the following definition.

Definition 0.2.2. Let $(E_i,\mathcal{E}_i)_{i\in I}$ be a family of measurable spaces. The **product σ-algebra** $\otimes_{i\in I}\mathcal{E}_i$ is the σ-algebra on $\prod_{i\in I}E_i$ generated by the rectangles $\prod_{i\in I}\Gamma_i$, elements of $\prod_{i\in I}\mathcal{E}_i$ such that $\Gamma_i = E_i$ except for a finite number of i.

Let $(E,\mathcal{E})$ be a measurable space and let T be a set. Denote $\otimes_{t\in T}\mathcal{E} = \mathcal{E}^{\otimes T}$ and $(E,\mathcal{E})^T = (E^T,\mathcal{E}^{\otimes T})$.

Thus, if $(\Omega,\mathcal{A},P,(X_t)_{t\in T})$ is a process taking values in $(E,\mathcal{E})$, X is measurable from $(\Omega,\mathcal{A})$ into $(E,\mathcal{E})^T$; its distribution is the image measure $X(P)$. The set of rectangles being closed under finite intersections, a probability on $(E,\mathcal{E})^T$ is characterized by its values on the rectangles (Vol. I, Corollary 3.1.12). This is equivalent to saying that the distribution $X(P)$ is characterized by the values of $P[(X_{t_1}, \ldots, X_{t_n}) \in \Gamma_1 \times \ldots \times \Gamma_n]$

for every $(t_1, \ldots, t_n) \in T^n$ and $(\Gamma_i)_{1\leqslant i\leqslant n} \in \mathcal{E}^n$ and $n \in \mathbb{N}$, or by the distributions of $(X_t)_{t\in J}$ for the finite subsets J of T.

Proposition 0.2.3. *Let $(\Omega,\mathcal{A},P,(X_t)_{t\in T})$ be a process taking values in $(E,\mathcal{E})$. The function X: $\omega \longmapsto (X_t(\omega))_{t\in T}$ is measurable from $(\Omega,\mathcal{A})$ into $(E,\mathcal{E})^T$. The image measure of P by X, $X(P)$, is called the* **distribution of the process.**

This distribution is characterized by the finite distribution functions of the process, i.e. *the distributions of $(X_t)_{t\in J}$ for finite subsets J of T.*

0.2.2. Equivalence Between Processes

Definition 0.2.4. (a) Two processes having the same state space, the same time space and the same distribution are said to be **equivalent**.

(b) Let $X = (X_t)_{t\in T}$ and $Y = (Y_t)_{t\in T}$ be two processes defined on the same probability space $(\Omega,\mathcal{A},P)$, which have the same state space and the same time space T. Y is a **modification** of X if, for every $t \in T$, X_t and Y_t are equal a.s.

The above deals with two equivalence relations; the second is more restrictive. Let us consider on E^T the coordinate functions γ_t: $(x_u)_{u \in T} \longmapsto x_t$. The σ-algebra $\mathcal{E}^{\otimes T}$ is the σ-algebra generated by the functions γ_t, $t \in T$. Let Q be a probability on $(E,\mathcal{E})^T$; the process $((E,\mathcal{E})^T, Q, (\gamma_t)_{t \in T})$ is certainly a process with distribution Q. It is called the **canonical process with distribution Q.** Hence, *every process is equivalent to a unique canonical process.*

Notes. (a) *Working with canonical processes is equivalent to working with distributions. In what follows a "process" will often be defined by its distribution. This amounts to considering an equivalence class of processes. A* **version** *of the process is an element of this equivalence class. The* **canonical version** *will often be used.*

If several processes are being studied at the same time, we may be interested in considering some of the versions defined on the same probability space, thus not always the canonical version.

(b) *In what follows, we shall often omit reference to $\mathcal{E}$, especially when dealing with the Borel σ-algebra of a topological space. If $\mathcal{E}$ is understood, $\mathcal{E}^{\otimes T}$ will be understood on E^T.*

0.2.3. Canonical Processes Having Given Finite Distribution Functions

An Infinite Sample From a Distribution. Let F be a distribution on $(E,\mathcal{E})$; assuming that, for $(t_1, \ldots, t_n) \in T^n$, $(X_{t_1}, \ldots, X_{t_n})$ is an n-sample from F, its finite distribution function is $F^{\otimes n}$. There is at most one probability, denoted $F^{\otimes T}$, on $(E,\mathcal{E})^T$ satisfying this condition. When $T = \mathbb{N}$, there is thus, up to equivalence, at most one sequence of independent r.v.'s with distribution F. The existence of such a sequence will be assumed (proved, for example, in Neveu [2]).

Theorem 0.2.5. *If $(E,\mathcal{E},F)$ is an arbitrary probability space, there exists a unique canonical sequence, denoted $\{E^{\mathbb{N}}, \mathcal{E}^{\otimes \mathbb{N}}, F^{\otimes \mathbb{N}}, (X_n)_{n \in \mathbb{N}}\}$, such that the sequence $(X_n)_{n \in \mathbb{N}}$ is a sequence of independent* r.v.'s *with distribution F.*

Denote $(E,\mathcal{E},F)^{\mathbb{N}} = (E^{\mathbb{N}}, \mathcal{E}^{\otimes \mathbb{N}}, F^{\otimes \mathbb{N}})$. Up to equivalence, a process is characterized by its finite distribution functions. Thus, to its family of finite distribution functions, corresponds a unique canonical process.

Conversely, given a family of distributions $(\mu_I)_{\{I, I \subset T, I \text{ finite}\}}$ where μ_I is a distribution on E^I, the possible existence of a canonical process, of which they are the finite distribution functions, results from various extension theorems. Theorem 0.2.5 is one such for arbitrary E, $T = \mathbb{N}$ and for samples.

Here is one connected with $E = \mathbb{R}^k$ or, more generally, with E a **Polish** space, i.e. a complete, separable metric space. We also assume this theorem. Its proof may be found in Neveu [2], Billingsley [4] Chapter 7, and Dellacherie–Meyer [1].

Definition 0.2.6. A Coherent Family of Finite Distribution Functions. If I, J are two finite subsets of T with $J \subset I$, let us denote by $\phi_{I,J}: E^I \longmapsto E^J$ the natural projection defined by $\phi_{I,J}\{(X_i)_{i \in I}\} = (X_j)_{j \in J}$. A family of probabilities π_I, indexed by the set of finite subsets of T where, for each I, π_I is a probability on E^I, is said to be **coherent** if, for $J \subset I$,

$$\phi_{I,J}[\pi_I] = \pi_J.$$

Kolmogorov's Theorem 0.2.7. *Assume E is a Polish space. To every finite subset I of T, associate a probability π_I on E^I. There exists a probability P on $(E, \mathcal{E})^T$ of which $\{\pi_I\}$ is the family of finite distribution functions if, and only if, the family $\{\pi_I\}$ is coherent. The probability P is then determined in a unique way by $\{\pi_I\}$.*

Examples. We shall verify that in the following cases the finite distribution functions are coherent.

(a) Let E be a Polish space. Theorem 0.2.5 may be extended to arbitrary T. If F is a probability on E, there exists a probability $F^{\otimes T}$ on $(E, \mathcal{E})^T$ such that the coordinates $(X_t)_{t \in T}$ form a **sample** from F. This term implies here that, for every finite subset I of T, the r.v.'s $\{X_t\}_{t \in I}$ are independent with distribution F.

(b) Real Gaussian processes.

Definition 0.2.8. A process taking values in $\mathbb{R}^k$ is **Gaussian** if all its finite distribution functions are Gaussian.

Proposition 0.2.9. *Let m be a function from T into $\mathbb{R}$ and Γ a function from T^2 into $\mathbb{R}$ such that, for every finite subset $\{t_1, \ldots, t_n\}$ of T, the matrix $\{\Gamma(t_i, t_j)\}_{1 \leq i,j \leq n}$ is symmetric and positive (Γ is* *said to be positive). Then there exists a Gaussian process, unique up to equivalence, the finite distribution functions of which are the distributions* $N_n[(m(t_i))_{1\leq i\leq n}; \{\Gamma(t_i,t_j)\}_{1\leq i,j\leq n}]$ *for* $\{t_1, ..., t_n\}$ *a finite subset of T. The functions m and* Γ *characterize the distribution of this Gaussian process. m is called its* **mean** *and* Γ *its* **covariance**.

(c) Processes with independent increments taking values in $\mathbb{R}^k$.

Definition 0.2.10. The time space is an interval T of $\mathbb{Z}$ or $\mathbb{R}_+$. A process $(\Omega,A,P,(X_t)_{t\in T})$ taking values in $\mathbb{R}^k$ is said to have **independent increments** (abbreviated by PII), if, for every sequence $t_1 < t_2 < ... < t_n$ of n points of T, the random vectors $(X_{t_{i+1}} X_{t_i})_{1\leq i\leq n-1}$ are independent.

For $T = \mathbb{Z}$ or $T = \mathbb{N}$, we speak instead of a **random walk.** Take $T = \mathbb{R}_+$ and $X_0 = 0$. Assume given a family of distributions $\{\mu_{s,t}; 0 \leq s \leq t\}$ on $\mathbb{R}^k$ such that for $t_1 < t_2 < t_3$, $\mu_{t_1,t_2} * \mu_{t_2,t_3} = \mu_{t_1,t_3}$. If we fix the distribution of $X_t - X_s$ to be $\mu_{s,t}$, we obtain, for $t_1 < t_2 < ... < t_n$, the distribution

$$(\mu_{0,t_1} \otimes \mu_{t_1,t_2} \otimes ... \otimes \mu_{t_{n-1},t_n})$$

of $(X_{t_1}, X_{t_2} - X_{t_1}, ..., X_{t_n} - X_{t_{n-1}})$. The distribution of $(X_{t_1}, ..., X_{t_n})$ is necessarily the image $\nu_{t_1...t_n}$ of this distribution by the function

$$(x_1, ..., x_n) \longmapsto (x_1, x_1 + x_2, ..., x_1 + ... + x_n).$$

We thus obtain a coherent family of finite distribution functions. From this we obtain a PII zero at 0, unique up to equivalence, characterized by the family $\{\mu_{s,t}; 0 \leq s \leq t\}$.

Particular Cases. Let a: $t \longmapsto a(t)$ be an increasing function from $\mathbb{R}_+$ into $\mathbb{R}$, continuous on the right and zero at 0.

Taking for $\mu_{s,t}$ the Poisson distribution with parameter $a(t) - a(s)$, $p(a(t) - a(s))$, we obtain a **Poisson process with intensity** a. Taking for $\mu_{s,t}$, the Gaussian distribution $N(0,a(t) - a(s))$, we obtain a centered Gaussian PII. Then the covariance of X_s and X_t is, for $s \leqslant t$,

$$E(X_s X_t) = E(X_s^2) + E(X_s(X_t - X_s)) = E(X_s^2).$$

Hence this is the centered (mean zero) Gaussian process with covariance $(s,t) \longmapsto a(s \wedge t)$.

Let $(\mu_t)_{t \leqslant 0}$ be a **convolution semigroup**, i.e., a family of distributions on $\mathbb{R}^k$ such that $\mu_{t+s} = \mu_t * \mu_s$ for every $(s,t) \in \mathbb{R}_+^2$. Set $\mu_{s,t} = \mu_{t-s}$. We obtain a PII $\{\Omega, A, P, (X_t)_{t \geqslant 0}\}$, zero at 0, unique up to equivalence, such that, for $0 \leqslant s \leqslant t$, $X_t - X_s$ has distribution μ_{t-s}. This is said to be the homogeneous PII associated with the convolution semigroup (μ_t). We can, for example, take $\mu_t = p(\lambda t)$ or $\mu_t = N(0,t)$. In the first case we have a **homogeneous Poisson process with intensity** $t \longmapsto \lambda t$. In the second case, we have a homogeneous Gaussian PII with covariance $(s,t) \longmapsto s \wedge t$; this process is called **Brownian motion.**

0.2.4. Notation

Denote the modulus in $\mathbb{C}$ and the Euclidean norm in $\mathbb{R}^k$ by $x \longmapsto |x|$. For real x, $x_+ = \sup(x,0)$ and $x_- = -\inf(x,0)$. For $x \in \mathbb{C}$, $\bar{x}$ is the conjugate of x. The scalar product in $\mathbb{R}^k$ is denoted by $<\cdot,\cdot>$. For $x \in \mathbb{R}^k$, x also designates the $k \times 1$ matrix of which it is the column vector.

Let $(\Omega, A, P, (X_t)_{t \in T})$ be a "real" process, i.e. taking values in $[-\infty,\infty]$. We shall denote $X = (X_t)_{t \in T}$, $|X| = (|X_t|)_{t \in T}$, $X^+ = (X_t^+)_{t \in T}$, $X^- = (X_t^-)_{t \in T}$.

X is said to be positive, or centered, or bounded, or pth power integrable.... if that is the case for the r.v's X_t for arbitrary t.

For $T = \mathbb{N}$ or $\mathbb{Z}$, $X = (X_n)$ is a sequence, and in general X is defined up to a modification, i.e. the r.v.'s X_n are equivalence classes for a.s. equality. Equality thus implies a.s. equality, which is not in general stated. For ordered T ($T = \mathbb{Z}$, $\mathbb{R}$, ...) and X a real process, denote $X_t^* = \sup_{s \leqslant t} |X_s|$, $\overline{X}_t = \sup_{s \leqslant t} X_s$, $\underline{X}_t = \inf_{s \leqslant t} X_s$. These are r.v.'s if T is countable; otherwise it depends on the regularity of the trajectories.

Complex processes X are also studied, taking values in $\mathbb{C}$, for which we shall again be able to speak of $|X|$; X will again be said to be centered, or bounded, or pth power integrable... if that is the case for X_t, for arbitrary t.

0.3. Convergence in Distribution

We state here some tools related to convergence in distribution since they will be used again and again in what follows (see [Vol. I, 3.4]).

0.3.1. Narrow Convergence of Probabilities in A Metric Space

Let E be a metric space equipped with its Borel σ-algebra (which will always be understood), $\mathcal{P}(E)$ the set of probabilities on E. Narrow convergence of a sequence of $\mathcal{P}(E)$ has been studied in [Vol. I, 3.4.3]. Denote $P(f)$ for $\int f\, dP$.

A sequence $(P_n) \subset \mathcal{P}$ converges narrowly to P $((P_n) \xrightarrow{n} P)$ if one of the following five equivalent properties holds.

1. For every bounded continuous f from E into $\mathbb{R}$ $(f \in \mathcal{C}_b(E))$ the sequence $(P_n(f))$ tends to $P(f)$.
2. For every bounded uniformly continuous f from E into $\mathbb{R}$, the sequence $(P_n(f))$ tends to $P(f)$.
3. For every closed set F of E: $\overline{\lim}\, P_n(F) \leqslant P(F)$.
4. For every bounded r.v. f on E, continuous P-a.s., the sequence $(P_n(f))$ tends to $P(f)$.
5. Property 4, replacing "f bounded" by "f equiintegrable for the family (P_n)" i.e.,

$$\lim_{a\to\infty} \sup_n P_n[|f| 1_{(|f|>a)}] = 0.$$

The equivalence of properties 1, 3 and 4 appears in Propositions 3.4.31 and 3.4.32 of Volume I. However the proof of "1 ⇒ 3" only uses "2 ⇒ 3", and the implication "1 ⇒ 2" is clear. Finally it is easy to deduce 5 from 4.

Proposition 3.4.32 of Volume I also establishes the following result: if E' is another metric space and if h is a measurable function from E into E', P-a.s. continuous then,

$$(P_n) \xrightarrow{n} P \quad \text{implies} \quad (h(P_n)) \xrightarrow{n} h(P).$$

0.3.2. Convergence in Distribution

Let $(\Omega,\mathcal{A},P)$ be a probability space and E a metric space equipped with the distance d, denoted (E,d). The sequence (X_n) of measurable functions from $(\Omega,\mathcal{A})$ into E converges in distribution to F, a distribution on E (or to X, measurable from $(\Omega,\mathcal{A})$ into E), if the sequence of distributions $(X_n(P))$ converges narrowly to F (or to $X(P)$). We denote this by

$$X_n \xrightarrow{\mathcal{D}(P)} F \quad \text{or} \quad X_n \xrightarrow{\mathcal{D}(P)} X$$

and $\mathcal{D}(P)$ is replaced by $\mathcal{D}$ if there is no doubt about P.

The sequence (X_n) converges in probability to X if the sequence of r.v.'s $(d(X_n,X))$ tends to 0 in probability. This is denoted by $X_n \xrightarrow{P} X$. Convergence in probability implies convergence in distribution: whereas convergence in distribution does not in general imply convergence in probability, except if the limit is a constant.

Let (X_n) and (Y_n) be two sequences of measurable functions on $(\Omega,\mathcal{A},P)$ taking values in the metric spaces E_1 and E_2. The properties $X_n \xrightarrow{\mathcal{D}} X$ and $Y_n \xrightarrow{\mathcal{D}} Y$ do not in general allow any conclusion to be drawn concerning (X_n,Y_n) (Vol. I, E.3.4.3). However the following proposition holds.

Proposition 0.3.11. *Let (X_n) and (Y_n) be two sequences of measurable functions on $(\Omega,\mathcal{A},P)$ taking values in two metric spaces (E_1,d_1) and (E_2,d_2). Assume $X_n \xrightarrow{\mathcal{D}} X$ and $Y_n \xrightarrow{P} a$, a being a constant in E_2. Then $(X_n,Y_n) \xrightarrow{\mathcal{D}} (X,a)$ on $E_1 \times E_2$ equipped with the product metric structure.*

Proof. Let ϕ be a uniformly continuous and bounded function from $E_1 \times E_2$ into $\mathbb{R}$. For $\varepsilon > 0$, there exists an η such that $d_2(y,a) < \eta$ implies, for all x, $|\phi(x,y) - \phi(x,a)| \leq \varepsilon$. Hence,

$$|E[\phi(X_n,Y_n) - \phi(X,a)]| \leq P[d_2(Y_n,a) \geq \eta]$$

$$+ \varepsilon + |E[\phi(X_n,a) - \phi(X,a)]|.$$

From which

$$\overline{\lim_{n\to\infty}} |E[\phi(X_n,Y_n)] - E[\phi(X,a)]| \leqslant \varepsilon.$$

This being true for all ε, the result is obtained.

Up till Chapter 7, we shall only speak of narrow convergence of random vectors. The following idea will then be used.

Definition 0.3.12. A sequence (X_n) of random vectors is said to be **tight**, or **bounded in probability** (or **mass preserving**), if we have

$$\lim_{a\to\infty} \overline{\lim_{n\to\infty}} P[|X_n| \geqslant a] = 0.$$

In particular if (X_n) converges in distribution on $\mathbb{R}^k$ it is tight.

Proposition 0.3.13. *Let (X_n) and (Y_n) be two random vectors of dimension k; assume that (X_n) is tight. Then*

$$Y_n \xrightarrow{P} 0 \quad \textit{implies} \quad X_nY_n \xrightarrow{P} 0.$$

Proof. For every $\varepsilon > 0$ and $\eta > 0$ we can find an a such that

$$\overline{\lim_{n\to\infty}} P[|X_n| \geqslant a] \leqslant \eta/2,$$

hence an n_0 such that for $n \geqslant n_0$, and $a > 0$

$$P[|X_nY_n| \geqslant \varepsilon] \leqslant P[|X_n| \geqslant a] + P\left[|Y_n| \geqslant \frac{\varepsilon}{a}\right] \leqslant \eta.$$

The following tool will prove useful to us.

Proposition 0.3.14. *Let (X_n) be a sequence of k-dimensional random vectors and let (Γ_n) be a sequence of Borel sets of $\mathbb{R}^k$. If*

$$1_{\Gamma_n} X_n \xrightarrow{\mathcal{D}} F \quad \textit{and} \quad P(\Gamma_n) \to 1,$$

then $X_n \xrightarrow{\mathcal{D}} F$.

Proof. Let ϕ be the Fourier transform of F, and let $u \in \mathbb{R}^k$,

$$E[\exp i<u,\mathbf{1}_{\Gamma_n} X_n>] = P[\Gamma_n^c] + E[\mathbf{1}_{\Gamma_n} \exp i<u,X_n>]$$

$$= E[\mathbf{1}_{\Gamma_n^c} (1 - \exp i<u,X_n>)] + E[\exp i<u,X_n>].$$

Since

$$|E[\mathbf{1}_{\Gamma_n^c} (1 - exp\ i<u,X_n>)]| \leqslant 2P(\Gamma_n^c) \to 0,$$

we have

$$E[\exp i<u,X_n>] \to \phi(u).$$

The result follows from this, and from Lévy's theorem (Vol. I, page 83).

Finally we have the following compactness result.

Proposition 0.3.15. *From every tight sequence (X_n) of random vectors a subsequence can be extracted which converges in distribution. If (X_n) does not converge in distribution, then two subsequences can be extracted which converge in distribution to different limits.*

Proof. Theorem 3.4.28 of Volume I is identical to this proposition by replacing narrow convergence of distributions with weak convergence. However with the help of Theorem 3.4.29 of Volume I, we see that a tight sequence, the distributions of which converge weakly, converges in distribution This completes the proof.

Bibliographic Notes

As in Volume I, we only assume very few probabilistic results in this book. However we continue to accept extension theorems such as Theorems 0.2.5 and 0.2.7.

They appear in many good books, for example, Neveu [2], Billingsley [4], Dellacherie-Meyer [1].

Convergence in distribution on metric spaces is studied in detail in Billingsley [2] and Parthasarathy.

Chapter 1
TIME SERIES

Objectives

In this chapter we study second order processes by exploiting the Hilbert space structure of square integrable variables. The ideas of least squares and regression dealt with in [Vol. I, 5] are taken up again. In this framework, we introduce sequences of Hilbert spaces increasing with time, which allow us, at each instant, to see what can be described with the help of the past observations of the process. Here we work only with linear combinations of these observations. This study is particularly important in the Gaussian case where orthogonality implies independence. The statistics of stationary time series are then introduced due to their importance in applications. This chapter is almost independent of the rest of the book.

1.1. Second Order Processes

1.1.1. The Spaces L^2 and $L^2_{\mathbb{C}}$

Let $(\Omega, \mathcal{A}, P)$ be a probability space. Consider $L^2_{\mathbb{C}}(\Omega, \mathcal{A}, P)$ (or $L^2_{\mathbb{C}}$ if there is no ambiguity), the vector space of equivalence classes for a.s. equality of square integrable complex r.v.'s (or measurable functions taking values in $\mathbb{C}$). $L^2_{\mathbb{C}}$ is a Hilbert space over $\mathbb{C}$ for the Hermitian product $< , >$, $(X,Y) \longmapsto <X,Y> = E(X\overline{Y})$. Denote by $\| \ \|_2$ the associated norm. The space L^2

of real r.v.'s of $L^2_{\mathbb{C}}$ is a Hilbert space over $\mathbb{R}$.

For X,Y in $L^2_{\mathbb{C}}$, the covariance Γ is defined by,

$$\Gamma(X,Y) = E[(X - E(X))\overline{(Y - E(Y))}\,].$$

This is a scalar product on the subspace of $L^2_{\mathbb{C}}$ of centered r.v.'s. The **variance** of X is $\sigma^2(X) = \Gamma(X,X)$; its **standard deviation**, $\sigma(X)$, is the positive square root of $\sigma^2(X)$. The correlation of X and Y is

$$\rho(X,Y) = \Gamma(X,Y)/\sigma(X)\sigma(Y).$$

X and Y are said to be **uncorrelated** if $\rho(X,Y)$ is zero. For X and Y centered, uncorrelated corresponds to orthogonal.

A second order complex random vector $(X_1, ..., X_n) = X$ is a sequence of n r.v.'s of $L^2_{\mathbb{C}}$. We associate with X its **mean** $E(X) = \{E(X_i);\ 1 \leqslant i \leqslant n\}$ (X is **centered** for $E(X) = 0$) and its **covariance matrix**

$$\Gamma(X) = \{\Gamma(X_i,X_j);\ 1 \leqslant i,j \leqslant n\}$$

$$= E((X - E(X))^t(\overline{X} - E(\overline{X}))).$$

The matrix $\Gamma(X)$ is Hermitian; it is, moreover, positive semi-definite since, for every $c = (c_1, ..., c_n)$ of $\mathbb{C}^n$,

$$\sum_{1 \leqslant i,j \leqslant n} c_i\overline{c}_j\Gamma(X_i,X_j) = E\left[\left|\sum_{i=1}^{n} c_i(X_i - E(X_i))\right|^2\right] \geqslant 0.$$

Hence $\Gamma(X)$ is diagonalizable in $\mathbb{C}^n$ (in $\mathbb{R}^n$ if X is real) and, if its rank is r, it has r strictly positive eigenvalues $\sigma_1^2, ..., \sigma_r^2$ and $n - r$ zero eigenvalues. In $\mathbb{C}^n$ equipped with the Hermitian product

$$<(z_1, ..., z_n), (z_1', ..., z_n')> = \sum_{i=1}^{n} z_i\overline{z}_i',$$

there is an orthonormal basis of eigenvectors $(v_1, ..., v_r, v_{r+1}, ..., v_n)$ associated with these eigenvalues. Let U be the matrix of which these vectors are the column vectors, U^{-1} is the conjugate ${}^t\overline{U}$ of ${}^tU$, and ${}^t\overline{U}\Gamma(X)U$ is diagonal. Then let $Y = {}^t\overline{U}X$: $Y_i = <X,v_i>$, $1 \leqslant i \leqslant n$, and if $E(X) = 0$:

$$\Gamma(Y) = E[Y^t\overline{Y}] = E[{}^t\overline{U}X^t\overline{X}U] = {}^t\overline{U}\Gamma(X)U.$$

Hence the random vector Y has its last $n - r$ components zero and the first r pairwise uncorrelated, with variances $\sigma_1^2, ..., \sigma_r^2$. Set $\varepsilon_i = Y_i/\sigma_i$ for $1 \leqslant i \leqslant r$.

Proposition 1.1.1. *Let X be a second order complex random vector of dimension n, with covariance matrix Γ. The matrix Γ is positive Hermitian. If its rank is r, we can find r centered complex* r.v.'s $\varepsilon_1, ..., \varepsilon_r$ *with variance* 1, *pairwise uncorrelated*, r *orthogonal and normed vectors $v_1, ..., v_r$ of $\mathbb{C}^n$ and $\sigma_1 > 0, ..., \sigma_r >$ 0 such that*

$$X = E(X) + \sum_{i=1}^{r} v_i \sigma_i \varepsilon_i.$$

If X is real, the vectors v_i can be taken in $\mathbb{R}^n$ and the r.v.'s ε_i *can be taken to be real.*

In [Vol. I, 5.2] we have given an almost identical proof in $\mathbb{R}^n$ for the study of Gaussian vectors. If X has distribution $N_n(m,\Gamma)$, the r.v.'s $\varepsilon_1, ..., \varepsilon_r$ are independent and distributed as $N(0,1)$.

1.1.2. Second Order Processes

By replacing a sequence of n observations by a process, the following definition is obtained.

Definition 1.1.2. A **second order process** is a process $(\Omega, A, P, (X_t)_{t \in T})$ taking values in $\mathbb{C}$, such that for all t, X_t is in $L^2_{\mathbb{C}}$. The mean m and the covariance Γ of this process are the functions,

$$t \longmapsto m(t) = E(X_t)$$

$$(t,s) \longmapsto \Gamma(t,s) = \Gamma(X_t, X_s).$$

The process is **centered** if m is zero. For $T = \mathbb{Z}$, we say that it is a **time series** or a **second order sequence**.

Examples of Time Series. (a) **Noisy signals.** A signal is emitted at time n, denote this by $m(n)$, $n \in \mathbb{N}$. It is received at time $n + d$, $d \geqslant 0$ by an observer, however at the time of reception it is mixed in with noise. The observer receives the superimposed signal plus noise, denoted by $X_n = m(n - d) + \varepsilon_n$. The simplest hypothesis on the sequence (ε_n) is that it is a sequence of independent and centered r.v.'s, for example, Gaussian r.v.'s; or possibly that they are centered square integrable uncorrelated r.v.'s.

Definition 1.1.3. We call **white noise** a real process (ε_t) such that the r.v.'s ε_t are centered, square integrable, have variance 1 and are pairwise uncorrelated.

Note. *To simplify the notation in what follows we impose* $\sigma^2(\varepsilon_t) = 1$. *The current terminology "white noise" does not require this condition.*

(b) **Periodic phenomenon with random amplitude, fixed period.** Consider a complex process, $X_n = Z \exp(in\lambda)$, of frequency λ, with fixed period $2\pi/\lambda$, and with amplitude $Z \in L^2_{\mathbb{C}}$, or there again a phenomenon which is a sum of harmonics of this type,

$$X_n = \sum_{j=1}^{k} a_j Z^j \exp(in\lambda_j)$$

with $(a_j) \in \mathbb{C}^k$, $(\lambda_j) \in ([0,2\pi[)^k$, (Z^j) a complex second order random vector the components of which are centered and pairwise orthogonal.

(c) **Autoregressive phenomenon of order p ($AR(p)$).** Assume that macroeconomic observations (prices, indices, stocks) depend linearly on p previous values. We then choose as a model,

$$X_n = a_1 X_{n-1} + \dots + a_p X_{n-p} + \varepsilon_n$$

where (ε_n) is a white noise, p representing the size of the "memory" of the phenomenon.

Let us assume that the r.v.'s (ε_n) are independent and identically distributed. For $p = 1$ and X_0 fixed, a Markov chain is obtained ([Vol. I, 6]) by considering $(X_n)_{n \geq 0}$. In the general case, assuming for example fixed values of $X_{-p}, \dots, X_{-1}$ (the initial values) and setting $Y_n = (X_n, X_{n+1}, \dots, X_{n+p-1})$, the sequence (Y_n) is a Markov chain taking values in $\mathbb{R}^p$.

Proposition 1.1.4. *The covariance* Γ *is a function from* T^2 *into* $\mathbb{C}$

of positive form i.e., *such that*

For $(s,t) \in T^2$: $\Gamma(t,s) = \overline{\Gamma(s,t)}$
For $k \in \mathbb{N}$, $(t_1, \ldots, t_k) \in T^k$ *and* $(c_1, \ldots, c_k) \in \mathbb{C}^k$,

$$\sum_{1 \leq i,j \leq k} c_i \bar{c}_j \Gamma(t_i, t_j) \geq 0.$$

Proposition 0.2.27 implies that to every function m from T into $\mathbb{R}$ and to every function Γ from T^2 into $\mathbb{R}$ of positive form there corresponds a Gaussian process with mean m and covariance Γ.

1.1.3. Prediction

A second order process X is observed. We can then consider as known the r.v.'s of $H^X_{\mathbb{C}}$, the Hilbert subspace of $L^2_{\mathbb{C}}$ generated by X. This amounts to considering as known all the linear combinations $\Sigma_{i=1}^k c_i X_{t_i}$ of k observations and their limits. This is the point of view adopted in this chapter. We could also consider as known all the measurable functions of the process, i.e. the σ-algebra $\mathcal{F}^X = \sigma(X_t;\ t \in T)$. This will be the point of view adopted in the following chapters.

Let Y be an r.v. in $L^2_{\mathbb{C}}$. We want to use X to attribute a value to Y, or **predict** Y by the process X; i.e. a **prediction** problem. We can then use $P^X(Y)$, the projection of Y on $H^X_{\mathbb{C}}$. We thus obtain in the least squares sense the best linear approximation of Y by X. This is the regression point of view ([Vol. 1.5.5]) which we shall use in this chapter. We can also look for the $\mathcal{F}^X$-measurable function closest to Y in the least squares sense, i.e. $E(Y|\mathcal{F}^X)$, the expectation of Y conditional on $\mathcal{F}^X$, which is the projection of Y on $K^X = L^2_{\mathbb{C}}(\bar{\mathcal{F}}^X)$, where $\bar{\mathcal{F}}^X$ is the completion of $\mathcal{F}^X$ in $\mathcal{A}$ ([Vol. I, 1.6.1]).

When the process X is real valued, H^X designates the Hilbert subspace of L^2 generated by $(X_t)_{t \in T}$,

$$H^X_{\mathbb{C}} = H^X \oplus iH^X.$$

What happens for a Gaussian process?

Proposition 1.1.5. *Let X be a Gaussian process. The Hilbert subspace H^X which it generates in L^2 contains only Gaussian* r.v.'s, *possibly constant (with zero variance).*

Proof. Let V be the vector space generated by X, i.e. the set of vectors $\Sigma_{i=1}^{k} c_i X_{t_i}$ for $k \in \mathbb{N}$, $(c_1, \ldots, c_k) \in \mathbb{R}^k$, $(t_1, \ldots, t_k) \in T^k$.

All the r.v.'s of V are Gaussian and H^X is the closure of V in L^2.

For all n, let X_n be an r.v. with distribution $N(m_n, \sigma_n^2)$; if (X_n) converges in quadratic mean to X with mean m and variance σ^2, (m_n) tends to m and (σ_n^2) to σ^2. From Lévy's theorem (Vol. I, Theorem 3.4.30), it follows that (X_n) tends in distribution to $N(m, \sigma^2)$; X is Gaussian.

Prediction. Let us assume that $T = \mathbb{Z}$ or $T = \mathbb{R}$. At time t we have observed the process $(X_s)_{s \leq t}$ which generates the Hilbert subspace H_t^X of $L^2_{\mathbb{C}}$ and the sub-σ-algebra $\mathcal{F}_t^X$ of $\mathcal{A}$. At each time t we set ourselves prediction problems. In particular, we can try to predict the future, i.e. to predict an r.v. in $H^X_{\mathbb{C}}$ with the help of $(X_s)_{s \leq t}$. We then use the **linear predictor** P_t^X (or P_t), the projection on H_t^X; or the conditional expectation E^t, the expectation conditional on $\mathcal{F}_t^X$. In the particular case where X is a Gaussian process, for all $Y \in H^X$, $P_t^X(Y) = E^t(Y)$. In fact, $Y - P_t^X(Y)$ is orthogonal to every r.v. Z of H_t^X, hence it is independent since the pair $(Y - P_t^X(Y), Z)$ is Gaussian: $Y - P_t^X(Y)$ is independent of H_t^X, hence also of $K_t^X = L^2_{\mathbb{C}}(\mathcal{F}_t^X)$. Thus if the linear prediction $P_{n-1}(X_n)$ is, for all n, $a_1 X_{n-1} + \ldots + a_p X_{n-p}$, $\varepsilon_n = X_n - P_{n-1}(X_n)$ is orthogonal to H^X_{n-1}. We have an $AR(p)$ sequence of which (ε_n) is a "white noise innovation." We shall come back to this in [1.3.4].

Denote $H^X_{\mathbb{C}} = H^X_\infty$ and $\mathcal{F}^X = \mathcal{F}^X_\infty$; and $H^X_{-\infty} = \cap_t H_t^X$, $\mathcal{F}^X_{-\infty} = \cap_t \mathcal{F}_t^X$. The notations P_t and E^t are extended to $t = \pm\infty$. Then, for every $Y \in L^2_{\mathbb{C}}$, we see that $(P_t(Y))$ tends to $P_\infty(Y)$ (resp. $P_{-\infty}(Y)$) if t tends to $+\infty$ (resp. $-\infty$), and that $(E^t(Y))$ tends to $E^\infty(Y)$ (resp. $E^{-\infty}(Y)$) if t tends to $+\infty$ (resp. $-\infty$) in the sense of convergence in quadratic mean. This results from the following lemma (in the case where $T = \mathbb{R}$, convergence for $t \to \pm\infty$ is equivalent to convergence for every subsequence (t_n) tending to $+\infty$ or $-\infty$).

Lemma 1.1.6. *Let H be a Hilbert space and $(H_n)_{n \in \mathbb{Z}}$ a nested sequence of subspaces of H. Denote by $H_{-\infty}$ the intersection $\cap H_n$ and by H_∞ the Hilbert subspace of H generated by $\cup H_n$. Then for all $x \in H$, we have*

$$P_\infty(x) = \lim_{n\to\infty} P_n(x) \quad \textit{and} \quad P_{-\infty}(x) = \lim_{n\to-\infty} P_n(x).$$

Proof. For arbitrary $n \in \mathbb{Z}$, $P_{n+1}(x) - P_n(x)$ is the projection of x on the orthogonal complement of H_n in H_{n+1} and, by Pythagoras' theorem, for every positive integer N we have (denoting the norm in H by $\|\cdot\|$),

$$\|P_N(x) - P_0(x)\|^2 = \sum_{n=0}^{N-1} \|P_{n+1}(x) - P_n(x)\|^2 \leq 4\|x\|^2$$

$$\|P_0(x) - P_{-N}(x)\|^2 = \sum_{n=-N}^{-1} \|P_{n+1}(x) - P_n(x)\|^2 \leq 4\|x\|^2.$$

Hence the series $\Sigma_{n\geq 0}\|P_{n+1}(x) - P_n(x)\|^2$ and $\Sigma_{n<0}\|P_{n+1}(x) - P_n(x)\|^2$ converge; the Cauchy sequences $(P_n(x))_{n\geq 0}$ and $(P_n(x))_{n<0}$ converge respectively to $x_\infty \in H_\infty$ and to $x_{-\infty} \in H_{-\infty}$. As $x - x_\infty$ (resp. $x - x_{-\infty}$) is orthogonal to H_∞ (resp. to $H_{-\infty}$) we obtain the result.

The study of the process $(E^t(Y))$, which is a **martingale**, will be covered for $T = \mathbb{Z}$ in [2] and for $T = \mathbb{R}_+$ in [7]. We shall show that it converges a.s. In this section we study linear predictions P_t^Y.

If Y is in $H_{-\infty}^X$, all the predictions $P_t(Y)$ of Y are equal to it. We can write (I being the identity transformation):

$$X_t = P_{-\infty}(X_t) + (I - P_{-\infty})(X_t) = Y_t + Z_t.$$

Definition 1.1.7. For $T = \mathbb{Z}$ or $T = \mathbb{R}$ the second order process $(X_t)_{t\in T}$ is said to be **regular** if $H_{-\infty}^X = \{0\}$, **singular** if $H_{-\infty}^X = H_{\mathbb{C}}^X$.

Proposition 1.1.8. Wold's Decomposition. *Every second order process is the sum of a regular process and a singular process. These processes are orthogonal and the decomposition is unique.*

Note. *For a white noise (ε_n), for arbitrary $n < m$, we have $P_n(\varepsilon_m) = 0$. A white noise is always regular.*

1.1.4. Processes with Orthogonal Increments

Definition 1.1.9. Let $T = \mathbb{Z}$ or $\mathbb{R}$. A centered second order process $(X_t)_{t\in T}$ has **orthogonal increments** if, for every $t < s$, $(X_s - X_t)$ and X_t are orthogonal.

Example. Let $X_n = \Sigma_{1 \leqslant p \leqslant n} \varepsilon_p$, be the sum of n r.v.'s. If (ε_n) is a white noise, X_n has orthogonal increments.

For a process with orthogonal increments prediction is simple: for $t < s$, $P_t(X_s) = X_t$. Thus

$$X_t \xrightarrow[t \to -\infty]{L^2_{\mathbb{C}}} X_{-\infty} = P_{-\infty}(X_0).$$

Assume that $T = \mathbf{Z}$ and let us set, for all n, $\Delta X_n = X_n - X_{n-1}$. For $n < m$,

$$E(|X_n - X_m|^2) = \sum_{n<p\leqslant m} E(|\Delta X_p|^2)$$
$$= E(|X_m|^2) - E(|X_n|^2).$$

If the series $\Sigma_{0 \leqslant p < \infty} E(|\Delta X_p|^2)$ converges, (X_n) is a Cauchy sequence if n tends to $+\infty$.

Theorem 1.1.10. *Let $(X_n)_{n \in \mathbf{Z}}$ be a time series with orthogonal increments.*

(a) $$X_n \xrightarrow[n \to -\infty]{L^2_{\mathbb{C}}} X_{-\infty} \quad \textit{and} \quad X_n = X_{-\infty} + \sum_{p \leqslant n} \Delta X_p.$$

(b) *If* $\sum_{p=0}^{\infty} E(|\Delta X_p|^2)$ *converges,* $X_n \xrightarrow[n \to \infty]{L^2_{\mathbb{C}}} X_\infty.$

1.1.5. Fourier Series

In this section we study Fourier series, a particular case of a time series with orthogonal increments.

Let us consider $\Omega = [-\pi,\pi[$, $A = B_{[-\pi,\pi[}$, $P = L/2\pi$ where L is Lebesgue measure. Let f_n be the function $x \longmapsto e^{inx}$ defined on $[-\pi,\pi[$. In $L^2_{\mathbb{C}}$, the sequence (f_n) is a sequence of orthogonal r.v.'s, all of them centered except f_0. Let (c_n) be a sequence of complex numbers. Set $S_N = \Sigma^N_{n=-N} c_n f_n$. The sequence (S_N) converges in quadratic mean if, and only if, $\Sigma^\infty_{n=-\infty} |c_n|^2 < \infty$, which we denote by $(c_n) \in \ell^2$.

In this case, let us designate the limit by S. We have $E(S\bar{f}_p) = \lim_{N\to\infty} E(S_N \bar{f}_p) = c_p$; (c_n) is the sequence of **Fourier coefficients** of S. **Parseval's relation** is written,

$$\lim_{N\to\infty} E(S_N^2) - E(S^2) = \sum_{-\infty}^{\infty} |c_p|^2,$$

$$\int_{-\pi}^{\pi} |S(x)|^2\, dx = 2\pi \sum_{-\infty}^{\infty} |c_p|^2.$$

Now let g be a function in $L^2_{\mathbb{C}}$, and $c_n = \langle g, f_{-n} \rangle$. The Fourier series of g is, by definition, $\Sigma_{-\infty}^{\infty} c_n f_n$; c_n is the Fourier coefficient of order n of g. We extend g to $\mathbb{R}$ as a function of period 2π, denoted again by g. Let D_N be the **Dirichlet kernel**, defined by,

$$D_N(z) = \frac{1}{2\pi} \frac{\sin[(2N+1)z/2]}{\sin(z/2)}.$$

Let $S_N(g) = \Sigma_{n=-N}^{N} c_n f_n$ be the Nth **Dirichlet transform** of g; then,

$$[S_N(g)](x) = \int_{-\pi}^{\pi} D_N(z) g(x + z) dz.$$

If g is twice continuously differentiable, the function

$$z \longmapsto \phi(x,z) = \frac{g(x+z) - g(x)}{2 \sin(z/2)}$$

is continuously differentiable and, ϕ' designating the derivative with respect to z,

$$S_N(g)(x) - g(x) = \left[-\phi(x,z) \frac{\cos \frac{2N+1}{2} z}{\frac{2N+1}{2}} \right]_{-\pi}^{\pi}$$

$$+ \int_{-\pi}^{\pi} \phi'(x,z) \frac{\cos \frac{2N+1}{2} z}{\frac{2N+1}{2} z} dx.$$

If g is of class C^2, $S_N(g)$ converges uniformly, hence, also in L^2, to g.

Let Φ_N be the **Fejer kernel** defined by

$$\Phi_N(z) = \frac{1}{2\pi N} \left(\frac{\sin(Nz/2)}{\sin(z/2)} \right)^2.$$

The Nth **Fejer transform** of g is $\sigma_N(g)$ with,

$$\sigma_N(g)(x) = \frac{S_0(g) + \ldots + S_{N-1}(g)}{N}(x).$$

$$= \int_{-\pi}^{\pi} \Phi_N(z) g(x + z) dz.$$

If g is continuous at x and bounded in modulus by $\|g\|$, let δ be the modulus of continuity at x associated with ε

$$|\sigma_N(g)(x) - g(x)| \leqslant \int_{-\delta}^{\delta} \Phi_N(z)|g(z + x) - g(x)|dz + 2\|g\| \int_{(-\delta,\delta)^c} \Phi_N(z)dz$$

$$\leqslant \varepsilon \int_{-\pi}^{\pi} \Phi_N(z)dz + 4\|g\| \frac{\pi - \delta}{2\pi N(\sin(\delta/2))^2}.$$

If g is continuous on $\mathbb{R}$, g is the uniform limit of its Fejer transforms, thus the uniform limit of trigonometric polynomials.

Let F be the subspace of $L^2_{\mathbb{C}}$ of functions g such that $S_N(g)$ tends to g in $L^k_{\mathbb{C}}$; F is closed. In fact, let $(g^k = \Sigma\, c^k_p f_p)$ be a Cauchy sequence,

$$E(|g^k - g^j|^2) = \sum_{p \in \mathbf{Z}} |c^k_p - c^j_p|^2.$$

Each sequence (c^k_p) is Cauchy and converges to c_p, and it can be verified that g^k converges to $g = \Sigma_n c_n f_n$.

The space F contains functions twice continuously differentiable on $\mathbb{R}$ of period 2π, hence the functions $x \longmapsto e^{inx}$ for $n \in \mathbb{Z}$. It is closed, hence contains continuous functions on $[-\pi,\pi]$. Let $f \in L^2_{\mathbb{C}}$ be orthogonal to F. $\int f\bar{g}\, dL$ is zero for every continuous function g on $[-\pi,\pi]$, hence for g an indicator function of a closed set (the limit of a decreasing sequence of continuous functions) and for g a bounded r.v. [Vol. I, 3.1.11 and 3.1.13]: f is zero.

Theorem 1.1.11. *To every function g in $L^2_{\mathbb{C}}([-\pi,\pi[, L/2\pi)$, (denoted here by $L^2_{\mathbb{C}}$), we associate the Fourier series $\Sigma_{-\infty}^{\infty} c_n e^{inx}$ with $c_n = (1/2\pi)\int_{-\pi}^{\pi} e^{-inx} g(x)dx$ the sequence $(\Sigma_{-N}^{N} c_n e^{inx})$ converges in $L^2_{\mathbb{C}}$ to g, denoted $(\Sigma_{-\infty}^{\infty} c_n e^{inx})$. Let ℓ^2 be the Hilbert space of sequences of complex numbers $(c_n)_{n\in\mathbf{Z}}$ such that $\Sigma_{-\infty}^{\infty}|c_n|^2$ converges, equipped with the Hermitian product $\langle(c_n),(d_n)\rangle = \Sigma_{-\infty}^{\infty} c_n \bar{d}_n$. The mapping $(c_n) \longmapsto \Sigma_{-\infty}^{\infty} c_n e^{inx}$ is an isometry of ℓ^2 onto $L^2_{\mathbb{C}}$.*

Convolution. Let f and g be two functions of $L^2_{\mathbb{C}}$ such that fg is in $L^2_{\mathbb{C}}$. If (c_n) and (d_n) are the Fourier coefficients of f and of g, then $(c * d)_n$ is the nth Fourier coefficient of fg, setting

$$(c * d)_n = \sum_{m \in \mathbb{Z}} c_{n-m} d_m .$$

By extending f on $\mathbb{R}$ as a function of period 2π, the function $f * g$ is defined by,

$$f * g(x) = \frac{1}{2\pi} \int_{-\pi}^{\pi} f(x - z) g(z) dz.$$

Then the Fourier coefficient of order n of $f * g$ is $c_n d_n$.

1.2. Spatial Processes with Orthogonal Increments

1.2.1. Spatial Poisson Processes

In order to describe the distribution of grasshoppers in a field, we can, for every measurable subset A of the field, study the random number $Z(A)$ of grasshoppers which are in A. A simple situation is the following: for arbitrary A, the r.v.'s follow a Poisson distribution and, to two disjoint measurable subsets A and B of the field, correspond the independent random numbers $Z(A)$ and $Z(B)$.

Recall that a Poisson distribution $p(\lambda)$ with parameter λ is the distribution which gives to the integer n the weight $e^{-\lambda}\lambda^n/n!$. We agree that a Poisson distribution with an infinite parameter is concentrated on $+\infty$. A transition probability Z from $(\Omega, \mathcal{A})$ into $(E, \mathcal{E})$ is a function from $\Omega \times E$ into $[0,1]$, such that, for every $\omega \in \Omega$, $Z(\omega, \cdot)$ is a probability and, for every A, $\omega \longmapsto Z(\omega, A)$ is an r.v. ([Vol. I, 6.3]).

Definition 1.1.12. Let $(E, \mathcal{E}, \mu)$ be a measure space, $(\Omega, \mathcal{A}, P)$ a probability space and Z a transition probability from $(\Omega, \mathcal{A})$ into $(E, \mathcal{E})$. We say that $\{\Omega, \mathcal{A}, P, Z\}$ is a Poisson process with intensity μ on $(E, \mathcal{E})$ if the following two conditions are satisfied:

(a) for every $A \in \mathcal{E}$, $\omega \longmapsto Z(\omega, A)$ is an r.v., denoted $Z(A)$, having distribution $p[\mu(A)]$

(b) for every finite family of disjoint elements $A_1, \ldots, A_k$ of $\mathcal{E}$, $Z(A_1), \ldots, Z(A_k)$ are independent r.v.'s.

Note. *This process has state space* $N \cup \{\infty\}$ *and time space* $\mathcal{E}$. *Its existence follows from Kolmogorov's theorem* 0.2.27.

Particular Case. Poisson Process on $\mathbb{R}_+$. The observation of successive breakdown times (T_n) amounts to observing random points on $]0,\infty[$ (if two simultaneous breakdowns cannot occur). In this case, Z can be described by its distribution function $N_t = Z([0,t])$. If T_0 is zero and if the r.v.'s $(T_n - T_{n-1} = \Delta T_n)_{n\geq 1}$ are independent and have an exponential density with parameter λ, Z is a Poisson process the intensity of which is λL (L is Lebesgue measure on $[0,\infty[$). In other words, for $s < t$, $N_t - N_s$ has distribution $p(\lambda(t - s))$ and is independent of the past ([Vol. I, E4.4.12]). We shall come back to this result in [6].

In a general way, a Poisson process N with intensity μ bounded on the compact sets of $\mathbb{R}_+$ may be characterized by $N_t = N[0,t]$ and μ by $\mu_t = \mu[0,t]$ with $t \geq 0$. Then $(N_t)_{t\geq 0}$ is a Poisson process with intensity $(\mu_t)_{t\geq 0}$ in the sense defined in [0.2.3].

Proposition 1.2.13. *Let* $(\Omega,\mathcal{A},P,Z)$ *be a Poisson process with intensity* μ *on* $(E,\mathcal{E})$.

(a) *If* A *and* B *have finite* μ*-measures, we have*

$$\sigma^2(Z(A)) = \mu(A), \quad \Gamma(Z(A),Z(B)) = \mu(A \cap B).$$

(b) *For every* r.v. $f \in L^1(E,\mathcal{E},\mu)$, $Z(f) = \int f\, dZ$ *is integrable, with mean* $\mu(f)$; *and, for* $f \in L^2(E,\mathcal{E},\mu)$, $Z(f)$ *is square integrable, with variance* $\mu(f^2)$. *For* f *and* g *in* $L^2(E,\mathcal{E},\mu)$, *we have*

$$\Gamma[Z(f), Z(g)] = \mu(fg).$$

Proof. (a) is easy and (b) holds for positive stepped f. Then use [Vol. I, 3.2.1].

Poisson processes and sequences of orthogonal r.v.'s. Orthogonal increments with base μ are generalized in [1.2.2].

1.2.2. Spatial Processes with Orthogonal Increments

Definition 1.2.14. Let $(E,\mathcal{E})$ be a measurable space and μ a σ-finite measure on $(E,\mathcal{E})$. We call a **spatial process with orthogonal increments** with base μ a second order process $Z = \{Z(A), A \in T\}$ with $T = \{A; A \in \mathcal{E}, \mu(A) < \infty\}$ which satisfies the following properties:

(a) $Z(\emptyset) = 0$
(b) for A and B disjoint in T: $Z(A \cup B) = Z(A) + Z(B)$
(c) for A and B in T: $\Gamma(Z(A),Z(B)) = \mu(A \cap B)$.

In particular if A and B are disjoint, $Z(A)$ and $Z(B)$ are uncorrelated.

Denote also $Z(A)$ by $\omega \longmapsto Z(\omega,A)$. The process is **centered** if $Z(A)$ is assumed centered, for every $A \in T$.

Examples. (a) Let Z be a spatial Poisson process with intensity μ. The process $\{Z(A); A \in T\}$ is a process with orthogonal increments with base μ.

(b) Let $(\lambda_n)_{n\in\mathbb{Z}}$ be a sequence of points of $(E,\mathcal{E})$ and $(X_n)_{n\in\mathbf{Z}}$ a sequence in $L^2_{\mathbb{C}}$ with orthogonal increments. Set $X_n - X_{n-1} = \Delta X_n$ and let δ be Dirac measure,

$$\mu = \sum_n \sigma^2(\Delta X_n)\delta_{\lambda_n} \quad \text{and} \quad Z(\omega,\cdot) = \sum_n \Delta X_n(\omega)\delta_{\lambda_n} .$$

In particular, for $E = \mathbf{N}$ or $\mathbf{Z}$ and $\lambda_n = n$, we obtain

$$Z(\omega,]-\infty,n]) = X_n(\omega) - X_{-\infty}(\omega).$$

(c) Spatial Gaussian Processes with Independent Increments. The function $(A,B) \longmapsto \mu(A \cap B)$ from T^2 into $\mathbb{R}$ is a positive form. From Proposition 0.2.7, there exists a centered Gaussian process of which it is the covariance. This process is a process with orthogonal increments with base μ. For $E = \mathbb{R}^n$ and μ Lebesgue measure, Z is called a **Brownian field**; for $E = \mathbb{R}$ and $Z([0,t]) = X_t$, the process $(X_t)_{t\geq 0}$ is a Brownian motion ([0.2.3]).

In Volume 1 we agreed to reserve the term of measure for positive measures. Staying faithful to this convention a signed measure is the difference between two measures. In examples (a) and (b), $Z(\omega,\cdot)$ is, for all ω, a signed measure.

This property is false in general and we shall avoid the classic expression of "random measure with orthogonal increments" which leads to confusion when Z is not (a.s.) a measure. For example, almost no trajectory ω of a Brownian motion is such that $Z(\omega,\cdot)$ is a measure (cf. [8.1.3]).

1.2.3. Stochastic Integrals

When $Z(\omega,\cdot)$ is, for all ω, a signed measure (examples (a) and (b)) it is natural to study, for an r.v. f defined on $(E,\mathcal{E})$, the integral

$$Z(f)(\omega) = \int_E f(s)Z(\omega,ds).$$

Denote then $Z(f) = \int f\, dZ$.

We are going to give a meaning to the r.v. $\int f\, dZ$ even in some cases where the $Z(\omega,\cdot)$ are not almost surely measures.

Let V be the vector space of step functions of $L^2_{\mathbb{C}}(E,\mathcal{E},\mu)$, dense in this space. For $f = \Sigma_{i=1}^n a_i 1_{A_i}$, $g = \Sigma_{j=1}^m b_j 1_{B_j}$ in V we have,

$$\int f\bar{g}\, d\mu = \sum_{i=1}^n \sum_{j=1}^m a_i \bar{b}_j \mu(A_i \cap B_j)$$

$$= \Gamma\left[\sum_{i=1}^n a_i Z(A_i),\ \sum_{j=1}^m b_j Z(B_j)\right].$$

Let $H^Z_{\mathbb{C}}$ be the Hilbert subspace of $L^2_{\mathbb{C}}(\Omega,\mathcal{A},P)$ generated by Z. It contains the r.v.'s $\Sigma_{i=1}^n a_i Z(A_i)$. Since V is dense in $L^2_{\mathbb{C}}$ we obtain:

Theorem 1.2.15. *Let Z be the spatial process with orthogonal increments of Definition 1.2.14. If Z is centered, there exists a unique isometry of $L^2_{\mathbb{C}}(E,\mathcal{E},\mu)$ onto $H^Z_{\mathbb{C}}$ which extends the mapping $1_A \longmapsto Z(A) = Z(1_A)$ defined for $A \in T$. This isometry is also denoted by Z.*

If Z takes real values, there exists a unique isometry from $L^2(E,\mathcal{E},\mu)$ into H^Z which extends the mapping $1_A \longmapsto Z(A)$.

For f and g in $L^2_{\mathbb{C}}(E,\mathcal{E},\mu)$, $Z(f)$ is centered and

$$\Gamma(Z(f),Z(g)) = \int f\bar{g}\, d\mu.$$

Examples. (a) For a Poisson process Z with intensity μ, $\bar{Z} = Z - \mu$ is centered and $\bar{Z}(f) = Z(f) - \mu(f) = \int f\, dZ - \int f\, d\mu$ from Proposition 1.2.13.

(b) In example (b) of [1.2.2] for $f \in L^2_{\mathbb{C}}(E,\mathcal{E},\mu)$ we obtain $Z(f) = \Sigma f(\lambda_n)\Delta X_n$, a series which converges in $L^2_{\mathbb{C}}(\Omega,\mathcal{A},P)$. In particular if $(E,\mathcal{E},\mu)$ is $\mathbf{Z}$, provided with the σ-algebra of all its subsets and with $\mu = \Sigma_{n\in\mathbf{Z}}\delta_n$, $L^2_{\mathbb{C}}$ is the space ℓ^2 of sequences $c = (c_n)$ such that $\Sigma_{n\in\mathbb{Z}}|c_n|^2$ converges. If

$$(\Omega,\mathcal{A},P) = \left([-\pi,\pi[,\ \mathcal{B}_{[-\pi,\pi[},\ \frac{1}{2\pi}L\right)$$

and $Z(x,\{n\}) = e^{inx}$, then the stochastic integral $Z(c)$ is the sum of the Fourier series $x \longmapsto \Sigma_{n\in\mathbb{Z}} c_n e^{inx}$ (Theorem 1.1.11). If $\Sigma_{n\in\mathbb{Z}}|c_n|$ converges, the series $Z(c)(x) = \Sigma_{n\in\mathbb{Z}} c_n e^{inx}$ converges uniformly at x and the function $Z(c)$ is defined for all x and continuous. On the other hand, the stochastic integral $Z(c)$ is defined only L-a.s. for arbitrary $c \in \ell^2$.

(c) For a Gaussian process with centered independent increments with base μ, for $f \in L^2(E,\mathcal{E},\mu)$, $Z(f)$ is in H^Z. Hence it is a Gaussian r.v. with distribution $\mathcal{N}(0,\mu(f))$.

Notation. We often denote $Z(f) = \int f\,dZ$, but we are dealing with an r.v. in $L^2_{\mathbb{C}}$. In general, there is not, for almost all ω, a signed measure $Z(\omega,\cdot)$ allowing $Z(f)$ to be defined as a normal integral. Nevertheless $Z(f)$ is called the **stochastic integral of f with respect to Z.**

1.2.4. Representation of a Second Order Process Using a Spatial Process with Orthogonal Increments

Recall Proposition 1.1.1, with the same notations and $v_i = (v_i^p)_{1\leqslant p\leqslant n}$. The sequence $Z = (\varepsilon_1 + \dots + \varepsilon_\ell)_{1\leqslant \ell\leqslant r}$ is a process with orthogonal increments on $\{1, \dots, r\}$ with base μ, $\mu(\ell) = 1$ for $1 \leqslant \ell \leqslant r$. For $1 \leqslant p,q \leqslant n$,

$$X_p = \sum_{i=1}^{r} v_i^p \sigma_i \varepsilon_i = \int v_{.}^p \sigma_{.}\, dZ,$$

$$\gamma(p,q) = \Gamma(X_p,X_q) = \sum_{i=1}^{r} v_i^p \overline{v_i^q} \sigma_i^2 = \int v_{.}^p \overline{v_{.}^q} (\sigma_{.})^2 d\mu.$$

The following theorem is a generalization of this result which can be recovered by taking in the theorem,

$$E = \{1, \dots, r\},\quad T = \{1, \dots, n\} \text{ and } a(p,i) = v_i^p \sigma_i.$$

Theorem 1.2.16 (Karhunen). *Let T be a space and $(E,\mathcal{E},\mu)$ a measure space. A function a from $T \times E$ into $\mathbb{C}$ is given such that, for all $t \in T$, $a(t,\cdot)$ is in $L^2_{\mathbb{C}}(E,\mathcal{E},\mu)$. Set*

$$\Gamma(s,t) = \int a(s,\lambda)\overline{a(t,\lambda)}\,d\mu(\lambda).$$

For every second order centered process $(\Omega, \mathcal{A}, P, (X_t)_{t\in T})$ *with covariance* Γ, *there exists a centered process* Z *with orthogonal increments with base* μ *such that,*

$$X_t = \int a(t,\lambda)dZ(\lambda).$$

The spaces $H_{\mathbb{C}}^X$ *and* $H_{\mathbb{C}}^Z$ *coincide if, and only if, the subspace generated by* $\{a(t,\cdot);\ t \in T\}$ *is dense in* $L_{\mathbb{C}}^2(E,\mathcal{E},\mu)$.

Proof. We have

$$E(|X_t - X_s|^2) = \int |a(t,\lambda) - a(s,\lambda)|^2 d\mu(\lambda).$$

Hence, if $a(t,\cdot) = a(s,\cdot)$ μ-a.s., the r.v.'s X_t and X_s are equal P-a.s.

Let K be the Hilbert subspace of $L_{\mathbb{C}}^2(E,\mathcal{E},\mu)$ generated by $\{a(t,\cdot);\ t \in T\}$. The function $a(t,\cdot) \longmapsto X_t$ has a unique extension to an isometry ψ from K into $H_{\mathbb{C}}^X$ due to the form of Γ. Let us assume $K = L_{\mathbb{C}}^2(E,\mathcal{E},\mu)$. Then for $A \in \mathcal{E}$ with finite μ-measure, $1_A \in L^2(E,\mathcal{E},\mu)$: let us denote $Z(A) = \psi(1_A)$. We thus define a process with orthogonal increments with base μ. However the mapping ψ is then the isometry from $L_{\mathbb{C}}^2(E,\mathcal{E},\mu)$ into $H_{\mathbb{C}}^Z$ which associates $Z(A)$ with 1_A. It is the stochastic integral,

$$X_t = \psi(a(t,\cdot)) = \int a(t,\lambda)dZ(\lambda),$$

from which $H_{\mathbb{C}}^X \subset H_{\mathbb{C}}^Z$. However we have $Z(A) \in H_{\mathbb{C}}^X$, for $\mu(A)$ finite. Thus $H_{\mathbb{C}}^X = H_{\mathbb{C}}^Z$. If K is a strict subspace of $L_{\mathbb{C}}^2(E,\mathcal{E},\mu)$, its orthogonal complement $K^\perp$ can be generated by a family $\{a(t',\cdot);\ t' \in T'\}$ of functions of $L^2(E,\mathcal{E},\mu)$, T' disjoint from T. We define a covariance on $T \cup T'$,

$$\Gamma(t,s) = \int a(t,\lambda)\overline{a(s,\lambda)}\, d\mu(\lambda)$$

which is zero if $t \in T$, $s \in T'$. Consider a centered process $\{\Omega', \mathcal{A}', P', (X_t)_{t\in T'}\}$ with covariance $\{\Gamma(t,s);\ (t,s) \in T'^2\}$. Consider also the process $\{\Omega \times \Omega', \mathcal{A} \otimes \mathcal{A}', P \otimes P', (Y_t)_{t\in T\cup T'}\}$ defined by

$$Y_t(\omega,\omega') = \begin{cases} X_t(\omega) & \text{for } t \in T \\ X_t(\omega') & \text{for } t \in T' \end{cases}.$$

For this process, we apply the first part of the proof and taking $t \in T$, by identifying $Y_t(\omega,\omega')$ and its projection $X_t(\omega)$ on $L^2(E,\mathcal{E},\mu)$,

$$X_t = \int a(t,\lambda)dZ(\lambda).$$

Here $H_{\mathbb{C}}^Z$ is isomorphic to $L_{\mathbb{C}}^2(E,\mathcal{E},\mu)$ and $H_{\mathbb{C}}^X$ is isomorphic to K. Hence $H_{\mathbb{C}}^X$ is a strict subspace of $H_{\mathbb{C}}^Z$.

Example. Let $T = \mathbb{Z}$ and $(c_n)_{n\in\mathbb{Z}}$ be a sequence in ℓ^2. Consider a centered time series (X_n), the covariance of which satisfies the relation,

$$\Gamma(n,m) = \sum_{p=-\infty}^{\infty} c_{n-p}\bar{c}_{m-p} \,.$$

Let us take $E = \mathbb{Z}$ and μ the measure which gives weight 1 to each point. To a process Z with orthogonal increments with base μ corresponds a white noise defined by $\varepsilon_n = Z(n)$. Hence there exists a white noise such that,

$$X_n = \sum_{p=-\infty}^{\infty} c_{n-p}\varepsilon_p \,.$$

In general $H_{\mathbb{C}}^X \subset H_{\mathbb{C}}^{\varepsilon}$, equality being equivalent to the family of sequences $\tilde{c}_n = (c_{n-k})_{k\in\mathbb{Z}}$ being dense in the space ℓ^2 of square summable sequences.

1.3. Stationary Second Order Processes

1.3.1. Stationarity

In order to study a distribution F, we have often used in Volume I the observation of an n-sample from this distribution, the repetition of identically distributed experiments being essential.

If a phenomenon evolves through time, $T = \mathbb{Z}$ or $T = \mathbb{R}$, we observe it for example between the times t and $t + h$ $(h > 0)$. However time does not repeat itself. If we want to repeat the experiment, we repeat an experiment of length h and the phenomenon is observed between s and $s + h$ $(s > t)$. If this new experiment has the same distribution as the first, we can speak about repetitions of identical experiments and hope to obtain statistical results.

Definition 1.3.17. A second order process $(X_t)_{t\in T}$ with $T = \mathbb{Z}$ or $\mathbb{R}$ is **stationary in the strict sense** if, for every $(t,s) \in T^2$ and $h > 0$, $\{X_u\}_{t\leqslant u\leqslant t+h}$ and $\{X_u\}_{s\leqslant u\leqslant s+h}$ have the same distribution. It is said to be **second order stationary** if its mean m is constant and if its covariance $\Gamma(t,s)$, for $(t,s) \in T^2$, depends only on $t - s$. This is then denoted by $\gamma(t - s)$: γ will be called, like Γ, the **covariance of the stationary process.** The term **autocovariance** is also used for γ, and the term **autocorrelation** for $\gamma/\gamma(0)$.

For a stationary second order time series, we shall often say, as an abbreviation, **stationary sequence.**

It is clear that strict stationarity implies second order stationarity. This latter stationarity scarcely has any physical meaning. However it is sufficient for mathematical developments using the Hilbert structure of $L^2_{\mathbb{C}}$. For Gaussian processes these two ideas coincide, since the covariance and mean characterize the distributions.

1.3.2. Spectral Representation

Let γ be the covariance of a stationary process. It is a **positive semidefinite** function, i.e. such that $(s,t) \longmapsto \gamma(t - s)$ is a positive form (Proposition 1.1.4). The form of positive semi-definite functions on $\mathbb{Z}$ or $\mathbb{R}$ is given by the following theorem. Here we denote $\int_{-\pi}^{\pi}$ for $\int_{[-\pi,\pi[}$.

Theorem 1.3.18. (1) (Hergoltz). *Let γ be a positive semi-definite function from $\mathbb{Z}$ into $\mathbb{C}$. There exists a bounded measure μ on $[-\pi,\pi[$ such that, for all $n \in \mathbb{Z}$*

$$\gamma(n) = \int_{-\pi}^{\pi} e^{in\lambda} d\mu(\lambda).$$

(2) (Bochner). *Let γ be a continuous and positive semi-definite function from $\mathbb{R}$ into $\mathbb{C}$. There exists a bounded measure μ on $\mathbb{R}$ such that, for all $t \in \mathbb{R}$*

$$\gamma(t) = \int_{-\infty}^{\infty} e^{it\lambda} d\mu(\lambda).$$

We shall not prove these theorems (their converse is clear). A proof can be found in Rudin [2]. They allow Karhunen's theorem to be applied. We write

$$\text{on } \mathbb{Z}: \quad \Gamma(n,n') = \gamma(n-n') = \int_{-\pi}^{\pi} e^{in\lambda}e^{-in'\lambda}d\mu(\lambda)$$

$$\text{on } \mathbb{R}: \quad \Gamma(t,t') = \gamma(t-t') = \int_{-\infty}^{\infty} e^{it\lambda}e^{-it'\lambda}d\mu(\lambda).$$

On $\mathbb{Z}$, we have seen in [1.1.5] that every continuous function from $[-\pi,\pi]$ into $\mathbb{R}$ is the uniform limit of its Fejer transforms.

The vector space generated by the functions $e^{in\cdot}: \lambda \longmapsto e^{in\lambda}$ is dense in $L^2_{\mathbb{C}}([-\pi,\pi[,\mu)$. If (X_n) is a second order stationary process with covariance γ, there is thus a unique isometry Z from $L^2_{\mathbb{C}}([-\pi,\pi[,\mu)$ into $H^X_{\mathbb{C}}$ such that $Z(e^{in\cdot}) = X_n$, as a result of Theorem 1.2.16 of Karhunen.

Theorem 1.3.19. *Let $X = (X_n)_{n\in\mathbb{Z}}$ be a centered stationary time series with covariance γ. There exists on $[-\pi,\pi[$ a process Z with orthogonal increments with base μ such that, for all n,*

$$X_n = \int_{-\pi}^{\pi} e^{in\lambda}dZ(\lambda).$$

The spaces $H^X_{\mathbb{C}}$ and $H^Z_{\mathbb{C}}$ coincide: Z is an isometry from $L^2_{\mathbb{C}}([-\pi,\pi[,\mu)$ onto $H^X_{\mathbb{C}}$.

The results on $\mathbb{R}$ are identical, by using the Fourier transform of continuous functions from $\mathbb{R}$ into $\mathbb{R}$ zero outside a compact set.

Theorem 1.3.20. *Let $X = (X_t)_{t\in\mathbb{R}}$ be a centered second order stationary process with covariance γ. There exists on $\mathbb{R}$ a process with orthogonal increments Z with base μ such that, for all t,*

$$X_t = \int_{-\infty}^{\infty} e^{it\lambda}dZ(\lambda).$$

The spaces $H^X_{\mathbb{C}}$ and $H^Z_{\mathbb{C}}$ coincide; Z is an isometry from $L^2_{\mathbb{C}}(\mathbb{R},\mu)$ onto $H^X_{\mathbb{C}}$.

Definition 1.3.21. In the above two statements the measure μ is the **spectral measure** of X and the process Z is its **spectral process.** If μ has a density with respect to Lebesgue measure this density is the **spectral density** of X.

Notes. (a) If a stationary sequence has a spectral density f in $L^2_{\mathbb{C}}([-\pi,\pi[,L)$, $\gamma(n)/2\pi$ is the Fourier coefficient of the index $-n$ of f; $2\pi f$ is the limit in $L^2_{\mathbb{C}}([-\pi,\pi[,L)$, of $\lambda \longmapsto \Sigma^N_{-N}\gamma(n)e^{-in\lambda}$ as

N tends to ∞.

(b) If the stationary sequence X is real, the covariance is real but the spectral process Z is not. The spectral measure is symmetric, and

$$\gamma(n) = \int_{-\pi}^{\pi} \cos n\lambda \; d\mu(\lambda) = \gamma(-n),$$

$$X(n) = \int_{-\pi}^{\pi} \cos n\lambda \; dZ_1(\lambda) - \int_{-\pi}^{\pi} \sin n\lambda \; dZ_2(\lambda)$$

with $Z = Z_1 + iZ_2$, Z_1 and Z_2 real spectral processes, and

$$\int_{-\pi}^{\pi} \sin n\lambda \; dZ_1(\lambda) = -\int_{-\pi}^{\pi} \cos n\lambda \; dZ_2(\lambda) = 0.$$

If g is an even continuous function from $[-\pi,\pi]$ into $\mathbb{R}$, its Fourier coefficients are real and g is the uniform limit of linear combinations with real coefficients of the functions $\lambda \longmapsto \cos n\lambda$ ($n \in \mathbb{N}$): $Z(g)$ is real. Likewise, if g is an odd continuous function, for $iZ(g)$. Hence if g is a continuous function: $Z(g) = Z(g^s) + Z(g^i)$, g^s is the symmetric part of g, $g = g^s + g^i$, and $Z(g^s)$, $Z(g^i)$ are orthogonal. Real continuous functions being dense in $L^2([-\pi,\pi[,\mu)$, for g in this space (extended by $g(\pi) = g(-\pi)$) we again have the following decomposition.

If the stationary sequence X is real, $H^X = H_1^X \oplus H_2^X$, with $H_1^X = Z(L_e^2(\mu))$, $H_2^X = Z(L_0^2(\mu))$, $L_e^2(\mu)$ (resp. $L_0^2(\mu)$) real subspaces of $L^2(\mu)$ generated by even (resp. odd) functions.

1.3.3. Stationary Time Series

Examples. (a) $\int_{-\pi}^{\pi} e^{i\lambda n} d\lambda$ equals 0, except for $n = 0$: a white noise is a stationary sequence with spectral density $1/2\pi$.

(b) (ε_n) being a white noise, let us set $X_n = \varepsilon_n - a\varepsilon_{n-1}$ for real a. Then $\sigma^2(X_n) = 1 + a^2$, $\Gamma(X_n,X_{n-1}) = -a$ and $\Gamma(X_n,X_{n+k}) = 0$ for $k > 1$. The sequence $X = (X_n)$ is second order stationary and its spectral density is

$$f(\lambda) = \frac{1}{2\pi}(1 + a^2) - \frac{a}{2\pi}(e^{i\lambda} + e^{-i\lambda}) = \frac{1}{2\pi}|1 - ae^{i\lambda}|^2.$$

(c) Let (λ_k) be a sequence of distinct, non-zero elements of $[-\pi,\pi[$, and let (a_k) be a positive summable sequence.

Let $(Z_k)_{k\in\mathbf{Z}}$ be a sequence of centered, pairwise orthogonal r.v.'s, Z_k with variance a_k. Then Z defined by $Z(A) = \Sigma Z_k 1_{\{\lambda_k \in A\}}$ is a process with orthogonal increments with base $\mu = \Sigma_k a_k \delta_{\lambda_k}$. This is the spectral process of the stationary sequence (X_n) defined by,

$$X_n = \sum_k Z_k \exp(in\lambda_k).$$

The covariance of this sequence is defined by $\gamma(n) = \Sigma a_k \exp(in\lambda_k)$. The phenomenon (X_n) is the superposition of sine waves of the given frequencies λ_k and random (and orthogonal) amplitudes with variances a_k. In the general case we can associate with a stationary sequence (X_n), with spectral process Z, sequences $(X_n^{(p)}) = X^{(p)}$ with

$$X_n^{(p)} = \sum_{-2^p \leq k < 2^p} Z\left[\left[\frac{k\pi}{2^p}, \frac{(k+1)\pi}{2^p}\right]\right] \exp(ink\pi/2^p)$$

which is a process of the above type. X can be considered as a limit of the sequence $X^{(p)}$ if $p \to \infty$ ... without trying to state here in which sense ... .

(d) Let (ε_n) be a white noise and $c = (c_n)$ an ℓ^2 sequence. Let us set

$$X_n = \sum_{p=-\infty}^{\infty} c_{n-p} \varepsilon_p .$$

Then $X = (X_n)$ is a centered stationary sequence. The Fourier series $\Sigma_{-\infty}^{\infty} c_{-p} e^{ipx}$ converges. Let S be its sum. From Parseval's relation applied to the functions $x \longmapsto e^{inx}S(x)$ and $x \longmapsto e^{imx}S(x)$,

$$\Gamma(X_n, X_m) = \gamma(n-m) = \sum_{p=-\infty}^{\infty} c_{n-p}\bar{c}_{m-p}$$
$$= \frac{1}{2\pi}\int_{-\pi}^{\pi} e^{inx}e^{-imx}|S(x)|^2 dx.$$

Hence the spectral density of X is the function

$$\lambda \longmapsto \frac{1}{2\pi}\left|\sum_{-\infty}^{\infty} c_{-p}e^{ip\lambda}\right|^2 .$$

Conversely, if X has a spectral density f, $\sqrt{f}$ is in $L^2([-\pi,\pi[)$ and can be developed in a Fourier series as $\sqrt{f(x)} = (1/\sqrt{2\pi})\Sigma_{-\infty}^{\infty} c_{-p} e^{ipx}$. Parseval's relation again gives,

$$\Gamma(X_n, X_m) = \sum_{-\infty}^{\infty} c_{n-p}\bar{c}_{m-p} .$$

This is within the framework of the example which follows Theorem 1.2.16.

Theorem 1.3.22. *Let X be a centered stationary sequence. The following two conditions are equivalent.*

1. *X has a spectral density f.*
2. *There exists a white noise (ε_n) with spectral process Z_ε and a sequence (c_n) in ℓ^2 such that*

$$f(\lambda) = \frac{1}{2\pi}\left|\sum_{-\infty}^{\infty} c_{-p}e^{i\lambda p}\right|^2 .$$

1.3.4. Regular Processes

We have just studied the general case of a spectral density $f = |\sqrt{f}|^2$. In fact there exists an infinity of factorizations of f, $f(\lambda) = |\sqrt{f(\lambda)}\exp(i\theta(\lambda))|^2$. Let H^2 be the space of functions in $L^2_{\mathbb{C}}([-\pi,\pi[,L)$ the Fourier development of which contains only negative exponentials. Giving a sequence $(d_n)_{n\geq 0}$ of ℓ^2 is equivalent to giving a function h in H^2 defined by

$$h(\lambda) = \sum_{n=0}^{\infty} d_n e^{-in\lambda}.$$

Let (X_n) then be a centered stationary time series the spectral density of which may be written $f = (1/2\pi)|h|^2$. There exists a white noise process (ε_n) such that for all n,

$$X_n = \sum_{p=0}^{\infty} d_p \varepsilon_{n-p} .$$

Using the notations of [1.1.3], for all $n > 0$, we have

$$H^X_{-n} \subset H^\varepsilon_{-n}, \quad P_{-n}(X_0) = \sum_{p=n}^{\infty} d_p \varepsilon_{-p}.$$

Since white noise is regular, $P^\varepsilon_n(X_0)$ tends in $L^2_{\mathbb{C}}$ to 0 if $n \to -\infty$; the same holds for $P^X_n(X_0)$ and the time series is regular.

We are now going to show that in fact we have $H^X_n = H^\varepsilon_n$ for all n, and that there is at most one possible factorization of $2\pi f$ of the form $|h|^2$, $h \in H^2$ (up to the choice of sign for h).

Theorem 1.3.22. *A centered stationary sequence X is a regular process if and only if, there exists a sequence $(d_n)_{n \geq 0}$ of ℓ^2 satisfying one of the following equivalent conditions:*

1. *X has spectral density*

$$\lambda \longmapsto \frac{1}{2\pi}\left|\sum_{p=0}^{\infty} d_p e^{-ip\lambda}\right|^2 .$$

2. *There exists a white noise (ε_n) such that, for all n. We have*

$$X_n = \sum_{p=0}^{\infty} d_p \varepsilon_{n-p}.$$

There then exists (up to a sign) only one sequence $(d_p)_{p \in \mathbf{N}}$ and one sequence $(\varepsilon_n)_{n \in \mathbf{N}}$ satisfying (1) and (2). This white noise is such that, for all n, $H_n^X = H_n^\varepsilon$. It is called the **innovation white noise** *of the regular stationary sequence X.*

We have $d_p = \Gamma(X_0, \varepsilon_{-p})$. For $n > 0$,

$$P_{-n}^X(X_0) = \sum_{p=n}^{\infty} d_p \varepsilon_{-p} .$$

Proof. We have proved that (1) and (2) are equivalent and imply regularity. It remains to prove the converse and also that, under (1) or (2), $H_n^X = H_n^\varepsilon$ and that up to signs the sequences $(d_p)_{p \in \mathbf{N}}$ and $(\varepsilon_p)_{p \in \mathbf{N}}$ are unique.

Recall the linear prediction of [1.1.3], denote $H_n = H_n^X$, $-\infty \leq n \leq \infty$. For all n, let $H_n \ominus H_{n-1}$ be the orthogonal complement of H_{n-1} in H_n. Its dimension is at most 1 (0 if X_n is in H_{n-1}). Let us consider on H_∞ the isometry T (translation in time) defined by $T(X_n) = X_{n+1}$, for each n. For $Y = \Sigma_{i=1}^k c_i X_{n_i}$ $(k \in \mathbb{N}, (n_1, \dots, n_k) \in \mathbb{Z}^k)$,

$$T(Y) = \sum_{i=1}^{k} c_i X_{n_i+1}.$$

We have $T^n(H_k) = H_{n+k}$. Hence

$$T^n(H_0 \ominus H_{-1}) = H_n \ominus H_{n-1}.$$

If $H_0 \ominus H_{-1}$ has dimension zero, $H_0 = H_{-1}$; all the H_n are equal to $H_{-\infty}$. We are dealing with a singular process.

If $H_0 \ominus H_{-1}$ has dimension 1, let ε_0 be a centered r.v. with variance 1 which generates $H_0 \ominus H_{-1}$, and $\varepsilon_n = T^n\varepsilon_0$. The sequence (ε_n) is a white noise (ε_n is a basis of $H_n \ominus H_{n-1}$, hence the sequence is orthogonal). For $n > 0$, $\{\varepsilon_0, \ldots, \varepsilon_{-n+1}\}$ is thus a basis of $H_0 \ominus H_{-n}$ and

$$X_0 = \sum_{-n<p\leqslant 0} \Gamma(X_0,\varepsilon_p)\varepsilon_p + P_{-1}(X_0).$$

Thus

$$X_0 = P_{-\infty}(X_0) + \sum_{p\leqslant 0} \Gamma(X_0,\varepsilon_p)\varepsilon_p.$$

Let $d_p = \Gamma(X_0,\varepsilon_{-p})$, $a^2 = E(P_{-\infty}(X_0))^2$

$$\Gamma(X_n,X_0) = \gamma(n) = a^2 + \sum_{p\geqslant 0} d_p \bar{d}_{p+n}$$

and X has the spectral measure

$$a^2\delta_{\{0\}} + \frac{1}{2\pi}\left|\sum_{p\geqslant 0} d_p e^{i\lambda p}\right|^2 d\lambda.$$

Consequences. (a) If X is regular, it has a spectral density. (b) If X has a spectral density, there are two cases:

$$X \text{ is regular and } E(X_0 - P_{-1}(X_0))^2 \neq 0$$

$$X \text{ is singular and } E(X_0 - P_{-1}(X_0))^2 = 0.$$

When X is regular we have $X_n = \Sigma_{p\geqslant 0} d_p \varepsilon_{n-p}$. The series Σd_p^2 is convergent (Theorem 1.1.10). Setting $h(\lambda) = \Sigma_{p\geqslant 0} d_p e^{-ip\lambda}$, we have $f(\lambda) = (1/2\pi)|h(\lambda)|^2$. Note that the only normed vectors of $H_0 \ominus H_{-1}$ are ε_0 and $-\varepsilon_0$.

Let us assume now that $X_n = \Sigma_{p\geqslant 0} d'_p \varepsilon'_{n-p}$ for a particular white noise (ε'_p) and a sequence (d'_p). We have $X_0 = d'_0\varepsilon'_0 + X'_0$ with X'_0 orthogonal to $H_0 \ominus H_{-1}$ and $d'_0 \neq 0$. Thus $d'_0\varepsilon'_0$ is the projection of X_0 on $H_0 \ominus H_{-1}$ as $d_0\varepsilon_0$. Thus $d_0\varepsilon_0 = d'_0\varepsilon'_0$, $E(\varepsilon_0^2) = E(\varepsilon_0'^2) = 1$. Following from this, we have $d_0^2 = d_0'^2$ and $\varepsilon_0 = \pm\varepsilon'_0$. Finally $d'_p = \Gamma(X_p,\varepsilon'_0)$ and $\varepsilon'_n = T^n\varepsilon'_0$. From which, up to the sign, the uniqueness of the sequences (d_p) and (ε_n) satisfying (1) and (2) follows. Up to the signs there only exists a single representation of X in the form $\Sigma_{k\geqslant 0} d_k \varepsilon_{n-k}$, and a single factorization of f in the form $(1/2\pi)|h|^2$, $h \in \mathcal{H}^2$.

Prediction by n Past Observations. Let X be a centered stationary process. Let us denote by $X_{0,-n}$ the projection of X_0 on $H_{-1} \ominus H_{-(n+1)}$, the best linear approximation of X_0 by

$(X_{-1}, \ldots, X_{-n})$. Let Γ_n be the covariance of $(X_1, \ldots, X_n)$. The r.v. $X_{0,-n}$ is centered. Let us denote by σ_n^2 the variance of $X_0 - X_{0,-n}$. We have

$$\lim_n \sigma_n^2 = \lim_n E[X_0 - X_{0,-n}]^2 = E[X_0 - P_{-1}^X(X_0)]^2.$$

Thus (σ_n^2) is a decreasing sequence. Its limit is zero if, and only if, the process is singular.

If the process is not singular, H_{-n-1}^X is, for all n, a strict subspace of H_{-n}^X. Thus the random vector $(X_{-1}, \ldots, X_{-n})$ is of rank n. We thus know how to calculate $X_{0,-n}$ [Vol. I, p. 139] and we have

$$\sigma_n^2 = E(X_0^2) - E(X_{0,-n}^2) = E(X_0^2) - {}^tU_n(\Gamma_n)^{-1}U_n$$

with ${}^tU_n = (E(X_0X_1), \ldots, E(X_0X_n))$.

We have

$$\Gamma_{n+1} = \begin{bmatrix} E(X_0^2) & {}^tU_n \\ U_n & \Gamma_n \end{bmatrix},$$

and

$$|\Gamma_{n+1}| = |\Gamma_n|(E(X_0^2) - {}^tU_n(\Gamma_n)^{-1}U_n).$$

From which $\sigma_n^2 = |\Gamma_{n+1}|/|\Gamma_n|$ and,

$$\begin{aligned} \text{Log } E(X_0 - P_{-1}^X(X_0))^2 &= \lim_n \frac{1}{n} \sum_{k=1}^n \text{Log } \sigma_k^2 \\ &= \lim_n \frac{\text{Log}|\Gamma_{n+1}| - \text{Log}|\Gamma_1|}{n}. \end{aligned}$$

Proposition 1.3.23. *Let X be a regular centered stationary sequence and Γ_n the covariance of $(X_1, \ldots, X_n)$. We have,*

$$\lim_{n\to\infty} \frac{1}{n} \text{Log}|\Gamma_n| = \text{Log } E(X_0 - P_{-1}^X(X_0))^2.$$

In order for a stationary sequence to be regular, it must have a spectral density (from Theorem 1.3.22). Here is an analytic criterion of regularity.

Theorem 1.3.24 (Szegö) - Regularity Criterion. (a) *Let X be a* *stationary sequence with spectral density f. We have*

$$E[X_0 - P^X_{-1}(X_0)]^2 = \exp\left[\frac{1}{2\pi}\int_{-\pi}^{\pi}\operatorname{Log}[2\pi f(\lambda)]d\lambda\right].$$

(b) *A stationary sequence is regular if, and only if, it has a spectral density f such that*

$$\int_{-\pi}^{\pi}\operatorname{Log} f(\lambda)d\lambda > -\infty.$$

Proof. Denote L^p for $L^p[(-\pi,\pi), (1/2\pi)L]$ and $\hat{f} = 2\pi f$. Assume X is centered. From Theorem 1.3.22, X is regular if, and only if, X has a spectral density f and $\hat{f} = |h|^2$, $h \in \mathcal{H}^2$. Let

$$d_0^2 = E[X_0 - P^X_{-1}(X_0)]^2.$$

Recall the notations which precede Proposition 1.3.23. The r.v. $X_{0,-n}$ is centered and, because of the isometry between $H^X_{\mathbb{C}}$ and $L^2_{\mathbb{C}}((-\pi,\pi),f.L)$, we have,

$$\sigma_n^2 = \inf\left\{\int_{-\pi}^{\pi}\left|1 - \sum_{k=1}^{n}\alpha_k e^{-i\lambda k}\right|^2 f(\lambda)d\lambda;\ (\alpha_k)_{1\leqslant k\leqslant n} \in \mathbb{C}^n\right\}$$

and the lower bound may be taken on the polynomials $\lambda \longmapsto \sum_{k=1}^n \alpha_k e^{-i\lambda k}$ with integral zero for L. Let $\mathcal{P}$ be the set of these polynomials, for arbitrary integer n. We have,

$$d_0^2 = E(X_0 - P^X_{-1}(X_0))^2 = \inf_{p\in\mathcal{P}} \int |1 - p(\lambda)|^2 f(\lambda)d\lambda.$$

Let A be the set of continuous functions obtained as uniform limits of elements of $\mathcal{P}$, then,

$$d_0^2 = \inf_{g\in A} \frac{1}{2\pi}\int_{-\pi}^{\pi}|1 - g(\lambda)|^2\hat{f}(\lambda)d\lambda.$$

Let $L_0^1 = \{g \in L^1,\ \int_{-\pi}^{\pi} g(\lambda)d\lambda = 0\}$. The following properties hold:

$$\text{if } \phi \in A,\ \operatorname{Re}\phi \in L_0^1 \text{ and } 1 - e^{\phi} \in A;$$

$$\text{if } (\phi,\psi) \in A^2,\quad \int_{-\pi}^{\pi}\phi(\lambda)\psi(\lambda)d\lambda = 0.$$

Since Log f is majorized by $f - 1$, it is quasi-integrable. Its integral is in $[-\infty,\infty[$. Let us denote

$$C = \exp \frac{1}{2\pi} \int_{-\pi}^{\pi} \text{Log}(\hat{f}(\lambda))d\lambda.$$

Let us assume first of all that Log $f \in L^1$. For $g \in L_0^1$, Jensen's inequality gives

$$\text{Log } C = \frac{1}{2\pi} \int_{-\pi}^{\pi} \text{Log}(e^{g(\lambda)}\hat{f}(\lambda))d\lambda$$

$$\leqslant \text{Log } \frac{1}{2\pi} \int_{-\pi}^{\pi} \hat{f}(\lambda)e^{g(\lambda)}d\lambda.$$

However,

$$C = \inf_{h \in L_0^1} \frac{1}{2\pi} \int_{-\pi}^{\pi} \hat{f}(\lambda)e^{h(\lambda)}d\lambda,$$

the infimum being attained for $h = 2\pi$ Log C − Log $\hat{f}$. Let us show that we can successively replace ($h \in L_0^1$) in the infimum by ($h \in L_0^1$ bounded) then by (h = Re ψ, $\psi \in A$). Let $h = h^+ - h^-$, $h^+ = h \vee 0$ is the bounded increasing limit of a sequence h_n^+, $h_n^+ \geqslant 0$. $h^- = -h \vee 0$ is the bounded increasing limit of $h_n^- \geqslant 0$.

Since $\int h^+(\lambda)d\lambda = \int h^-(\lambda)d\lambda$, there exists a sequence $m(n)$ of integers and a sequence c_n tending to 1 such that

$$\int h_n^+(\lambda)d\lambda = \int c_n h_{m(n)}^-(\lambda)d\lambda.$$

Let us set $k_n^- = c_n h_{m(n)}^-$; $\exp(h^+ - h_n^+ - h^- + k_n^-)$ is majorized by 1 on $h \leqslant 0$, and by e^h on $h > 0$. We obtain,

$$C = \inf_{g \in L_0^1,\ g \text{ bounded}} \frac{1}{2\pi} \int_{-\pi}^{\pi} \hat{f}(\lambda)e^{g(\lambda)}d\lambda.$$

Let f be a function such that $|f| \leqslant M$. It is the limit of a sequence of continuous functions bounded by M since

$$f(x) = \lim_{\sigma \to 0} \frac{1}{\sigma\sqrt{2\pi}} \int \exp\left[-\frac{(x-y)^2}{2\sigma^2}\right] f(y)dy.$$

from which, using the diagonal process, it is the limit of a sequence of trigonometric polynomials bounded by $2M$. Hence

$$C = \inf_{g \in A_0} \frac{1}{2\pi} \int_{-\pi}^{\pi} \hat{f}(\lambda)\exp(\text{Re } g(\lambda))d\lambda.$$

This relation remains true if Log f is not integrable (apply the formula to $\hat{f} + a$, for all $a > 0$).

Let $g \in A$. Let us apply the preceding relation to $\hat{f} = |1 - g|^2$,

$$\exp\left[\frac{1}{2\pi}\int_{-\pi}^{\pi} \operatorname{Log}|1 - g(\lambda)|^2 d\lambda\right]$$

$$= \inf_{h\in A} \frac{1}{2\pi}\int_{-\pi}^{\pi} |1 - g(\lambda)|^2 e^{2\mathrm{Re}\, h(\lambda)} d\lambda$$

$$\geqslant \inf_{h\in A} \left|\frac{1}{2\pi}\int_{-\pi}^{\pi} (1 - g(\lambda))\, e^{h(\lambda)} d\lambda\right|^2.$$

However $e^h - 1$ is in A and

$$\frac{1}{2\pi}\int_{-\pi}^{\pi} (1 - g(\lambda))(e^{h(\lambda)} - 1 + 1) d\lambda = 1;$$

$$\exp\left[\frac{1}{2\pi}\int_{-\pi}^{\pi} \operatorname{Log}(|1 - g(\lambda)|^2) d\lambda\right] \geqslant 1.$$

Now let us set $G(\lambda) = \operatorname{Log}|1 - g(\lambda)|^2$ and

$$\gamma = \frac{1}{2\pi}\int_{-\pi}^{\pi} G(\lambda) d\lambda \geqslant 0.$$

We have

$$\frac{1}{2\pi}\int_{-\pi}^{\pi} |1 - g(\lambda)|^2 \hat{f}(\lambda) d\lambda$$

$$\geqslant \frac{1}{2\pi}\int_{-\pi}^{\pi} \exp[G(\lambda) - \gamma]\hat{f}(\lambda) d\lambda$$

$$\geqslant \frac{1}{2\pi}\inf\left\{\int_{-\pi}^{\pi} \exp[\phi(\lambda)]\hat{f}(\lambda) d\lambda; \quad \phi \in L_0^1\right\}$$

$$= C.$$

Hence $C \leqslant d_0^2$.

Conversely, note that $g \in A$ implies $1 - e^g \in A$ and $|e^g|^2 = e^{2\mathrm{Re}\, g}$. From which,

$$d_0^2 = \inf_{g\in A} \frac{1}{2\pi}\int_{-\pi}^{\pi} |1 - g(\lambda)|^2 \hat{f}(\lambda) d\lambda$$

$$\leqslant \inf_{g\in A} \frac{1}{2\pi}\int_{-\pi}^{\pi} e^{2\mathrm{Re}\, g(\lambda)} \hat{f}(\lambda) d\lambda = C.$$

1.3.5. Linear Filters

Let X be a stationary sequence with spectral measure μ and spectral process Z_X and let T be the translation which associates with X, $T(X) = (Y_n)_{n \in \mathbf{Z}}$ with $Y_n = X_{n+1}$. We have

$$Y_n = \int_{-\pi}^{\pi} e^{i\lambda(n+1)} dZ_X(\lambda)$$

thus the spectral measure of $T(X)$ is also μ, its spectral process is $Z_{T(X)} = e^{i\lambda} Z_X$.

We can also study weighted moving averages of the form

$$Y_n = a_{-k} X_{n-k} + \dots + a_0 X_n + \dots + a_h X_{n+h}$$

which clearly define a new stationary sequence. If $k = h$ and $a_j = 1/(2k+1)$, $|j| \leqslant k$, such a sequence is said to be a **smoothing average** of X.

Using the representation of X, we see that,

$$Y_n = \int_{-\pi}^{\pi} e^{i\lambda n} h(\lambda) dZ_X(\lambda)$$

with $h(\lambda) = a_{-k} e^{-ik\lambda} + \dots + a_h e^{ih\lambda}$.

The following result generalizes these examples.

Theorem 1.3.25. *Let X be a stationary time series with spectral measure μ and spectral process Z_X. Let $h \in L^2_{\mathbb{C}}([-\pi,\pi[,\mu)$. The time series Y defined by*

$$Y_n = \int_{-\pi}^{\pi} e^{in\lambda} h(\lambda) Z_X(d\lambda)$$

is stationary. Its spectral measure is $|h|^2 \cdot \mu$, its spectral process is defined, for every Borel set A in $[-\pi,\pi[$, by

$$Z_Y(A) = Z_X(h \cdot 1_A).$$

We say that Y is obtained from X by the filter with response function h (or that Y is filtered from X by h).

In general, the information given by Y is weaker than that given by X: $H^Y_{\mathbb{C}} \subset H^X_{\mathbb{C}}$. The equality $H^Y_{\mathbb{C}} = H^X_{\mathbb{C}}$ is satisfied if, and only if, $\mu(h = 0) = 0$. Then, X is obtained from Y by the filter with response function $1/h$.

Proof. It is again sufficient to apply Karhunen's theorem since the covariance of Y is

$$\gamma_Y(n) = \int_{-\pi}^{\pi} e^{i\lambda n}|h(\lambda)|^2 d\mu(\lambda).$$

The space $H_{\mathbb{C}}^Y$ coincides with $H_{\mathbb{C}}^Z = H_{\mathbb{C}}^X$ if, and only if, the Hilbert subspace K generated by the functions $\lambda \longmapsto e^{i\lambda n}h(\lambda)$, for $n \in \mathbb{Z}$, is dense in $L_{\mathbb{C}}^2(\mu)$. Assume $\mu(h = 0) = 0$. Let $f \in L_{\mathbb{C}}^2(\mu)$ be orthogonal to K. We have, for all $n \in \mathbb{Z}$,

$$\int_{-\pi}^{\pi} e^{i\lambda n} f(\lambda)h(\lambda)d\mu(\lambda) = \int_{-\pi}^{\pi} e^{i\lambda n}(f/\bar{h})(\lambda)|h(\lambda)|^2 d\mu(\lambda)$$
$$= 0.$$

Thus f is zero in $L^2(|h|^2 \cdot \mu)$, hence zero a.s. Conversely, if μ charges $\{h = 0\}$, all the functions $\lambda \longmapsto e^{i\lambda n}h(\lambda)$ are zero on this set, and the same holds for all the functions of K. K is a strict subspace of $L_{\mathbb{C}}^2(\mu)$.

If μ does not charge $\{h = 0\}$, i.e., if $H_{\mathbb{C}}^X = H_{\mathbb{C}}^Y$, the function $1/h$ is in $L^2(|h|^2 \cdot \mu)$ and X is obtained from Y by the filter with response function $1/h$.

Examples. (a) Let $B \in \mathcal{B}_{[-\pi,\pi[}$ and let $h = 1_B$: then Y is the **band pass filter** B. The process Y only allows frequencies of Z lying in B to get through. The frequencies lying in B^c have been "filtered." If μ charges B^c, $H_{\mathbb{C}}^Y$ is strictly included in $H_{\mathbb{C}}^X$.

(b) Theorems 1.3.21 and 1.3.22 imply that if X has a spectral density $f = (1/2\pi)|h|^2$, there exists a white noise ε such that X is filtered by the response function h. If h can be taken in the space H^2, we can take for ε the innovation white noise ($H_n^\varepsilon = H_n^X$ for all n).

If the stationary sequence X is observed, at time n, only H_n^X is observable. A filtered process Y will only be observable at each instant for $Y_n \in H_n^X$. By stationarity this is equivalent to $Y_0 \in H_0^X$.

Definition 1.3.26. A filter which transforms X to Y is said to be **adapted** (or **causal**) if, for all n, we have: $Y_n \in H_n^X$ (or if $Y_0 \in H_0^X$).

Note simply that, for a regular process X, we have $dZ_X(\lambda) = h(\lambda)dZ_\varepsilon(\lambda)$ where ε is the innovation white noise of X and $h \in H^2$. Let $g \in L_{\mathbb{C}}^2((-\pi,\pi), f \cdot L)$ be the response function of the filter giving Y. If $hg \in H^2$, then the filter transforming ε to

Y is causal.

For example, if $L(h = 0)$ is zero, $H^X = H^\varepsilon$ and ε is deduced from X by the filter with response function $1/h$. For f bounded and $1/h$ in H^2, this filter is adapted.

Proposition 1.3.27. *A regular process X admits an infinite autoregressive representation, ε being its innovation white noise,*

$$\varepsilon_n = \sum_{k=0}^{\infty} d_k X_{n-k},$$

if, and only if, the spectral density f can be written $f = |h|^2/2\pi$, where we have in $L^2([-\pi,\pi], f\cdot L)$,

$$\frac{1}{h}(\lambda) = \sum_{k=0}^{\infty} d_k e^{-ik\lambda}, \quad \textit{with } d_0 \neq 0.$$

Proof. Filtering X by $1/h$, we obtain ε,

$$\varepsilon_n = \sum_{k=0}^{\infty} d_k X_{n-k}.$$

Example. Let $a \in \mathbb{C}$ and $X_n = \varepsilon_n - a\varepsilon_{n-1}$, (ε_n) white noise,

$$h(\lambda) = 1 - ae^{-i\lambda}.$$

For $|a| < 1$,

$$\frac{1}{h}(\lambda) = \sum_{n=0}^{\infty} a^n e^{-in\lambda};$$

(ε_n) is the innovation, $d_n = a^n$. For $|a| > 1$, (ε_n) is no longer the innovation. We have,

$$\frac{1}{h}(\lambda) = -\frac{1}{ae^{-i\lambda}}\left[1 - \frac{1}{a}e^{i\lambda}\right]^{-1} = \sum_{n=0}^{\infty} \frac{1}{a^{n+1}} e^{i(n+1)\lambda}$$

and the representation of ε_n is in terms of functions of the variables X_{n+k}, $k > 0$. Finally if $|a| = 1$, $H_0^X = H_0^\varepsilon$, however we can no longer write ε as a series in X, ε_0 is here only the limit of a sequence of elements of H_0^X.

Proposition 1.3.28. Composition of Filters. *Let X be a stationary sequence with spectral measure μ. For $h \in L^2_{\mathbb{C}}(\mu)$ and $g \in L^2_{\mathbb{C}}(h\cdot\mu)$, hg is in $L^2_{\mathbb{C}}(\mu)$ and the filter Z with response function hg is obtained by filtering X by the response function h, then by filtering the series Y thus obtained by the response function g.*

Proof. It is sufficient to apply Theorem 1.3.25,

$$Y_n = \int_{-\pi}^{\pi} e^{in\lambda} h(\lambda) dZ_X(\lambda)$$

$$Z_n = \int_{-\pi}^{\pi} e^{in\lambda} g(\lambda) dZ_Y(\lambda) = \int_{-\pi}^{\pi} e^{in\lambda} h(\lambda) g(\lambda) dZ_X(\lambda).$$

1.3.6. ARMA Processes

As a generalization of regression, used for example in economics, we have autoregressive models of order p, AR(p), met in [1.1.2]. Let us recall their definition.

Definition 1.3.29. A sequence (X_n) of complex r.v.'s is an **autoregressive process of order** p, (AR(p)), if it is a centered second order stationary sequence and if there exists a white noise (ε_n) and $(a_0, a_1, ..., a_p) \in \mathbb{C}^{p+1}$ such that, for all n,

$$\sum_{k=0}^{p} a_k X_{n-k} = \varepsilon_n.$$

For a regular process (X_n), we have obtained a representation $X_n = \sum_{k=0}^{\infty} b_k \varepsilon_{n-k}$ with the innovation white noise (ε_n). Assume that the projection of X_0 on H^X_{-q-1} is zero. This implies that X_n and X_m are only correlated for $|n - m| \leq q$; and since $H^X_{-q-1} = H^\varepsilon_{-q-1}$, X_n and ε_{n-k} are only correlated for $|k| \leq q$. From Theorem 1.3.22, we have $b_k = \Gamma(X_n, \varepsilon_{n-k})$. From which,

$$X_n = \sum_{k=0}^{q} b_k \varepsilon_{n-k}.$$

Definition 1.3.30. A sequence (X_n) of r.v.'s is a **moving average of order** q (MA(q)) if there exists a white noise (ε_n) and $(b_0, ..., b_q) \in \mathbb{C}^{q+1}$ such that

$$X_n = \sum_{k=0}^{q} b_k \varepsilon_{n-k}.$$

These two types of sequences are part of the class of stationary "ARMA" sequences.

Definition 1.3.31. A centered second order stationary sequence (X_n) is an ARMA(p,q) **process** if there exists a white noise (ε_n) and $(a_0, ..., a_p, b_0, ..., b_q) \in \mathbb{C}^{p+q+2}$ such that, for all n,

$$\sum_{k=0}^{p} a_k X_{n-k} = \sum_{j=0}^{q} b_j \varepsilon_{n-j}.$$

Note that the definition does not require (ε_n) to be the innovation noise.

Proposition 1.3.32. Spectral Processes of ARMA Processes. *Let (X_n) be an ARMA process associated with the coefficients $a_0, \ldots, a_p, b_0, \ldots, b_q$. Set $P(z) = \Sigma_{k=0}^{p} a_k z^k$, $Q(z) = \Sigma_{j=0}^{q} b_j z^j$. Let μ be the spectral measure of (X_n). Let us assume that there does not exist a $\lambda \in [-\pi,\pi[$ simultaneously satisfying $\mu(\{\lambda\}) > 0$ and $P(e^{-i\lambda}) = 0$. Then μ is absolutely continuous with respect to Lebesgue measure, and has density*

$$\lambda \longmapsto \frac{1}{2\pi}\left|\frac{Q}{P}(e^{-i\lambda})\right|^2.$$

Proof. From Theorem 1.3.22, $Y_n = \Sigma_{j=0}^{q} b_j \varepsilon_{n-j}$ has spectral density $\lambda \longmapsto (1/2\pi)|Q(e^{-i\lambda})|^2$. If μ is the spectral measure of (X_n), the spectral measure of (Y_n) also has the density $\lambda \longmapsto |P(e^{-i\lambda})|^2$ with respect to μ. The function $\lambda \longmapsto |P(e^{-i\lambda})|^2$ is μ-a.s. nonzero. Hence the equality of the measures

$$|P(e^{-i\lambda})|^2 \mu(d\lambda) \quad \text{and} \quad \frac{1}{2\pi}|Q(e^{-i\lambda})|^2 d\lambda$$

implies the proposition. Moreover, $H_{\mathbb{C}}^X = H_{\mathbb{C}}^Y$ and X is the process Y filtered by the response function $\lambda \longmapsto 1/P(e^{-i\lambda})$. Since Y is the white noise ε filtered by $\lambda \longmapsto Q(e^{-i\lambda})$, X is ε filtered by $\lambda \longmapsto Q/P(e^{-i\lambda})$. Finally, if Q is not the zero polynomial, the measure L does not charge the set of zeros of $\lambda \longmapsto Q(e^{-i\lambda})$. Hence $H_{\mathbb{C}}^X = H_{\mathbb{C}}^Y = H_{\mathbb{C}}^{\varepsilon}$, and ε is X filtered by $\lambda \longmapsto P/Q(e^{-i\lambda})$.

Notes. (a) Several ARMA representations may correspond to the same stationary sequence. However, within the framework of Proposition 1.3.32, if X is nonzero, the white noise ε such that the relation of Definition 1.3.31 is satisfied with coefficients $a_0, \ldots, a_p, b_0, \ldots, b_q$ is unique.

(b) Let us assume that there exists a unique λ_0 such that $\mu(\lambda_0) > 0$ and $P(e^{i\lambda_0}) = 0$. Let A be a centered r.v. with variance $\mu(\lambda_0)$. If X is an ARMA process associated with $(a_0, \ldots, a_p, b_0, \ldots, b_q)$ with spectral measure μ, the sequence $\tilde{X}_n = (X_n - Ae^{in\lambda_0})$ is an ARMA process associated with the same coefficients and with spectral measure $\tilde{\mu} - \mu(\lambda_0)\delta_{\lambda_0}$, hence $\tilde{\mu}$ has density.

$$\lambda \longmapsto \frac{1}{2\pi}\left|\frac{Q}{P}(e^{-i\lambda})\right|^2.$$

(X and A are chosen to be orthogonal.)

Regular ARMA Representation. Consider a stationary sequence X with spectral density

$$\lambda \longmapsto \frac{1}{2\pi}\left|\frac{Q}{P}(e^{-i\lambda})\right|^2,$$

where P and Q are two polynomials with no common factors. P does not have a root of modulus 1 and f is bounded; h: $\lambda \longmapsto Q/P(e^{-i\lambda})$ is in H^2. P and Q can be taken to have no root in the unit open disc since

$$|e^{-i\lambda} - \alpha| = |\alpha|\left|e^{-i\lambda} - \frac{1}{\bar{\alpha}}\right|.$$

If Q does not have a root of modulus 1, $1/h$ is in H^2. It can be seen that, even if Q has roots of modulus 1, $1/h$ is the limit in $L^2([-\pi,\pi], f\cdot L)$ of polynomials in $e^{-i\lambda}$. The innovation ε is thus X filtered by $1/h$ and filtering X by P and ε by Q, we have an ARMA representation of X using its innovation white noise.

Finally, from Proposition 1.3.27, X admits an infinite autoregressive representation when P/Q is in H^2, hence when Q has no zeros in the unit disc.

Theorem 1.3.33. *Let P and Q be two polynomials*

$$P(z) = \sum_{k=0}^{p} a_k z^k, \quad Q(z) = \sum_{j=0}^{q} b_j z^j.$$

Assume that P and Q are relatively prime, P and Q only having zeros with modulus greater than 1.

(a) *Let $X = (X_n)$ be a stationary sequence with spectral density*

$$\lambda \longmapsto \frac{1}{2\pi}\left|\frac{Q}{P}(e^{-i\lambda})\right|^2.$$

Then X has an extra ARMA *representation using its innovation white noise $\varepsilon = (\varepsilon_n)$ of the form*

$$\sum_{k=0}^{p} a_k X_{n-k} = \sum_{k=0}^{q} b_j \varepsilon_{n-j}.$$

(b) *The process X admits an infinite autoregressive representation if and only if Q does not have any zeros in the* *closed unit disc.*

Example. Take $a \in \mathbb{C}$ with modulus different from 1, $P(z) = 1 - az$, $Q(z) = 1$:

$$\frac{Q}{P}(e^{-i\lambda}) = \frac{1}{1 - ae^{-i\lambda}} = h(\lambda)$$

is a function in $L^2_{\mathbb{C}}$ and there exists an ARMA process X with spectral density $(1/2\pi)|h|^2$ and a white noise ε such that:

$$X_n - aX_{n-1} = \varepsilon_n .$$

For $|a| < 1$, ε is the innovation white noise:

$$h(\lambda) = \sum_{n=0}^{\infty} a^n e^{-in\lambda} \quad \text{and} \quad X_n = \sum_{p=0}^{\infty} a^p \varepsilon_{n-p} .$$

For $|a| > 1$, ε is not the innovation white noise.

We then have:

$$\bar{a}X_n - X_{n-1} = \eta_n, \ (\eta_n) \text{ being the innovation.}$$

1.4. Time Series Statistics

1.4.1. Estimation of the Mean of a Second Order Stationary Sequence

n observations $X_1, \ldots, X_n$ are taken of a stationary sequence X with spectral measure μ, in order to estimate its characteristics. If m is the mean of X, an unbiased estimator of m is the empirical or sample mean $\bar{X}_n = (X_1 + \ldots + X_n)/n$.

In order to study these empirical estimators, let us consider a linear combination $\Sigma_{p=1}^n a_p X_p$, for $(a_p)_{1 \leq p \leq n} \in \mathbb{C}^n$. By using Z, the spectral process of $X - m$, it follows:

$$\sigma^2\left[\sum_{p=1}^{n} a_p X_p\right] = E\left[\sum_{p=1}^{n} a_p(X_p - m)\right]^2$$

$$= \int_{-\pi}^{\pi} \left|\sum_{p=1}^{n} a_p e^{iup}\right|^2 d\mu(u).$$

For example

$$\sigma^2(\overline{X}_n) = \int_{-\pi}^{\pi} \left|\frac{1}{n} \sum_{p=1}^{n} e^{iup}\right|^2 \mu(du).$$

We have

$$\left|\frac{1}{n} \sum_{p=1}^{n} e^{iup}\right|^2 = \frac{2\pi}{n} \Phi_n(u),$$

where Φ_n is the Fejer kernel defined in [1.1.5]. For all $\delta > 0$:

$$\mu(\{0\}) \leq \sigma^2(\overline{X}_n) \leq \mu([-\delta,\delta]) + \frac{2(\pi - \delta)}{n^2(\sin(\delta/2))^2} .$$

Passing to the limit in n, then in δ, we see that $\sigma^2(\overline{X}_n)$ converges to $\mu(\{0\})$. The empirical estimator of the mean m converges in quadratic mean if, and only if, the spectral measure does not charge 0. If μ charges 0, consider $Z_0 = Z(\{0\})$, the centered r.v. with variance $\mu(\{0\})$, and $Y = (Y_n)$ the stationary sequence with spectral process $A \longmapsto Z(A) - Z_0 1_A(0)$. The sequence Y is orthogonal to Z_0 and from the preceding result, $\overline{Y}_n \xrightarrow{L^2} m$, and thus $\overline{X}_n \xrightarrow{L^2} (m + Z_0)$. The sequence X has a random component Z_0 which cannot be detected while observing only a trajectory; from the statistical point of view, if $Z_0(\omega) = z$, since only the trajectory ω is observed, the model X studied here cannot be distinguished from the sequence $(Y_n + z)$. It is possible to distinguish the two models by observing several trajectories and by estimating the variance of the limit of $(\overline{X}_n)$. In what follows, we are only interested in the observation of a single trajectory; hence we assume that μ does not charge 0.

In particular, if the sequence has a spectral density f, then,

$$\sigma^2(\overline{X}_n) = \frac{2\pi}{n} \int_{-\pi}^{\pi} \Phi_n(u) f(u) du.$$

Thus, from [1.1.5], if f is continuous at 0,

$$\sigma^2(\overline{X}_n) = \frac{2\pi}{n} f(0) + o\left[\frac{1}{n}\right].$$

The condition $n\sigma^2(\overline{X}_n) \to 2\pi f(0)$ implies that $\overline{X}_n \xrightarrow{a.s.} m$ from the following lemma.

Lemma 1.4.34. *Let* (X_n) *be a sequence of centered complex* r.v.'s, *the variances of which form a bounded sequence. Let* $\overline{X}_n$ *be the sample mean at time n. Assume the sequence* $(nE(|\overline{X}_n|^2))$ *is bounded. Then we have the law of large numbers*:

$$\overline{X}_n \xrightarrow{\text{a.s.}} 0.$$

Proof. Let $M = \sup\{nE(|\overline{X}_n|)^2\}$. Applying Tchebyschev's inequality:

$$P[|\overline{X}_{n^2}| \geqslant \varepsilon n^{-1/4}] \leqslant \frac{\sqrt{n}}{\varepsilon^2} E(|\overline{X}_{n^2}|^2).$$

The term on the right is majorized by $M\varepsilon^{-2}n^{-3/2}$. The series $\Sigma P(|\overline{X}_{n^2}|) \geqslant \varepsilon n^{-1/4})$ converges. By the Borel-Cantelli lemma, we have a.s. from a certain point onwards $|\overline{X}_{n^2}| < \varepsilon n^{-1/4}$. This is true for all $\varepsilon > 0$ and $\overline{X}_{n^2}$ tends a.s. to 0. Let

$$Z_n = \sup_{n^2+1 \leqslant k < (n+1)^2} |\overline{X}_k| \quad \text{and} \quad K = \sup_n E(|X_n|^2)$$

$$= \sup_{n,m} E(|X_n||X_m|):$$

$$Z_n \leqslant \frac{1}{n^2}[|X_{n^2+1}| + |X_{n^2+2}| + \dots + |X_{(n+1)^2}|] + |\overline{X}_{n^2}|$$

$$E(Z_n^2) \leqslant \frac{1}{2}\frac{K}{n^4} 2n(2n+1) + \frac{M}{n^2} + \frac{2\sqrt{MK}}{n^3} 2n.$$

From which

$$P[Z_n \geqslant \varepsilon n^{-1/4}] \leqslant \frac{\sqrt{n}}{\varepsilon^2} E[Z_n^2],$$

and, as above, $(Z_n) \xrightarrow{\text{a.s.}} 0$. From which: $(\overline{X}_n) \xrightarrow{\text{a.s.}} 0$.

Let us assume that the sequence X is Gaussian; $\sqrt{n}(\overline{X}_n - m)$ is a centered Gaussian r.v., the variance of which tends to $2\pi f(0)$ and $\sqrt{n}(\overline{X}_n - m) \xrightarrow{\mathcal{D}} N(0, 2\pi f(0))$.

Theorem 1.4.35. (1) *The empirical estimator* $\overline{X}_n = (X_1+\dots+X_n)/n$ *of the mean* m *of a stationary sequence with spectral measure* μ *converges in quadratic mean to* m *if, and only if,* μ *does not charge* 0.

(2) *If* X *has a spectral density* f *continuous at* 0:

$$\sigma^2(\overline{X}_n) = \frac{2\pi}{n} f(0) + o\left[\frac{1}{n}\right];$$

$$\overline{X}_n \xrightarrow{\text{a.s.}} m.$$

If moreover X *is Gaussian, we have:*

$$\sqrt{n}(\overline{X}_n - m) \xrightarrow{\mathcal{D}} N(0, 2\pi f(0)).$$

1.4.2. Estimation of the Covariance of a Stationary Gaussian Sequence

Consider a real, second order sequence (X_n), stationary in the strict sense. Let m be its mean, γ its covariance.

In order to estimate $\gamma(p) = \gamma(-p)$, for $p \in \mathbb{N}$, it is natural to use the **empirical** (or **sample**) **covariances**,

$$c_n(p) = \frac{1}{n} \sum_{k=1}^{n-p} (X_k - m)(X_{k+p} - m) \quad \text{if } m \text{ is known;}$$

$$\overline{c}_n(p) = \frac{1}{n} \sum_{k=1}^{n-p} (X_k - \overline{X}_n)(X_{k+p} - \overline{X}_n) \quad \text{if } m \text{ is unknown.}$$

If X is centered and if the r.v.'s (X_n) have moments of order 4, for all $p \in \mathbb{N}$, the sequence $(X_n X_{n+p})$ is stationary with mean $\gamma(p)$. Convergence in quadratic mean of $c_n(p)$ to $\gamma(p)$ is assured if this sequence has a spectral measure which does not charge 0.

If X is centered, Gaussian, we calculate (cf. [Vol. I, E.5.2.2]):

$$\begin{aligned} E(X_{n_1}X_{n_2}X_{n_3}X_{n_4}) &= E(X_{n_1}X_{n_2})E(X_{n_3}X_{n_4}) \\ &+ E(X_{n_1}X_{n_3})E(X_{n_2}X_{n_4}) \\ &+ E(X_{n_1}X_{n_4})E(X_{n_2}X_{n_3}). \end{aligned}$$

The covariance γ_p of $(X_n X_{n+p})$ is thus given by,

$$\begin{aligned} \gamma_p(k) &= E[X_n X_{n+p} X_{n+k} X_{n+p+k}] \\ &\quad - E(X_n X_{n+p})E(X_{n+k} X_{n+p+k}) \\ &= \gamma^2(k) + \gamma(k+p)\gamma(k-p). \end{aligned}$$

Assume X is centered Gaussian, with spectral density f. Then,

$$\gamma_p(k) = \left[\int_{-\pi}^{\pi} e^{ik\lambda} f(\lambda)d\lambda\right]^2$$

$$+ \left[\int_{-\pi}^{\pi} e^{ik\lambda} e^{ip\lambda} f(\lambda)d\lambda\right]\left[\int_{-\pi}^{\pi} e^{ik\lambda} e^{-ip\lambda} f(\lambda)d\lambda\right];$$

$$\gamma_p(k) = \int_{-\pi}^{\pi} e^{ik\lambda} g_p(\lambda)d\lambda,$$

with $g_p = 2\pi[f*f + (f_p f)*(f_{-p}f)]$ denoting by f_p the function $\lambda \longmapsto e^{ip\lambda}$ (see [1.1.5] for the notation $*$). Hence $(X_n X_{n+p})_{n\in\mathbb{Z}}$ is a stationary sequence, with mean $\gamma(p)$, and with density g_p. If f is continuous, g_p is continuous at 0. By Theorem 1.4.33:

$$\frac{1}{n-p}\sum_{k=1}^{n-p} X_k X_{k+p} \xrightarrow[n\to\infty]{\text{a.s.}} \gamma(p).$$

Theorem 1.4.36. *For a stationary centered Gaussian sequence with covariance γ having a continuous spectral density, we have, for all $p \in \mathbb{N}$,*

$$c_n(p) \xrightarrow[n\to\infty]{\text{a.s.}} \gamma(p).$$

We shall obtain a central limit theorem relative to the sequence $(c_n(p))$ in [1.4.4].

1.4.3. Likelihood of a Stationary Gaussian Sequence

If (X_n) is a stationary Gaussian sequence with spectral density f, $X^{(n)} = (X_1, ..., X_n)$ is a centered Gaussian vector with covariance $2\pi T_n(f)$ where $T_n(f)$ is the Toeplitz matrix of f with the following definition.

Definition 1.4.37. Let h be an integrable function on $(-\pi,\pi)$. The nth **Toeplitz matrix** of h, denoted $T_n(h)$, is the $n \times n$ matrix,

$$T_n(h) = \left\{\frac{1}{2\pi}\int_{-\pi}^{\pi} e^{i\lambda(j-k)} h(\lambda)d\lambda\right\}_{1\leq j,k\leq n}.$$

The matrix $T_n(h)$ is Hermitian; if h is even, it is symmetric. If h is positive $T_n(h)$ is positive, since, for $y = (y_1, ..., y_n)$ in $\mathbb{R}^n$, we have,

$$\langle T_n(h)y,y\rangle = \frac{1}{2\pi}\int_{-\pi}^{\pi}\left|\sum_{j=1}^{n} y_j e^{i\lambda j}\right|^2 h(\lambda)d\lambda.$$

For $0 \leqslant m \leqslant h \leqslant M \leqslant \infty$, we have,

$$m|y|^2 \leqslant \langle T_n(h)y,y\rangle \leqslant M|y|^2.$$

Taking $m > 0$, $T_n(h)$ is invertible and

$$\langle [T_n(h)]^{-1}y,y\rangle \leqslant m^{-1}|y|^2.$$

Thus, if the spectral density f satisfies $0 < m \leqslant f \leqslant M$, the eigenvalues of $T_n(f)$ are all in $[m,M]$, and $X^{(n)}$ has density

$$x^{(n)} = (x_1, \ldots, x_n)$$

$$\longmapsto (2\pi)^{-n}\frac{1}{(|T_n(f)|)^{1/2}}\exp\left[-\frac{1}{4\pi}{}^t x^{(n)}(T_n(f))^{-1}x^{(n)}\right].$$

1.4.4. Estimation of the Spectral Density of a Stationary Sequence

Let X be a real centered sequence, second order stationary, with covariance γ. Assume that X has a spectral density f. Since $\gamma(p)$ is the Fourier coefficient of order $-p$ of $2\pi f$, we have,

$$f(\lambda) = \frac{1}{2\pi}\sum_{p\in\mathbb{Z}}\gamma(p)e^{-ip\lambda}.$$

With the help of the observations $(X_1, \ldots, X_n)$, $\gamma(p)$ is estimated by the sample covariance $c_n(|p|)$, zero for $|p| > n$. From which follows the idea of estimating $f(\lambda)$ by

$$I_n(\lambda) = \frac{1}{2\pi}\sum_{p\in\mathbb{Z}}c_n(|p|)e^{-ip\lambda}$$

$$= \frac{1}{2\pi n}\left\{\sum_{k=1}^{n}X_k^2 + \sum_{p=1}^{n}\sum_{k=1}^{n-p}X_{k+p}X_k(e^{-ip\lambda} + e^{ip\lambda})\right\}$$

$$= \frac{1}{2\pi n}\left\{\sum_{k=1}^{n}X_k^2 + \sum_{k=1}^{n}\sum_{h=k+1}^{n}X_kX_h 2\cos(h-k)\lambda\right\}$$

$$= \frac{1}{2\pi n}\sum_{k=1}^{n}\sum_{h=1}^{n}X_kX_h\cos(h-k)\lambda.$$

Definition 1.4.38. The **spectrogram** (or **empirical spectral density**) of a second order stationary sequence is the family of r.v.'s, $\{I_n(\lambda); n \geqslant 1, \lambda \in [-\pi,\pi[\}$ with

$$I_n(\lambda) = \frac{1}{2\pi n}\left|\sum_{p=1}^{n} X_p e^{-ip\lambda}\right|^2.$$

Let us calculate $E[I_n(\lambda)]$. Still using the isometry Z, we have,

$$E\left[\left|\sum_{p=1}^{n} X_p e^{-ip\lambda}\right|^2\right] = \int_{-\pi}^{\pi}\left|\sum_{p=1}^{n} e^{i(u-\lambda)p}\right|^2 f(u)du$$

$$= \int_{-\pi}^{\pi}\left|\sum_{p=1}^{n} e^{iup}\right|^2 f(u+\lambda)du$$

$$= \int_{-\pi}^{\pi}\left[\frac{\sin\frac{n}{2}u}{\sin\frac{u}{2}}\right]^2 f(u+\lambda)du.$$

Hence $E(I_n(\lambda))$ is the nth Fejer transform of f by λ; and, if f is bounded and continuous in λ:

$$\lim_{n\to\infty} E(I_n(\lambda)) = f(\lambda).$$

If f is continuous on $[-\pi,\pi]$, the limit is uniform in λ.

However, to estimate $f(\lambda)$, the only quality of $I_n(\lambda)$ is its asymptotic unbiasedness. It is not a consistent estimator, since $E(I_n(\lambda) - f(\lambda))^2$ does not tend to zero even if f is a very regular function. $I_n(\lambda)$ cannot then be used as an estimator of $f(\lambda)$ without further precautions.

On the other hand, if X is Gaussian and if g is continuous from $[-\pi,\pi]$ into $\mathbb{R}$, $\int_{-\pi}^{\pi} g(\lambda)I_n(\lambda)d\lambda$ is a good estimator of $\int_{-\pi}^{\pi} g(\lambda)f(\lambda)d\lambda$.

The following theorem shows this.

Theorem 1.4.39. *For a stationary centered Gaussian sequence having a continuous spectral density, the sequence of measures $(I_n \cdot L)$ converges narrowly almost surely to the spectral measure.*

Proof. From Theorem 1.4.36, for all p, the sequence $(c_n(-p))$ of Fourier coefficients of order p of $2\pi I_n$ converges a.s. to $\gamma(-p)$, the Fourier coefficient of order p of $2\pi f$. The result then follows from the following lemma.

Lemma 1.4.40. *Let (f_n) be a sequence of positive functions of $L^2([-\pi,\pi[,L)$ and let $f \in L^2([-\pi,\pi[,L)$ be positive. Let $\hat{f}_n^p$ (resp. $\hat{f}^p$) be the Fourier coefficient of order p of f_n (resp. of f).*

If, for every $p \in \mathbb{Z}$, $(\hat{f}_n^p)$ tends to $\hat{f}^p$, then the sequence $(f_n \cdot L)$ converges narrowly to $f \cdot L$.

Proof. Taking $p = 0$, we obtain

$$\int_{-\pi}^{\pi} f_n dL \xrightarrow{n\to\infty} \int_{-\pi}^{\pi} f\, dL .$$

The sequence of measures $(f_n \cdot L)$ is thus bounded. Let μ be an closure point of this sequence. For any $p \in \mathbb{Z}$, we have:

$$\int_{-\pi}^{\pi} e^{ipx} f_n(x)dx \xrightarrow{n\to\infty} \int_{-\pi}^{\pi} e^{ipx} d\mu(x)$$

$$= \int_{-\pi}^{\pi} e^{ipx} f(x)dx.$$

Thus $f \cdot L$ is the unique closure point of $(f_n \cdot L)$; it is its limit [Vol. I, Theorem 3.4.27].

We shall now study a central limit theorem for the random variables

$$I_n(\phi) = \int_{-\pi}^{\pi} I_n(\lambda)\phi(\lambda)d\lambda.$$

Theorem 1.4.41. Central Limit Theorem for Spectrograms. *Let f be the spectral density of a stationary centered Gaussian sequence X. Assume f is strictly positive, with period 2π, and differentiable with bounded derivative on $\mathbb{R}$. Let $\phi^1, \ldots, \phi^k$ be bounded* r.v.'s *on $(-\pi,\pi)$, then the random vector*

$$\sqrt{n}\left(I_n(\phi^j) - \int_{-\pi}^{\pi} f(\lambda)\phi^j(\lambda)d\lambda\right)_{1\leq j\leq k}$$

tends in distribution to $N(0,\Gamma)$, where Γ is the matrix with general term

$$\Gamma_{ij} = 4\pi \int_{-\pi}^{\pi} f^2(\lambda)\phi^i(\lambda)\phi^j(\lambda)d\lambda.$$

In particular, if $J_1, \ldots, J_k$ are disjoint intervals, the variables $\sqrt{n}\int_{-\pi}^{\pi} I_n(\lambda)1_{J_j}(\lambda)d\lambda$ are asymptotically independent.

Proof. Let ϕ be a bounded function on $(-\pi,\pi)$. Denote $I(\phi) = \int_{-\pi}^{\pi}\phi(\lambda)f(\lambda)d\lambda$ and ϕ_k the Fourier coefficient of order $-k$ of

ϕ whilst $\gamma(k)$ is that of $2\pi f$:

$$I(\phi) = \sum_{k=-\infty}^{\infty} \gamma(k)\phi_k .$$

(a) **Study of the Expectation of $\sqrt{n}[I_n(\phi) - I(\phi)]$.** We have

$$I_n(\phi) = \frac{1}{2\pi} \sum_{p\in\mathbb{Z}} c_n(|p|)\int_{-\pi}^{\pi} \phi(\lambda)e^{ip\lambda}d\lambda$$

$$= \sum_{p\in\mathbb{Z}} c_n(|p|)\phi_p.$$

Hence

$$E[I_n(\phi)] = \sum_{|k|\leqslant n} \left[1 - \frac{|k|}{n}\right] \gamma(k)\phi_k;$$

$$\sqrt{n}(E[I_n(\phi)] - I(\phi)) \leqslant \frac{1}{\sqrt{n}} \left|\sum_{|k|\leqslant n} |k|\gamma(k)\phi_k\right|$$

$$+ \sqrt{n} \left|\sum_{|k|>n} \gamma(k)\phi_k\right|.$$

Let us assume f is differentiable and f' is bounded:

$$\int_{-\pi}^{\pi} e^{i\lambda k}f(\lambda)d\lambda = \left[\frac{e^{i\lambda k}f(\lambda)}{ik}\right]_{-\pi}^{\pi}$$

$$- \frac{1}{ik}\int_{-\pi}^{\pi} e^{i\lambda k}f'(\lambda)d\lambda,$$

and the Fourier coefficient of order $-k$ of $2\pi f'$ is $-ik\gamma(k)$. From the hypothesis:

$$\sum_{k=-\infty}^{\infty} k^2\gamma^2(k) = 2\pi\int_{-\pi}^{\pi}|f'(\lambda)|^2 d\lambda < \infty;$$

$$\left|\sum_{|k|\leqslant n} k\gamma(k)\phi_k\right| \leqslant \left[\sum_{k=-\infty}^{\infty} k^2\gamma^2(k)\right]^{1/2} \left[\sum_{k=-\infty}^{\infty} \phi_k^2\right]^{1/2} < \infty;$$

$$\left|\sum_{|k|>n} \gamma(k)\phi_k\right| \leqslant \sum_{|k|>n} \frac{|k|}{n}|\gamma(k)||\phi_k|.$$

$$\leqslant \frac{1}{n}\left[\sum_{|k|>n} k^2\gamma^2(k) \sum_{|k|>n} |\phi_k|^2\right]^{1/2} .$$

Hence $\sqrt{n}E(I_n(\phi) - I(\phi))$ tends to 0, if $n \to \infty$, and to show the theorem it is sufficient to replace $I(\phi)$ by $E[I_n(\phi)]$.

(b) **Study of the Variance of $\sqrt{n}\ I_n(\phi)$.** The following study uses only the continuity of f. We are going to show that $n\sigma^2(I_n(\phi))$ tends to $4\pi\int_{-\pi}^{\pi}\phi^2(\lambda)f^2(\lambda)d\lambda$. The functions f and $\lambda \longmapsto I_n(\lambda)$ being even, if ϕ is odd, we have

$$I_n(\phi) = \int_{-\pi}^{\pi} \phi(\lambda)f^2(\lambda)d\lambda = 0.$$

Hence, in what follows, ϕ can be assumed to be even, i.e. the Fourier coefficients are real and $\phi_k = \phi_{-k}$. In the general case we deduce the property by taking

$$\phi(x) = \frac{\phi(x) + \phi(-x)}{2} + \frac{\phi(x) - \phi(-x)}{2},$$

the decomposition of ϕ into the sum of an even and odd function. We have,

$$I_n(\phi) = \frac{1}{n}\sum_{k=1}^{n}\sum_{h=1}^{n} \phi_{h-k}X_kX_h ,$$

from which (following [1.4.2]):

$$
\begin{aligned}
n\sigma^2[I_n(\phi)] &= \frac{1}{n}\sum_{1\leq h,k,\ell,m\leq n} \phi_{k-h}\phi_{\ell-m}\Gamma(X_kX_h, X_\ell X_m) \\
&= \frac{1}{n}\sum_{1\leq h,k,\ell,m\leq n} \phi_{k-h}\phi_{\ell-m}(\gamma(k-\ell)\gamma(h-m) \\
&\quad + \gamma(k-m)\gamma(h-\ell)) \\
(*)\qquad &= \frac{2}{n}\sum_{1\leq h,k,\ell,m\leq n} \phi_{k-h}\phi_{\ell-m}\gamma(k-\ell)\gamma(h-m) \\
&= \frac{2}{n}\sum_{1\leq h,k,\ell,m\leq n} \phi_{k-h}\phi_{\ell-m} \\
&\quad \cdot \int_{[-\pi,\pi]^2} e^{i(u(k-\ell)+v(h-m))}f(u)f(v)dudv \\
&= \frac{2}{n}\int_{[-\pi,\pi]^2} \left|\sum_{1\leq k,h\leq n} \phi_{k-h}2e^{i(uk+vh)}\right|^2 \\
&\quad \cdot f(u)f(v)dudv.
\end{aligned}
$$

Assume first of all that f is a trigonometric polynomial of degree d. Then $\gamma(p)$ is zero for $|p| > d$. Take $n > d$ and set $p = k - \ell$, $q = h - m$ in (*):

$$\frac{n}{2}\sigma^2[I_n(\phi)] = \sum_{|p|\leqslant d, |q|\leqslant d} \gamma(p)\gamma(q)\psi_n(p,q)$$

with

$$\psi_n(p,q) = \frac{1}{n}\sum_{\ell=1}^{n-|p|}\sum_{m=1}^{n-|q|} \phi_{\ell-m}\phi_{\ell-m+p-q} = \psi_n(q,p).$$

Take $q \geqslant p \geqslant 0$ and sum successively over three zones of values with $\ell - m = r$,

$$\psi_n(p,q) = \frac{1}{n}\Bigg[(n-q)\sum_{r=0}^{q-p} \phi_{r+p-q}\phi_r$$

$$+ \sum_{r=q-p+1}^{n-p} (n-p-r)\phi_{r+p-q}\phi_r$$

$$+ \sum_{r=q-n}^{-1} (n-q+r)\phi_{r+p-q}\phi_r\Bigg].$$

Let $(\phi^2)_k$ be the kth Fourier coefficient of ϕ^2, $(\phi^2)_{p-q} = \Sigma_{r=-\infty}^{\infty}\phi_{p-q+r}\phi_r$. However, $\Sigma_{r=-\infty}^{\infty}(\phi^2)_r < \infty$, hence

$$\sum_{r=-\infty}^{\infty} |\phi_r\phi_{r+p-q}| < \infty$$

and, by Lebesgue's theorem,

$$\lim_{n\to\infty} \psi_n(p,q) = (\phi^2)_{p-q}\ .$$

Hence,

$$\lim_{n\to\infty} \frac{1}{2} n\sigma^2[I_n(\phi)] = \sum_{|q|\leqslant d, |p|\leqslant d} (\phi^2)_{p-q}\gamma(p)\gamma(q)$$

$$= \sum_{|q|\leqslant d}\left[\sum_{|p|\leqslant d} (\phi^2)_{p-q}\gamma(p)\right]\gamma(q)$$

$$= \sum_{|q|\leqslant d}\left(\int_{-\pi}^{\pi}\phi^2(\lambda)f(\lambda)e^{-i\lambda q}d\lambda\right)\gamma(q)$$

$$= 2\pi\int_{-\pi}^{\pi}\phi^2(\lambda)f^2(\lambda)d\lambda.$$

Let us now assume that f is continuous on $\mathbb{R}$. For all $\varepsilon > 0$, we can find trigonometric polynomials g_ε^-, g_ε^+ such that $g_\varepsilon^- \leqslant f \leqslant g_\varepsilon^+$ and $g_\varepsilon^+ - g_\varepsilon^- \leqslant \varepsilon$. Denoting by $\sigma^2 g$ the calculated variance for a stationary centered Gaussian sequence with spectral density g, we have

$$\sigma^2_{g_\varepsilon^-}[I_n(\phi)] \leqslant \sigma_f^2[I_n(\phi)] \leqslant \sigma^2_{g_\varepsilon^+}[I_n(\phi)];$$

$n\{\sigma^2_{g_\varepsilon^+}[I_n(\phi)] - \sigma^2_{g_\varepsilon^-}[I_n(\phi)]\}$ tends to

$$4\pi \int_{-\pi}^{\pi} \phi^2(\lambda)(g_\varepsilon^+ - g_\varepsilon^-)^2(\lambda)d\lambda$$

when $n \to \infty$. All the closure points of $n\sigma^2[I_n(\phi)]$ differ from $4\pi \int_{-\pi}^{\pi}\phi^2(\lambda)f^2(\lambda)d\lambda$ by less than $4\pi\varepsilon \int_{-\pi}^{\pi}\phi^2(\lambda)d\lambda$. This is true for all ε,

$$n\sigma^2[I_n(\phi)] \xrightarrow[n\to\infty]{} 4\pi \int_{-\pi}^{\pi} \phi^2(\lambda)f^2(\lambda)d\lambda.$$

(c) Asymptotic Normality of $(\sqrt{n}(I_n(\phi^j) - E[I_n(\phi^j)]))_{1\leqslant j\leqslant k}$.

Let $u = (u^j)_{1\leqslant j\leqslant k}$ be in $\mathbb{R}^k$ and let $\phi = \Sigma_{j=1}^k u^j\phi^j$. Our aim is to show

$$J_n = \sqrt{n}(I_n(\phi) - E[I_n(\phi)]) \xrightarrow{\mathcal{D}} N(0, \Gamma(\phi)),$$

with

$$\Gamma(\phi) = \sum_{1\leqslant i,j\leqslant k} u^j u^{j'} 4\pi \int_{-\pi}^{\pi} \phi^j(\lambda)\phi^{j'}(\lambda)f^2(\lambda)d\lambda$$

$$= 4\pi \int_{-\pi}^{\pi} |\phi(\lambda)|^2 f^2(\lambda)d\lambda.$$

We have assumed f to be continuous and strictly positive. Hence f is minorized by a strictly positive constant. Following [1.4.3], the covariance matrix $\Gamma_n = 2\pi T_n(f)$ of $(X_1, ..., X_n) = X^{(n)}$ is invertible. We have

$$J_n = \frac{1}{\sqrt{n}}(\langle T_n(\phi)X^{(n)},X^{(n)}\rangle - E[\langle T_n(\phi)X^{(n)},X^{(n)}\rangle]).$$

However [Vol. I, 5.1.2] there exists a matrix, denoted $\Gamma_n^{1/2}$ having an inverse $\Gamma_n^{-1/2}$ such that Γ_n is equal to ${}^t\Gamma_n^{1/2}\Gamma_n^{1/2}$, and such that $Y^{(n)} = \Gamma_n^{-1/2}X^{(n)}$ is an n-sample from $N(0,1)$. Let us denote $Y^{(n)} = (Y_{j,n})_{1\leqslant j\leqslant n}$, $A_n = {}^t\Gamma_n^{1/2}T_n(\phi)\Gamma_n^{1/2}$, and let $\lambda_{j,n}$ be, for $j = 1, ..., n$, the eigenvalues of A_n. We have

$$J_n = \frac{1}{\sqrt{n}}(\langle A_nY^{(n)},Y^{(n)}\rangle - E[\langle A_nY^{(n)},Y^{(n)}\rangle])$$

$$= \frac{1}{\sqrt{n}}\sum_{j=1}^{n} \lambda_{j,n}(Y_{j,n}^2 - 1);$$

$$\sigma^2(J_n) = \frac{2}{n}\sum_{j=1}^{n} \lambda_{j,n}^2.$$

The characteristic function of J_n is given by

$$E[\exp iuJ_n] = \prod_{j=1}^{n} \frac{\exp(-i\frac{u}{\sqrt{n}}\lambda_{j,n})}{\left[1 - 2i\frac{u}{\sqrt{n}}\lambda_{j,n}\right]^{1/2}}$$

(since the Fourier transform of $\chi^2(1)$, the distribution of $Y_{j,n}^2$, is $u \longmapsto 1/(1 - 2iu)^{1/2}$). From which

$$\text{Log } E[\exp iuJ_n] = \sum_{j=1}^{n}\left(-\frac{1}{2}\text{Log}\left[1 - 2i\frac{u}{\sqrt{n}}\lambda_{j,n}\right] - i\frac{u}{\sqrt{n}}\lambda_{j,n}\right)$$

$$= -\frac{2u^2}{n}\left[\sum_{j=1}^{n}\lambda_{j,n}^2\right] + R_n .$$

However

$$\sigma^2(J_n) = \frac{4}{n}\sum_{j=1}^{n}\lambda_{j,n}^2 \xrightarrow[n\to\infty]{} 4\pi\int_{-\pi}^{\pi}\phi^2(\lambda)f^2(\lambda)d\lambda,$$

from part (b) of the proof. Now

$$\left|\text{Log}(1 - x) + x + \frac{x^2}{2}\right|$$

$$= \left|\frac{x^3}{3} + \cdots + \frac{x^n}{n} + \cdots\right| \leqslant \frac{|x|^3}{1 - |x|}$$

and

$$R_n \leqslant \frac{2}{n} u^2 \left[\sum_{j=1}^{n} \lambda_{j,n}^2 \right] \sup_{1 \leqslant j \leqslant n} \frac{2|u/\sqrt{n}|\,|\lambda_{j,n}|}{1 - 2|u/\sqrt{n}|\,|\lambda_{j,n}|}.$$

Finally,

$$\begin{aligned}\sup_{1 \leqslant j \leqslant n} |\lambda_{j,n}| &= \sup_{|y| \leqslant 1} |<A_n y, y>| \\ &= \sup_{|y| \leqslant 1} <T_n(\phi)\Gamma_n^{1/2} y, \Gamma_n^{1/2} y> \\ &\leqslant \sup|\phi| \sup_{|y| \leqslant 1} |\Gamma_n^{1/2} y|^2 \\ &= \sup|\phi| \sup_{|y| \leqslant 1} <\Gamma_n y, y> \leqslant \sup|\phi| \sup|f|.\end{aligned}$$

Hence R_n tends to 0 if $n \to \infty$. The theorem is proved.

Corollary 1.4.42. Central Limit theorem for Sample Covariances. *If the spectral density f of a stationary Gaussian sequence is differentiable, with $f > 0$, $f(-\pi) = f(\pi)$ and f' bounded, we have for every sequence of integers $(p_1, p_2, \ldots, p_k)$*

$$\sqrt{n}(c_n(p_j) - \gamma(p_j))_{1 \leqslant j \leqslant k} \xrightarrow{\mathcal{D}} N_k(0, \Gamma)$$

with

$$\Gamma = \left\{ 8\pi \int_{-\pi}^{\pi} \cos(p_i\lambda)\cos(p_j\lambda) f^2(\lambda)\, d\lambda \right\}_{1 \leqslant i,j \leqslant k}.$$

Proof. It is sufficient to notice that

$$c_n(p) = \int_{-\pi}^{\pi} (\cos p\lambda) I_n(\lambda)\, d\lambda$$

and to apply the preceding theorem.

Bibliographic Notes

First of all let us give mention of some works dealing with time series. Chatfield gives a first approach to some statistical aspects. An intuitive and comprehensive introduction is given in Grenander-Rosenblatt. The basic work dealing with applications of ARMA models, notably to engineering and to econometrics, is Box and Jenkins.

A more mathematical and condensed treatment, centered on ARMA models is to be found in Azencott and Dacunha-Castelle. Koopmans is an introduction to the particular theme of the estimation of spectral densities.

Hannan's book is the fundamental work concerned with the multidimensional extension of time series. Let us also mention Anderson, Brillinger (closer to data analysis).

For other aspects an original and abstract presentation will be found in Grenander. The probabilistic part of second order processes is detailed in Rozanov and Yaglom. A deep study of stationary Gaussian processes is Ibraguimov-Rozanov.

The remainder of Fourier analysis may be found in Fomine-Kolmogorov, and Rudin [1]. Prediction theory is developed in Rudin [2] and in Hoffman, which inspired our presentation of Szegö's theorem.

Chapter 2
MARTINGALES IN DISCRETE TIME

Objectives

With martingales in discrete time, we tackle all the original ideas of the theory of processes. The following chapters will therefore be either applications of this, or extensions to continuous time. They will be technically more difficult, but will be based on the same ideas.

Martingales (resp. submartingales) are sequences of r.v.'s which tend to be constant (resp. to increase). This idea is clarified using the concept of a compensator. We then establish some inequalities and some convergence theorems for submartingales, as important in probability as are the related theorems for increasing sequences of $\mathbb{R}$.

First application: a method for the study of absolute continuity of one process with respect to another and for approximating Radon Nikodym derivatives and dissimilarities. This will be the key to Chapter 3, where we deal with the principles of the statistics of processes.

Second application: some asymptotic theorems, laws of large numbers and central limit theorems, useful in more general frameworks than that of sums of independent identically distributed r.v.'s covered in Volume I.

2.1. Some Examples

2.1.1. Time Series

Let us return to the study of a sequence $(X_n)_{n\in\mathbb{Z}}$ of r.v.'s of $L^2(\Omega,\mathcal{A},P)$ dealt with in [1]. For $\mathcal{F}_n = \sigma(X_p; p \leqslant n)$ and $Y \in L^2(\Omega,\mathcal{A},P)$; the best approximation of Y by a function of the observations prior to n (i.e. by an $\mathcal{F}_n$-measurable function) is $E^n(Y) = E(Y|\mathcal{F}_n)$. The σ-algebras $(\mathcal{F}_n)$ form an increasing sequence and, for $p > 0$, we have:

$$E^n[E^{n+p}(Y)] = E^n(Y).$$

The sequence $E^n(Y)$ converges, in $L^2(\Omega,\mathcal{A},P)$, to $E^\infty(Y)$ if n tends to $+\infty$ and to $E^{-\infty}(Y)$ if n tends to $-\infty$; the same result holds for the linear predictions studied in [1.1.3]. We are going to be able to show here that there is also a.s. convergence of $E^n(Y)$.

The sequence (X_n) is a **martingale** when, for $p > 0$, we have: $E^n(X_{n+p}) = X_n$. It then has orthogonal increments. In fact,

$$E^{n-1}[\Delta X_n] = E^{n-1}[X_n - X_{n-1}] = 0$$

and ΔX_n is orthogonal to the r.v.'s X_p, for $p \leqslant n - 1$.

Here again Theorem 1.1.8 is going to be able to be extended to a.s. convergence.

2.1.2. Likelihoods

Consider two probabilities P and Q on $(\Omega,\mathcal{A})$. If $\mathcal{F}$ is a sub-σ-algebra of $\mathcal{A}$, denote by $P_\mathcal{F}$ and $Q_\mathcal{F}$, the traces of P and Q on $\mathcal{F}$. Take a partition of Ω by disjoint events $A_1, ..., A_n$, with union Ω, and let $\mathcal{F} = \sigma(A_1, ..., A_n)$. Assume $P(A_1) > 0$ for $1 \leqslant i \leqslant n$. Then $Q_\mathcal{F}$ is absolutely continuous with respect to $P_\mathcal{F}$, and its likelihood is,

$$\frac{dQ_\mathcal{F}}{dP_\mathcal{F}} = \sum_{i=1}^{n} \frac{Q(A_i)}{P(A_i)} 1_{A_i}.$$

Thus, taking $\Omega = [0,1[$,

$$\mathcal{F}_n = \sigma\left\{\left[\frac{k}{2^n}, \frac{k+1}{2^n}\right[; \ 0 \leqslant k < 2^n\right\}$$

and for P Lebesgue measure L, we have,

$$\frac{dQ_{\mathcal{F}_n}}{dP_{\mathcal{F}_n}} = \sum_{k=0}^{2^n} 2^n Q\left(\left[\frac{k}{2^n}, \frac{k+1}{2^n}\right[\right) 1_{[k/2^n,(k+1)/2^n[} \;.$$

In the last example, the Borel σ-algebra is $\vee \mathcal{F}_n$ and the family of σ-algebras $(\mathcal{F}_n)$ is increasing. In the general case, if the σ-algebra $\mathcal{A}$ is **separable**, i.e. generated by a countable family of events (B_p), we take $\mathcal{F}_n = \sigma(B_1, \ldots, B_n)$ and a partition $(A_1, \ldots, A_{u(n)})$ which generates the same σ-algebra. If Q is absolutely continuous with respect to P ($Q << P$), we can set

$$L_n = \frac{dQ_{\mathcal{F}_n}}{dP_{\mathcal{F}_n}} = \sum_{i=1}^{u(n)} \frac{Q(A_i)}{P(A_i)} 1_{A_i}$$

the ratio $Q(A_i)/P(A_i)$ being taken zero for $P(A_i) = Q(A_i) = 0$.

Let $\Gamma \in \mathcal{F}_n$, so $\Gamma \in \mathcal{F}_{n+1}$, $\mathcal{F}_n \subset \mathcal{F}_{n+1}$, and we have,

$$Q(\Gamma) = E[L_n 1_\Gamma] = E[L_{n+1} 1_\Gamma]$$

since L_n is $\mathcal{F}_n$-measurable, this means $E(L_{n+1} | \mathcal{F}_n) = L_n$. Does the sequence (L_n) converge to dQ/dP?

Without assuming $Q << P$, it could happen that, for each n, we have (as in the example of $([0,1[,L))$ $Q_{\mathcal{F}_n} << P_{\mathcal{F}_n}$, which again allows the calculation of L_n made above. Does the sequence (L_n) converge? Can we deduce tht $Q << P$ and the value of dQ/dP? The answer in general, is no otherwise every measure on $[0,1[$ would be absolutely continuous with respect to L. However the answer will help us considerably in the investigation of likelihood.

2.2. Martingales

2.2.1. Filtration

A sequence of observations is made in order to understand a random phenomenon $(\Omega, \mathcal{A}, P)$. The set of observed events up till time $n \in T$, with $T = \mathbb{N}$ or $\mathbb{Z}$, constitutes a σ-algebra $\mathcal{F}_n$: the sequence of these sub σ-algebras of $\mathcal{A}$, $\mathbb{F} = (\mathcal{F}_n)$, is called a **filtration**. It is an increasing sequence ($\mathcal{F}_n \subset \mathcal{F}_{n+1}$) since, at time $n + 1$, the events prior to n have been observed. Let $\mathcal{F}_\infty$

$= \vee \mathcal{F}_n$, and $\mathcal{F}_{-\infty} = \cap \mathcal{F}_n$ if $T = \mathbb{Z}$. Let Y be a quasi-integrable r.v. on $(\Omega, \mathcal{A}, P)$. At time n, its value is predicted by $E(Y| \mathcal{F}_n)$ which we denote in this chapter by $E^n(Y)$.

Let (X_n) be a sequence of variables. To say that X_n is observed at time n implies that it is $\mathcal{F}_n$-measurable. At time $n - 1$, we can predict the value of all the $\mathcal{F}_{n-1}$-measurable r.v.'s.

Definition 2.2.1. Given a filtration $\mathbb{F} = (\mathcal{F}_n)$ on $(\Omega, \mathcal{A}, P)$, a sequence (X_n) of measurable functions taking values in $(E, \mathcal{E})$ is said to be **$\mathbb{F}$-adapted** if, for every n, X_n is measurable from $(\Omega, \mathcal{F}_n)$ into $(E, \mathcal{E})$. It is **$\mathbb{F}$-predictable** if, for every n, X_n is $\mathcal{F}_{n-1}$ measurable (for $T = \mathbb{N}$, $\mathcal{F}_{-1}$ is taken to be trivial).

2.2.2. Tendency to Increase or Decrease

A fundamental theorem on the convergence of real sequences is the following: a sequence of real numbers increasing and bounded above (resp. decreasing and bounded below) converges. The study of increasing sequences (X_n) (for almost all ω, $(X_n(\omega))$ is an increasing sequence of real numbers) follows easily from this... but is too restrictive. On the other hand, many sequences "tend to increase." Let us clarify this: for a sequence of integrable r.v.'s (X_n), let us denote $\Delta X_n = X_n - X_{n-1}$. At time $n - 1$, ΔX_n is predicted by $E^{n-1}(\Delta X_n) = \widehat{\Delta X}_n$. For example, if $X_n = \Sigma_{i=1}^n \Delta X_i$ is the sum of n independent integrable r.v.'s, and $\mathcal{F}_n = \sigma(X_1, ..., X_n)$, then $E^{n-1}(\Delta X_n) = E(\Delta X_n)$. A sequence will tend to increase if the predicted increments $\widehat{\Delta X}_n$ are positive. We shall see that the convergence theorems for these sequences having a tendency to increase or decrease are analogous, and as important, as the theorems which deal with sequences of real numbers. Here is the more orthodox vocabulary for "having a tendency to ."

According to the French dictionary *le Petit Robert* a martingale is "every betting system more or less scientific (probability theory)." If X_n is the player's winnings after n games the sequence (X_n) tends to increase (or to be constant, or decrease) if the game is favorable (fair or unfair). It is this example which has led to the following definitions. However, do not pin your hope on finding a martingale in the

Robert sense. For an r.v. X, we denote $X^+ = \sup(X,0)$ and $X^- = -\inf(X,0)$.

Definition 2.2.2. On $(\Omega,\mathcal{A},P)$ equipped with a filtration $\mathbb{F} = (\mathcal{F}_n)_{n\in T}$ with $T = \mathbb{N}$ or $\mathbb{Z}$, a sequence $(X_n)_{n\in T}$ of r.v.'s (defined up to a.s. equality) adapted to $\mathbb{F}$ is:

(a) an $\mathbb{F}$-**submartingale** if, for every $n \in T$:

$$E(X_n^+) < \infty, \quad E^n(X_{n+1}) \geq X_n;$$

(b) an $\mathbb{F}$-**supermartingale** if, for every $n \in T$:

$$E(X_n^-) < \infty, \quad E^n(X_{n+1}) \leq X_n;$$

(c) an $\mathbb{F}$-**martingale** if, for every $n \in T$:

$$E(|X_n|) < \infty, \quad E^n(X_{n+1}) = X_n.$$

Notes. (a) $\mathbb{F}$ *is not quoted if it is clear. If* $\mathbb{F}$ *is not stated, we are dealing with* $\mathbb{F} = \{\mathcal{F}_n\} = \{\sigma(X_p;\ p \leq n)\}$.

(b) *For* $p > 1$, *it is easily seen by recurrence that we have* $E^n(X_{n+p}) \geq X_n$ *(resp.* $\leq$ *or* $=$*) for a submartingale (resp. supermartingale or martingale).*

(c) *The sequence* (X_n) *is said to be* **integrable** (*or* **pth power integrable, or centered**), *if each of its terms is. If* X *is a martingale,* $E(X_n)$ *does not depend on* n, *we denote it* $E(X)$.

(d) *If* X *is a supermartingale,* $-X$ *is a submartingale. Hence we are satisfied with statements relative to submartingales.*

2.2.3. Compensating an Adapted Sequence of Integrable r.v.'s

We have defined above $\Delta\tilde{X}_n = E^{n-1}(\Delta X_n)$. To say that (X_n) is a submartingale means that $(\Delta\tilde{X}_n)$ is positive. Take $T = \mathbb{N}$ and set $\tilde{X}_n = \Delta\tilde{X}_1 + \ldots + \Delta\tilde{X}_n$; then $(X_n - \tilde{X}_n)$ is a martingale since $\tilde{X}_n$ is $\mathcal{F}_{n-1}$ measurable and

$$E^{n-1}(X_n - \tilde{X}_n) = E^{n-1}(\Delta X_n) + (X_{n-1} - \tilde{X}_n) = X_{n-1} - \tilde{X}_{n-1}$$

Proposition 2.2.3. *To every* $\mathbb{F}$*-adapted sequence* $X = (X_n)_{n>0}$, *of integrable* r.v.'s *we can associate a unique predictable sequence* $\tilde{X} = (\tilde{X}_n)_{n\geq 0}$, *zero at* 0, *such that* $X - \tilde{X} = (X_n - \tilde{X}_n)$ *is an*

$\mathbb{F}$*-martingale:* $\breve{X}$ *is the compensator of* X. *The sequence* X *is a martingale (resp. sub or supermartingale) if* $\tilde{X}$ *is zero (resp. increasing or decreasing).*

Only the uniqueness remains to be shown. However, if $\hat{X}$ and $\breve{X}$ are compensators of X we have,

$$E^{n-1}(\hat{X}_n - \breve{X}_n) = \hat{X}_n - \breve{X}_n = \hat{X}_{n-1} - \breve{X}_{n-1}$$
$$= \dots = \hat{X}_0 - \breve{X}_0 = 0.$$

Particular Case. Let (M_n) be a square integrable martingale, i.e. in L^2. For each n,

$$0 \leqslant E^{n-1}(\Delta M_n)^2 = E^{n-1}(M_n^2 - 2M_nM_{n-1} + M_{n-1}^2)$$
$$= E^{n-1}(M_n^2) - 2M_{n-1}E^{n-1}(M_n) + M_{n-1}^2$$
$$= E^{n-1}(\Delta M_n^2).$$

Hence the sequence (M_n^2) is a submartingale. For $T = \mathbb{N}$, setting

$$A_n = \sum_{p=1}^{n} E^{p-1}(\Delta M_p^2) = \sum_{p=1}^{n} E^{p-1}(\Delta M_p)^2,$$

$(M_n^2 - A_n)$ is a martingale.

Examples. (a) Let $(\xi_n)_{n\geqslant 1}$ be a sequence of independent measurable functions taking values in a measurable space $(E,\mathcal{E})$ with the same distribution μ. For an r.v. f on $(E,\mathcal{E})$, define

$$N_n(f) = \sum_{p=1}^{n} f(\xi_p).$$

Take $\mathcal{F}_n = \sigma(\xi_1, \dots, \xi_n)$; when $f \in L^1(\mu)$, $(N_n(f) - n\mu(f))_{n\geqslant 1}$ is a centered martingale. When $f \in L^2(\mu)$, $([N_n(f) - n\mu(f)]^2 - n\sigma^2(f))_{n\geqslant 1}$ is a centered martingale. In these formulas we have denoted,

$$\mu(f) = \int f\, d\mu, \quad \sigma^2(f) = \int f^2\, d\mu - [\mu(f)]^2.$$

(b) Let $(\xi_n)_{n\geqslant 1}$ be a Markov chain taking values in $(E,\mathcal{E})$, with transition probability π: $E[\xi_{n+1} \in \cdot \mid \xi_0, \dots, \xi_n] = \pi(\xi_n, \cdot)$

[Vol. I, 6.3]. Let $\mathcal{F}_n = \sigma\{\xi_0, \ldots, \xi_n\}$. For F, a bounded r.v. on $(E,\mathcal{E})^2$, set $\pi F(x) = \int \pi(x,dy)F(x,y)$. Then, $\Sigma_{p=1}^n F(\xi_{p-1}, \xi_p)$ is compensated by

$$\sum_{p=1}^{n} \pi F(\xi_{p-1}).$$

2.2.4. Operations and Decompositions for Submartingales

(a) If X and Y are two martingales, $\alpha X + \beta Y$ is also a martingale (α and β real numbers).

(b) If X and Y are two submartingales, $[\sup(X_n,Y_n)]_{n\in T} = \sup(X,Y)$ is also a submartingale. In particular this is the case for $\sup(X,a)$ for every constant a.

(c) Let X be a martingale and Φ a *convex* function from $\mathbb{R}$ into $\mathbb{R}$, such that, $\Phi(X_n)^+$ is integrable for every n. Then $\Phi(X) = (\Phi(X_n))_{n\in T}$ is a submartingale. The result remains true if X is a submartingale and if Φ is *increasing and convex.*

The proofs of (a) and (b) are evident. For (c) we use Jensen's inequality [Vol. I, 6.2],

$$E^n(\Phi(X_{n+1})) \geq \Phi(E^n(X_{n+1})) \geq \Phi(X_n).$$

Example. Let $p \geq 1$. If X is a pth power integrable martingale, $|X|^p$ is a submartingale. If X is a positive, pth power integrable, submartingale X^p is a submartingale.

(d) **Doob's Decomposition 2.2.4.** Let $X = (X_n)_{n\in \mathbb{Z}}$ be an integrable submartingale satisfying $\sup_{n<0} E(X_n^-) < \infty$. There exists an integrable martingale $M = (M_n)$ and a predictable integrable increasing sequence $A = (A_n)$ for which we have: $X = M + A$. The sequence A is unique if we impose on it $\lim_{n\to-\infty} A_n = 0$ (or $A_0^1 = 0$ if $T = \mathbb{N}$). Moreover if X is positive, we have for all n, $E(A_n) \leq E(X_n)$.

Proof. For $n < m$, we write

$$E\left[\sum_{n<p\leq m} \Delta\tilde{X}_p\right] = E[X_m - X_n] \leq E(X_m) + \sup_{n\leq m} E(X_n^-).$$

The r.v.'s $(\Delta\tilde{X}_p)$ are positive. If n decreases to $-\infty$, the sequence $\Sigma_{n<p\leq m}\Delta\tilde{X}_p$ thus increases to an integral r.v. A_m;

$$E(A_m) \leqslant E(X_m) + \sup_{p \leqslant m} E(X_p^-).$$

Setting $M = X - A$, we calculate $\Delta M_n = \Delta X_n - \Delta A_n = \Delta X_n - \Delta X_n$, and M is a martingale. The difference of two processes A which would agree would be a predictable martingale, hence constant, tending to 0 if n tends to $-\infty$: it would be zero.

(e) **Krickeberg's Decomposition 2.2.5.** Let $X = (X_n)_{n \in \mathbb{Z}}$ be a martingale satisfying $\sup E(|X_n|) < \infty$. It is the difference of two positive martingales.

Proof. The process $|X|$ is a positive submartingale: its Doob decomposition is written $|X| = M + A$. The increasing sequence (A_n) converges a.s. to the integrable r.v. A_∞. In fact, $E(A_n)$ is majorized by $\sup E(|X_n|) - E(M)$. Setting $Y_n = M_n + E^n(A_\infty)$, (Y_n) is a positive martingale which majorizes $|X|$. Hence $Y - X$ is also a positive martingale.

2.3. Stopping

2.3.1. Stopping Times

When a gambler plays in a casino, he must decide when to stop. This decision cannot take into account future results of the roulette. If X_n is his total gain after n plays, he will be able to stop, for example, either because he is ruined and has lost his initial fortune a, or because he judges that he has won enough, having won $b > 0$; in other words, he stops at time

$$\nu_{a,b}(\omega) = \inf\{n;\ X_n(\omega) \geqslant b \ \text{ or } \ X_n(\omega) \leqslant -a\}.$$

Definition 2.3.6. A stopping time ν **(adapted to $\mathbb{F}$)** is a function ν from Ω into $T \cup \{\infty\}$, such that, for all n, the event $\{\nu = n\}$ is in $\mathcal{F}_n$. The **σ-algebra $\mathcal{F}_\nu$ of events prior to ν** is the set of $A \in \mathcal{F}_\infty$ such that for every $n < \infty$, $A \cap \{\nu = n\}$ is in $\mathcal{F}_n$.

Properties 2.3.7. (a) ν, *a function from Ω into $T \cup \{\infty\}$, is a stopping time if, and only if, for every n, $\{\nu \leqslant n\}$ is in $\mathcal{F}_n$.*

(b) $\mathcal{F}_\nu$ *is a σ-algebra. For $n \leqslant \infty$,*

$$\mathcal{F}_\nu \cap \{\nu = n\} = \mathcal{F}_n \cap \{\nu = n\}.$$

(c) *Let* $X = (X_n)$ *be a sequence of measurable functions taking values in* $(E,\mathcal{E})$, *adapted to* $\mathbb{F}$. *For* $\Gamma \in \mathcal{E}$, *the time of entry into* Γ, $T_\Gamma = \inf\{n;\ X_n \in \Gamma\}$, *is a stopping time (convention: the* inf *of an empty set of* $\mathbb{R}$ *is* $+\infty$).

(d) ν *constant, equal to* $n \in T$, *is a stopping time.*

(e) *Let* ν_1, ν_2 *be stopping times*; $\sup(\nu_1,\nu_2) = \nu_1 \vee \nu_2$ *and* $\inf(\nu_1,\nu_2) = \nu_1 \wedge \nu_2$ *are stopping times, and* $\{\nu_1 < \nu_2\}$ *and* $\{\nu_1 = \nu_2\}$ *are in* $\mathcal{F}_{\nu_1} \cap \mathcal{F}_{\nu_2}$. *For* $\nu_1 \leq \nu_2$, *we have* $\mathcal{F}_{\nu_1} \subset \mathcal{F}_{\nu_2}$.

Proof. It is easy. Let us show for example (e) by using (a): $\nu_1 \vee \nu_2$ and $\nu_1 \wedge \nu_2$ are stopping times since,

$$\{\nu_1 \vee \nu_2 \leq n\} = \{\nu_1 \leq n\} \cap \{\nu_2 \leq n\};$$

$$\{\nu_1 \wedge \nu_2 \leq n\} = \{\nu_1 \leq n\} \cup \{\nu_2 \leq n\}.$$

Moreover $\{\nu_1 < \nu_2\}$ and $\{\nu_1 = \nu_2\}$ are in $\mathcal{F}_{\nu_1} \cap \mathcal{F}_{\nu_2}$ since,

$$\{\nu_1 < \nu_2\} \cap \{\nu_1 = n\} = \{\nu_2 \leq n\}^c \cap \{\nu_1 = n\}$$

$$\{\nu_1 < \nu_2\} \cap \{\nu_2 = n\} = \{\nu_1 < n\} \cap \{\nu_2 = n\}$$

$$\{\nu_1 = \nu_2\} \cap \{\nu_1 = n\} = \{\nu_2 = n\} \cap \{\nu_1 = n\}.$$

For $A \in \mathcal{F}_{\nu_1}$ and $\nu_1 \leq \nu_2$, we write

$$A \cap \{\nu_2 = n\} = \bigcup_{p \leq n} \{A \cap (\nu_1 = p)\} \cap \{\nu_2 = n\}$$

and A is in $\mathcal{F}_{\nu_2}$. If ν is a stopping time, we denote in this chapter by E^ν, the expectation conditional on $\mathcal{F}_\nu$.

2.3.2. Stopping a Submartingale

Let us consider an $\mathbb{F}$-adapted sequence $X = (X_n)$ and a stopping time ν. The sequence X stopped at time ν is $X^\nu = (X_{n\wedge\nu}) = (X_n^\nu)$. The event $\{\nu \geq n\}$ complementary to $\cup_{p<n}\{\nu = p\}$ is in $\mathcal{F}_{n-1}$. Assuming that X is a submartingale,

$$\begin{aligned} E^{n-1}(X_n^\nu) &= E^{n-1}(X_n 1_{(\nu\geq n)}) + E^{n-1}(X_\nu 1_{(\nu<n)}) \\ &= 1_{(\nu\geq n)} E^{n-1}(X_n) + 1_{(\nu<n)} X_\nu \end{aligned}$$

$$\geq 1_{(\nu \geq n)} X_{n-1} + 1_{(\nu < n)} X_\nu = X^\nu_{n-1}$$

(with equality if X is a martingale). Moreover, if (X_n) is an integrable sequence, we have,

$$\Delta \tilde{X}^\nu_n = E^{n-1}(\Delta X^\nu_n) = 1_{(\nu \geq n)} E^{n-1}(\Delta X_n)$$
$$= 1_{(\nu \geq n)} \Delta \tilde{X}_n \; .$$

This proves part (a) of the following theorem.

Stopping Theorem 2.3.8. (a) *If X is an $\mathbb{F}$-(sub) martingale and ν is an $\mathbb{F}$-stopping time, X^ν is an $\mathbb{F}$-(sub) martingale. If moreover $\sup_{n<0} E(X^-_n)$ is finite and if $X = M + A$ is the Doob decomposition of X, that of X^ν is $M^\nu + A^\nu$.*

(b) *If X is an $\mathbb{F}$-submartingale, and ν_1 and ν_2 two stopping times satisfying $\nu_1 \leq \nu_2 \leq n$ for a certain n; we have:*

$$E^{\nu_1}(X_{\nu_2}) \geq X_{\nu_1}$$

(equality if X is a martingale).

Proof of (b). For $\nu \leq n$ and $\Gamma \in F_\nu$,

$$E[X_n 1_\Gamma] = \sum_{p \leq n} E[X_n 1_{\Gamma \cap (\nu = p)}]$$
$$= \sum_{p \leq n} E[E^p(X_n) 1_{\Gamma \cap (\nu = p)}]$$
$$\geq \sum_{p \leq n} E[X_p 1_{\Gamma \cap (\nu = p)}] = E[X_\nu 1_\Gamma].$$

Hence $E^\nu(X_n) \geq X_\nu$ (with equality for a martingale). We then apply the preceding to $\nu = \nu_1$ and to X^{ν_2}.

Here is a first consequence of the stopping theorem.

Wald's Theorem 2.3.9. *Consider a sequence $(Y_n)_{n \geq 1}$ of $\mathbb{F}$-adapted* r.v.'s, *with the same distribution with mean μ. Assume that, for all n, Y_n is independent of F_{n-1}. Set $S_n = Y_1 + \ldots + Y_n$. For every integrable stopping time ν, we have, $E[S_\nu] = \mu E(\nu)$.*

If moreover the r.v.'s *Y have a finite variance σ^2 and are centered, we have, for every integrable stopping time ν, $E[S^2_\nu] = \sigma^2 E(\nu)$.*

Proof. $(S_n - n\mu)$ is a martingale, hence,

$$E[S_{n\wedge\nu}] = \mu E[n \wedge \nu].$$

If the r.v.'s Y are positive, we pass to the limit by the Beppo-Levi theorem. If not, the theorem is true for the sequence $(|Y_n|)$, and $(S_{n\wedge\nu})$ is majorized by $\Sigma_{n\leqslant\nu}|Y_n|$ which is integrable: we can pass to the limit. For the proof of the second half of this theorem, see [2.6.1].

The theorem applies to the stopping of the gambler. When the r.v.'s Y are not (a.s.) zero, for $a > 0$ and $b > 0$, $\nu_{a,b} = \inf\{n; S_n \leqslant -a, S_n \geqslant b\}$ is integrable. To show this, refer back to Lemma 2.5.12 of Volume I. We shall come back to gambler's ruin in [5.3.3].

2.3.3. Inequalities

Kolmogorov's Theorem 2.3.10. (1) *Let* $X = (X_n)$ *be a submartingale,* $\lambda > 0$ *and, for all* n, $\bar{X}_n = \sup_{k\leqslant n} X_k$. *Then*

(a) $$P[\bar{X}_n \geqslant \lambda] \leqslant \frac{1}{\lambda} E(X_n^+ 1_{(\bar{X}_n \geqslant \lambda)}) \leqslant \frac{1}{\lambda} E(X_n^+).$$

(b) *If we assume the submartingale* X *to be positive, then for* $p \in]1,\infty[$ ($\|\cdot\|_p$ *being the norm of* L^p),

$$\|\bar{X}_n\|_p \leqslant \frac{p}{p-1} \|X_n\|_p .$$

(2) *Let* (X_n) *be a positive supermartingale. We have, for every constant* $\lambda > 0$ *and for all* n,

$$P\left[\sup_{k\geqslant n} X_k \geqslant \lambda\right] \leqslant \frac{1}{\lambda} E(X_n).$$

Proof. (1) (a) Let $N < n$ and let $\nu = \inf\{k; k \geqslant N, X_k \geqslant \lambda\}$; X^+ is a submartingale and,

$$X^+_{\nu\wedge n} \leqslant E(X_n^+ | F_{\nu\wedge n}),$$

$$\lambda P[\nu \leqslant n] = \lambda P\left[\sup_{N\leqslant k\leqslant n} X_k \geqslant \lambda\right] \leqslant E[X^+_{\nu\wedge n} 1_{(\nu\leqslant n)}]$$

$$\leqslant E[X_n^+ 1_{(\nu\leqslant n)}].$$

We obtain (a) by letting N tend to $-\infty$.

(b) $$\|\overline{X}_n\|_p^p = p\int_0^\infty \lambda^{p-1} P(\overline{X}_n \geqslant \lambda)\,d\lambda$$

$$\leqslant p\int_0^\infty \lambda^{p-1} E(X_n 1_{(\overline{X}_n \geqslant \lambda)})\,d\lambda\ .$$

By Fubini's theorem,

$$\|\overline{X}_n\|_p^p \leqslant \frac{p}{1-p}\, E[X_n(\overline{X}_n)^{p-1}].$$

We then use Hölder's inequality,

$$E[X_n(\overline{X}_n)^{p-1}] \leqslant \|X_n\|_p \|\overline{X}_n\|_p^{p-1}$$

from which the result follows.

(2) Consider $\nu = \inf\{k;\ k \geqslant n,\ X_k \geqslant \lambda\}$ and N, $n < N < \infty$:

$$E[X_n] \geqslant E[X_{\nu\wedge N}] \geqslant E[X_{\nu\wedge N} 1_{(\nu\leqslant N)}]$$

$$\geqslant \lambda P[\nu \leqslant N].$$

Letting N tend to ∞, we obtain the stated result.

Corollary 2.3.11. *Let M be a martingale, $\lambda > 0$ and $p \geqslant 1$. Then $|M|^p$ is a positive submartingale. Denoting $M_n^* = \sup_{k\leqslant n}|M_k|$, we have,*

$$P[M_n^* \geqslant \lambda] \leqslant \frac{1}{\lambda^p}\, E[|M_n|^p],$$

$$\|M_n^*\|_p \leqslant \frac{p}{p-1}\|M_n\|_p, \quad \text{if } p > 1.$$

Note finally that $(M_{n+m} - M_n)_{m\geqslant 0}$ is a martingale adapted to $(\mathcal{F}_{n+m})_{m\geqslant 0}$. Hence,

$$P\left[\sup_{0\leqslant k\leqslant m} |M_{n+k} - M_n|^p \geqslant \lambda\right] \leqslant \frac{1}{\lambda} E(|M_{n+m} - M_n|^p);$$

$$P\left[\sup_{n\leqslant i,j\leqslant m} |M_i - M_j|^p \geqslant 2\lambda\right] \leqslant \frac{1}{\lambda} E(|M_{n+m} - M_n|^p).$$

These inequalities applied to a sequence (Y_n) of independent r.v.'s having the same distribution with mean μ and to the martingale $(Y_1 + \dots + Y_n - n\mu)$ are Kolmogorov's classic inequalities.

2.4. Convergence of a Submartingale

In what follows $T = \mathbb{Z}$, $\mathbb{F}$ is given and $X = (X_n)_{n\in\mathbb{Z}}$ is a submartingale. The case $T = \mathbb{N}$ certainly follows by taking $X_n = X_0$ and $F_n = F_0$ for $n < 0$. The condition essential to convergence will be the existence of a uniform bound for the integral of X_n^2 (L^2 case) or of X_n^+ (a.s. case).

2.4.1. Convergence in L^2

Theorem 2.4.12. *Let X be a martingale or a positive submartingale. Assume* $\sup E(X_n^2) < \infty$. *Then the sequence (X_n) converges in L^2 if n tends to $\pm\infty$.*

Proof. We use the Doob decomposition $X^2 = M + A$ of the submartingale X^2. For $n < m$,

$$E(X_n - X_m)^2 = E(X_n^2) + E(X_m^2) - 2E(X_n E^n(X_m))$$

$$\leq E(X_m^2) - E(X_n^2) = E(A_m - A_n).$$

However $E(A_m) \leq E(X_m^2) \leq \sup E(X_m^2) < \infty$. Hence the increasing sequence (A_n) tends, in L^1 and a.s., to an r.v. A_∞ when n tends to $+\infty$. It decreases to 0 in L^1 and a.s. if n tends to $-\infty$. Hence (X_n) is a Cauchy sequence in L^2, if n tends to $\pm\infty$.

2.4.2. Almost Sure Convergence

Theorem 2.4.13. *Let X be a submartingale. Assume* $\sup E(X_n^+) < \infty$. *Then X converges* a.s. *if n tends to $+\infty$, to an integrable* r.v. $X_{+\infty}$. *It converges* a.s., *if n tends to $-\infty$ to an* r.v. $X_{-\infty}$ *taking values in $[-\infty,\infty[$. $X_{-\infty}$ is integrable if we also assume that* $\sup E(X_n^-)$ *is finite (or* $\sup E(|X_n|) < \infty$).

Proof. (a) Let us assume the conditions are valid; then we use the notations of [2.4.1]. Assume also that $X \geqslant 0$. The martingale $M = X^2 - A$ converges in L^1, if n tends to $\pm\infty$, to an r.v. $M_{\pm\infty}$ (since X^2 and A converge). For every $\lambda > 0$, we have, for $n < m$ (from [2.3.3]),

$$P\left[\sup_{n \leqslant i,j \leqslant m} |M_i - M_j| \geqslant 2\lambda\right] \leqslant \frac{1}{\lambda} E(|M_m - M_n|).$$

Letting m tend to $+\infty$ or n tend to $-\infty$,

$$P\left[\sup_{n \leqslant i,j} |M_i - M_j| \geqslant 2\lambda\right] \leqslant \frac{1}{\lambda} \sup_{m>n} E(|M_m - M_n|),$$

$$P\left[\sup_{i,j \leqslant m} |M_i - M_j| \geqslant 2\lambda\right] \leqslant \frac{1}{\lambda} \sup_{n<m} E(|M_m - M_n|).$$

The sequences

$$\left\{\sup_{n \leqslant i,j} |M_i - M_j|\right\}_{n \geqslant 0}$$

and

$$\left\{\sup_{i,j \leqslant -n} |M_i - M_j|\right\}_{n \geqslant 0}$$

are decreasing, they have a limit a.s. However they tend to 0 in probability, since (M_n) is Cauchy in L^1. Hence they also tend a.s. to 0. The sequence $(M_n(\omega))$ is a Cauchy sequence if n tends to $\pm\infty$ for almost all ω: it converges a.s. to $M_{\pm\infty}$. Since A converges a.s., the same holds for the sequence X^2, hence also for the positive sequence X.

(b) If X is a positive martingale, e^{-X} is a positive bounded submartingale: hence it converges a.s. if n tends to $\pm\infty$. Thus (X_n) converges a.s. to the positive r.v.'s $X_{\pm\infty}$.

However applying Fatou's theorem,

$$E[X_{\pm\infty}] \leqslant \varliminf_{n \to +\infty} E(X_n) = E(X_0).$$

The r.v.'s $X_{\pm\infty}$ are thus integrable (and therefore finite a.s.).

(c) Let X be a martingale, for which $\sup E(|X_n|)$ is finite. Applying the Krickeberg decomposition and (b), it converges a.s. to an integrable r.v. if n tends to $\pm\infty$.

(d) Let X be a submartingale for which $\sup E(|X_n|)$ is finite. We use the Doob decomposition $X = M + A$, and for $m < n$ the inequality,

$$E(A_n) - E(A_m) = E(X_n) - E(X_m)$$

$$\leqslant E(X_n) + \sup_{p\leqslant n} E(X_p^-) \leqslant 2 \sup E(|X_n|)$$

When $m \to -\infty$, the sequence (A_m) thus decreases to 0 in L^1. Hence the sequence (A_n) converges to A_∞, a.s. and in L^1, if n tends to $+\infty$. Now $\sup E(|M_n|) \leqslant 3 \sup E(|X_n|)$. We apply (c), and M converges to integrable r.v.'s if n tends to $\pm\infty$. Hence the same holds for X.

(e) Let X be a submartingale satisfying $\sup E(X_n^+) < \infty$. We then have: $|X_n| = -X_n + 2X_n^+$. Hence,

$$\sup_{n\geqslant 0} E(|X_n|) \leqslant 2 \sup_{n\geqslant 0} E(X_n^+) - E(X_0) < \infty$$

and, if $n \to +\infty$, the sequence $(X_n)_{n\geqslant 0}$ tends a.s. to an integrable r.v. $X_{+\infty}$.

Let us study the case $n \to -\infty$. For every rational number q, $(\sup (X,q))$ is a submartingale which satisfies condition (d). Hence it converges a.s., if n tends to $-\infty$, to a finite r.v. $X^q_{-\infty}$,

$$\left\{\overline{\lim}_{n\to-\infty} X_n \neq \underline{\lim}_{n\to-\infty} X_n\right\} = \bigcup_{q\in \mathbb{Q}} \left\{\overline{\lim}_{n\to-\infty} X_n > q \geqslant \underline{\lim}_{n\to-\infty} X_n\right\}$$

$$= \bigcup_{q\in \mathbb{Q}} \left\{\overline{\lim}_{n\to-\infty} \sup(X_n,q) > q \right.$$

$$\left.\geqslant \underline{\lim}_{n\to-\infty} \sup(X_n,q)\right\}$$

and hence

$$P\left\{\overline{\lim}_{n\to-\infty} X_n \neq \underline{\lim}_{n\to-\infty} X_n\right\} = 0,$$

$(X_{-n})_{n\geqslant 0}$ converges a.s. to an r.v. $X_{-\infty}$ which cannot be equal to $+\infty$, since then $X^q_{-\infty}$ would equal $+\infty$ for all q.

2.4.3. Convergence in L^p for $p > 1$

Theorem 2.4.14. *If X is a martingale or a positive submartingale,* pth *power integrable for $p \in]1,\infty[$, then (X_n) converges in L^p to $X_{-\infty}$ if n tends to $-\infty$. If $\sup\|X_n\|_p$ is finite, then (X_n) converges in L^p to $X_{+\infty}$ if n tends to $+\infty$.*

Proof. Use the inequalities of [2.3.3]. If X_0^p is integrable, $2\sup_{n\leq 0}|X_n|^p$ is integrable and majorizes the a.s. convergent sequence $\{|X_n - X_{-\infty}|^p\}_{n\geq 0}$. If $\sup\|X_n\|_p$ is infinite, $2\sup_n|X_n|^p$ is integrable and majorizes the a.s. convergent sequence $\{|X_n - X_\infty|^p\}_{n\geq 0}$.

2.4.4. Convergence in L^1

If X is a (sub) martingale which converges a.s. to $X_{\pm\infty}$ when n tends to $\pm\infty$, is there also convergence in L^1? This is important in order to be able to extend the (sub)martingale property to the times $\pm\infty$. In fact, if a sequence of integrable r.v.'s (Y_n) converges in L^1 to Y_∞, and if $\mathcal{B}$ is a sub-σ-algebra of $\mathcal{A}$, $(E(Y_n|\mathcal{B}))$ converges in L^1 to $(E(Y_\infty|\mathcal{B}))$ since conditional expectation diminishes the L^1 norm. Thus, if X is a (sub)martingale which converges in L^1 to $X_{+\infty}$ when n tends to $+\infty$, then,

$$E^n(X_{+\infty}) = \lim_{m\to\infty} E^n(X_m) = (\geq)\ X_n.$$

Let X be a (sub)martingale which converges in L^1 to $X_{-\infty}$ if n tends to $-\infty$. Then, for $\Gamma \in \mathcal{F}_{-\infty}$ and $m < n$,

$$E[1_\Gamma X_n] = E(1_\Gamma E^m(X_n))$$
$$= (\leq)\ E(1_\Gamma X_m).$$

Letting m tend to $-\infty$

$$E(1_\Gamma X_n) = E(1_\Gamma E^{-\infty}(X_n)) = (\leq)\ E(1_\Gamma X_{-\infty})$$

$$E^{-\infty}(X_n) = (\leq)\ X_{-\infty}.$$

In order to deduce convergence in L^1 from a.s. convergence, the basic tool is equi-integrability.

Definition 2.4.15. A sequence of r.v.'s $(X_n)_{n\geq 0}$ is said to be **equi-integrable**, if

$$\lim_{a\to\infty} \overline{\lim_{n}}\, E(|X_n| 1_{(|X_n|>a)}) = 0.$$

Notes. (a) $E(|X_m|) \leq a + E(|X_n|1_{\{|X_n|>a\}})$. *Thus, for such a sequence,* $\overline{\lim}_n E(|X_n|)$ *is bounded and* $\sup_n E(|X_n|)$ *is finite.*

(b) *Let* Φ *be a function* $\mathbb{R} \to \mathbb{R}^+$ *such that*

$$\lim_{|x|\to\infty} \frac{\phi(x)}{|x|} = \infty.$$

Then $\sup E(\phi(X_n)) < \infty$ *implies that the sequence* (X_n) *is equiintegrable. In fact*

$$E(|X_n|1_{\{|X_n|>a\}} \leq \frac{a}{\phi(a)} E(\phi(|X_n|)1_{\{|X_n|>a\}}$$

$$\leq \frac{a}{\phi(a)} \sup_n E(\phi(X_n)).$$

Proposition 2.4.16. *An equiintegrable sequence which converges* a.s. *converges also in* L^1.

Proof. Let (X_n) be an equiintegrable sequence which converges a.s. to X_∞. Let $\varepsilon > 0$ and let a be such that

$$\sup \int_{\{|X_n|>a\}} |X_n| dP$$

is majorized by ε.

To an r.v. X, we can associate the r.v. $\tau_a(X)$ said to be **truncated** at a defined by

$$\tau_a(X) = X1_{(|X|\leq a)} + a1_{(X\geq a)} - a1_{(X\leq -a)}.$$

Then for n and $m \geq 0$,

$$\|X_m - X_n\|_1 \leq \|\tau_a(X_m) - \tau_a(X_n)\|_1$$

$$+ \|X_m - \tau_a(X_m)\|_1 + \|X_n - \tau_a(X_n)\|_1.$$

The sequence $\tau_a(X_n)$ converges in L^1 to $\tau_a(X_\infty)$ (since it converges a.s. and is bounded), and $\|X_n - \tau_a(X_n)\|_1$ is majorized by $\int_{\{|X_n|>a\}}|X_n|dP$. Hence,

$$\overline{\lim_{n,m\to\infty}} \|X_m - X_n\|_1 \leq 2\varepsilon.$$

This is true for all $\varepsilon > 0$, and (X_n) converges in L^1.

Theorem 2.4.17. (a) *An equiintegrable (sub)martingale* X *converges* a.s. *and in* L^1 *to* $X_{+\infty}$ *if* n *tends to* $+\infty$, *and we have for all* n,

$$X_n = (\leqslant)\ E^n(X_\infty).$$

(b) *A (sub)martingale* X *such that* $\sup_{n\to-\infty} E(X_n)$ *is finite, converges* a.s. *and in* L^1 to $X_{-\infty}$ *if* n *tends to* $-\infty$, *and we have for all* n,

$$X_{-\infty} = (\leqslant)\ E^{-\infty}(X_n).$$

Proof. (a) follows from the above, the equiintegrability implying that $\sup E(|X_n|)$ is finite.

(b) As (X_n) is a submartingale, the sequence $E(X_n)$ is decreasing. As (X_n^+) is a submartingale, we have,

$$E(|X_n|) = 2E(X_n^+) - E(X_n) \leqslant 2\ E(X_0^+) - E(X_n)$$

and

$$\sup_{n\leqslant 0} E(|X_n|) \leqslant 2\ E(X_0^+) - \lim \downarrow E(X_n) < \infty.$$

Thus under the hypothesis of the theorem (which is equivalent to $\sup_{n\leqslant 0} E(|X_n|) < \infty$) the submartingale $(X_n)_{n\leqslant 0}$ converges a.s. for $n \to -\infty$. Let us show that the sequence $(X_{-n})_{n\geqslant 0}$ is equiintegrable. Let $\varepsilon > 0$ and let k be such that $0 \leqslant E(X_k) - \lim_{n\to-\infty} E(X_n) \leqslant \varepsilon$. Let $a > 0$ and $n < k$,

$$E(|X_n| \mathbf{1}_{(|X_n|\leqslant a)}) = E\left[X_n(\mathbf{1}_{(X_n\geqslant a)} + \mathbf{1}_{(X_n>-a)} - 1)\right]$$

$$\leqslant E\left[E^n(X_k)(\mathbf{1}_{(X_n\geqslant a)} + \mathbf{1}_{(X_n>-a)})\right] - E(X_n)$$

$$\leqslant E\left[X_k(\mathbf{1}_{(X_n\geqslant a)} + \mathbf{1}_{(X_n>-a)} - 1)\right] + \varepsilon$$

$$\leqslant E\left[X_k \mathbf{1}_{(|X_n|\geqslant a)}\right] + \varepsilon.$$

The sequence $(\mathbf{1}_{(|X_n|\geqslant a)} X_k)$ tend a.s. and in L^1 (since it is majorized by $|X_k|$) to

$$\mathbf{1}_{(|X_{-\infty}|\geq a)}X_k \text{ when } n \text{ tends to } -\infty.$$

Hence,

$$\overline{\lim_{n\to-\infty}} E(|X_n|\mathbf{1}_{(|X_n|\geq a)}) \leqslant E\left[\mathbf{1}_{(|X_{-\infty}|\geq a)}X_k\right] + \varepsilon$$

and

$$\lim_{a\to\infty} \overline{\lim_{n\to-\infty}} E\left[|X_n|\mathbf{1}_{(|X_n|\geq a)}\right] \leqslant \varepsilon.$$

This is true for all $\varepsilon > 0$, from which the result follows: $(X_n)_{n\geq 0}$ is equiintegrable.

Corollary 2.4.18. *Let $Y \in L^1$, $E^n(Y)$ tends to $E^{\pm\infty}(Y)$ when n tends to $\pm\infty$,* a.s. *and in L^2. Hence, for Y $\mathcal{F}^{\infty}$-measurable, $E^n(Y)$ tends to Y, if n tends to ∞,* a.s. *and in L^1 (in L^p for $Y \in L^p$).*

Proof. Let us show that the sequence $(E^n(Y))$ is equiintegrable. It is sufficient to prove it for $|Y|$, hence we can assume $Y \geqslant 0$. Then, for a and $b > 0$,

$$\begin{aligned}
E[E^n(Y)\mathbf{1}_{(E^n(Y)\geq a)}] &= E[Y\mathbf{1}_{(E^n(Y)\geq a)}] \\
&\leqslant E[Y\mathbf{1}_{(Y\geq a)}] + bP[E^n(Y) \geqslant a] \\
&\leqslant b\,\frac{E(E^n(Y))}{a} + E(Y\mathbf{1}_{(Y\geq b)}) \\
&\leqslant \frac{bE(Y)}{a} + E(Y\mathbf{1}_{(Y\geq b)}).
\end{aligned}$$

Taking $a = b^2$, we obtain the majorant

$$\frac{E(Y)}{\sqrt{a}} + E(Y\mathbf{1}_{(Y\geq\sqrt{a})}),$$

which tends to 0 if a tends to ∞.

Notes. *We often use this result in the following situation: $X = (X_n)_{n\in\mathbf{N}}$ is a sequence of* r.v.'s, *ϕ an* r.v. *defined on $(\mathbb{R},\mathcal{B}_{\mathbb{R}})^{\mathbf{N}}$ such that $\phi(X)$ is integrable. Then for all $\varepsilon > 0$, there exists an n and an* r.v. *$\psi(X_1, ..., X_n)$ such that $E(|\phi(X) - \psi(X_1, ..., X_n)|) \leqslant \varepsilon$.*

If $X = (X_n)_{n\in\mathbb{Z}}$ is a martingale which converges in L^1 for $n \to \pm\infty$, we can extend it to a sequence $(X_n)_{-\infty\leqslant n\leqslant\infty}$ adapted to the filtration $\hat{\mathbb{F}} = (F_n)_{-\infty\leqslant n\leqslant\infty}$. We then have a stopping theorem for all stopping times adapted to $\hat{\mathbb{F}}$.

Theorem 2.4.19. *Let $(X_n)_{n\in\mathbb{Z}}$ be an equiintegrable $\mathbb{F}$-martingale and $X_{\pm\infty}$ its limits for $n \to \pm\infty$. Let $\hat{\mathbb{F}} = (F_n)_{-\infty\leqslant n\leqslant\infty}$.*

(a) *If ν_1 and ν_2 are two $\hat{\mathbb{F}}$-stopping times and $\nu_1 \leqslant \nu_2$, then,*

$$E(X_{\nu_2} \mid F_{\nu_1}) = X_{\nu_1}.$$

(b) *If ν is an $\hat{\mathbb{F}}$ stopping time, X_ν is integrable and, for all n,*

$$E(X_\nu \mid F_n) = X_{\nu\wedge n}\ .$$

Proof. (a) From Theorem 2.3.8, $(X_{\nu\wedge n})_{n\in\mathbf{Z}}$ is a martingale adapted to $(F_{\nu\wedge n})$; it converges a.s. to X_ν if $n \to \infty$. Let us show that it is equiintegrable. Apply Theorem 2.3.8 to the submartingale $|X_n|$,

$$E(1_{(|X_{\nu\wedge n}|\geqslant a)}|X_{\nu\wedge n}|) \leqslant E[1_{(|X_{\nu\wedge n}|\geqslant a)}|X_n|]$$

$$\leqslant E[1_{(|X_{\nu\wedge n}|\geqslant a)}|X_\infty|]\ ;$$

$$\overline{\lim_{n\to\infty}}\ E(1_{(|X_{\nu\wedge n}|\geqslant a)}|X_{\nu\wedge n}|) \leqslant E[1_{(|X_\nu|\geqslant a)}|X_\infty|]$$

and

$$\lim_{a\to 0}\ \overline{\lim_{n\to\infty}}\ E(1_{(|X_{\nu\wedge n}|\geqslant a)}|X_{\nu\wedge n}|) = 0.$$

The sequence $(X_{\nu\wedge n})$ converges in L^1 to X_ν, if $n \to \infty$. Fix N an integer; let $n > N$,

$$E\left[X_{\nu_2\wedge n} \mid F_{\nu_1\wedge N}\right] = X_{\nu_1\wedge N}\ .$$

If n tends to ∞, we obtain

$$E\left[X_{\nu_2} \mid F_{\nu_1\wedge N}\right] = X_{\nu_1\wedge N}.$$

We then let N tend to ∞ and we obtain result (a) of the theorem, as a result of Corollary 2.4.18.

(b) We write

$$E(X_\infty - X_\nu \mid \mathcal{F}_{n\vee\nu}) = X_{n\vee\nu} - X_\nu$$

$$= \mathbf{1}_{(\nu\leq n)}(X_n - X_{n\wedge\nu}).$$

This r.v. is $\mathcal{F}_n$-measurable, hence:

$$E(X_\infty - X_\nu \mid \mathcal{F}_n) = X_n - X_{n\wedge\nu}$$

and

$$E(X_\nu \mid \mathcal{F}_n) = X_{\nu\wedge n}.$$

From which result (b) follows.

2.5. Likelihoods

2.5.1. Approximation of Radon-Nikodym Derivatives

Here we are concerned with answering the question posed in [2.1.2]. The ideas in this paragraph will be useful in all chapters for the statistics of processes.

Theorem 2.5.10. *Let $\mathbb{F} = (\mathcal{F}_n)$ be a filtration on Ω and let $\mathcal{F}_\infty = \bigvee \mathcal{F}_n$. We are given two probabilities P and Q on $(\Omega, \mathcal{F}_\infty)$, and we assume that, for all n, the trace of Q on $\mathcal{F}_n$ is absolutely continuous with respect to that of P: $Q = L_nP$ on $(\Omega, \mathcal{F}_n)$, where L_n is $\mathcal{F}_n$-measurable.*

(a) *(L_n) converges $(P + Q)$-a.s. to an r.v. L_∞, and $Q = \mathbf{1}_{(L_\infty=\infty)}Q + L_\infty P$ with $P(L_\infty = \infty) = 0$.*

(b) *Each of the following conditions is necessary and sufficient for Q to be absolutely continuous with respect to P: $Q(L_\infty = \infty) = 0$; or $E_P(L_\infty) = 1$. We then have $Q = L_\infty P$, and $L_n = E^n(L_\infty)$ converges in $L^1(P)$ to L_∞.*

(c) *Each of the following conditions is necessary and sufficient for Q to be singular with respect to P: $Q(L_\infty = \infty) = 1$; or $E_P(L_\infty) = 0$.*

Proof. Denote by P_n and Q_n the traces of P and Q on $\mathcal{F}_n$. Q_n is dominated by $P_n + Q_n$ and thus there exists an

$\mathcal{F}_n$-,measurable U_n, such that

$$Q_n = U_n \cdot \frac{P_n + Q_n}{2} .$$

The sequence $(U_n)_{n \geqslant 0}$ is a martingale adapted to $\mathbb{F}$ on $(\Omega, \mathcal{F}_\infty, (P+Q)/2)$. In fact, for all $\Gamma \in \mathcal{F}_n$

$$\int_\Gamma U_{n+1}\, d\left[\frac{P+Q}{2}\right] = \int_\Gamma dQ_{n+1} = Q(\Gamma)$$

$$= \int_\Gamma U_n\, d\left[\frac{P+Q}{2}\right].$$

Since, for all $A \in \mathcal{F}_n$,

$$Q_n(A) \leqslant 2\, \frac{P_n(A) + Q_n(A)}{2} ,$$

we have $0 \leqslant U_n \leqslant 2$, and thus the sequence $(U_n)_{n \geqslant 0}$ converges $(P + Q)$-a.s. and in $L^1(P + Q)$. Let U_∞ be its limit. For every $\Gamma \in \mathcal{F}_\infty$ we have

$$\int_\Gamma U_\infty\, d\left[\frac{P+Q}{2}\right] = \lim_{n \to \infty} \int_\Gamma U_n\, d\left[\frac{P+Q}{2}\right] = Q(\Gamma),$$

from which $Q = U_\infty(P+Q)/2$, giving $(2 - U_\infty)Q = U_\infty P$ and $P(U_\infty = 2) = 0$. If $Q(U_\infty = 2) = 0$, then Q is absolutely continuous with respect to P, with density $U_\infty/(2 - U_\infty)$. Set $L_n = U_n/2-U_n$ and $L_\infty = U_\infty/(2 - U_\infty)$; (L_n) tends $(P + Q)$-a.s. to L_∞ and $P(L_\infty = \infty) = 0$. We have,

$$Q_n = L_n P_n, \quad \frac{2}{1 + L_\infty} Q = \frac{2L_\infty}{1 + L_\infty} P.$$

From which the Lebesgue decomposition of Q with respect to P follows,

$$Q = 1_{(L_\infty = \infty)} Q + L_\infty P$$

with $P(L_\infty = \infty) = 0$, $Q(L_\infty = 0) = 0$.

Consequence. We have the answers to [2.1.2] and a method of approximating probability densities on a space $(\Omega, \mathcal{A})$ when $\mathcal{A}$ is separable.

2.5.2. Formulations Using Hellinger Distance or Kullback Information

We have defined **Hellinger distance** H^2 in [Vol I, 6.4]. If μ dominates P and Q and if $P = p\mu$ and $Q = q\mu$,

$$H^2(P,Q) = \frac{1}{2}\int(\sqrt{p} - \sqrt{q})^2 d\mu = 1 - \int \sqrt{pq}\, d\mu$$

which we also write as

$$H^2(P,Q) = \frac{1}{2}\int(\sqrt{P} - \sqrt{Q})^2 = 1 - \int\sqrt{PQ}\,.$$

By considering the traces of P and Q on $(\Omega, \mathcal{F}_n)$: $H^2(P_n,Q_n) = 1 - E_P(\sqrt{L_n})$. On $(\Omega, \mathcal{F}_\infty, P)$, the sequence $(\sqrt{L_n})$ is equiintegrable for $E(L_n) = 1$ for all n (see the observation following Definition 2.4.15). Hence $(\sqrt{L_n})$ converges a.s. and in L^1 to $\sqrt{L_\infty}$. As $(\sqrt{L_n})$ is a supermartingale, $E(\sqrt{L_n})$ decreases to $E(\sqrt{L_\infty})$. But,

$$\int \sqrt{PQ} = \int \sqrt{U_\infty(2-U_\infty)}\; d\left(\frac{P+Q}{2}\right) = E_P(\sqrt{L_\infty}).$$

Theorem 2.5.21. *In the framework of Theorem 2.5.20, the sequence $E_P(\sqrt{L_n})$ decreases to $E_P(\sqrt{L_\infty}) = \int\sqrt{PQ}$ (and $E_P(\sqrt{L_n} - \sqrt{L_\infty})^2$ tends to 0). The measures P and Q are perpendicular if, and only if, this limit $\int\sqrt{PQ}$ is zero.*

Consequence. From [2.2.1], we obtain

$$\int \sqrt{PQ} = \lim_n \sum_{i=1}^{u(n)} \sqrt{P(A_i)Q(A_i)}\ .$$

Let μ_1 and μ_2 be two bounded measures, μ_1 absolutely continuous with respect to μ_2, $\mu_1 = f\mu_2$. An extension to measures of the definition [Vol. I, 6.4.2] of the **Kullback information** of μ_1 and μ_2, is given by

$$K(\mu_1,\mu_2) = \int(f \text{ Log } f + 1 - f)d\mu_2;$$

K takes values in $[0,\infty]$. If μ_1 is not absolutely continuous with respect to μ_2, we set $K(\mu_1,\mu_2) = \infty$.

With the preceding notations, $x \longmapsto x \text{ Log } x + 1 - x$ is convex and positive, thus $L_n \text{Log } L_n + 1 - L_n = Z_n$ is a sub-martingale for P converging a.s.:

$$K(Q_n,P_n) = E_Q(\text{Log } L_n) = E_P(L_n \text{ Log } L_n) = E_P(Z_n)$$

is an increasing sequence. If its limit is finite it follows from note (b) of [2.4.4] that (L_n) is equiintegrable for P and $E_P(L_n) = 1$ tends to $E_P(L_\infty)$ and $Q \ll P$. Fatou's lemma gives: $K(Q,P) = E_P[Z_\infty] \leq \lim_n K(Q_n,P_n)$; but we know [Vol. I, 6.4.2] that $K(Q,P) \geq K(Q_n P_n)$. Then: $K(Q,P) = \lim_n K(Q_n,P_n)$. If $\lim_n K(Q_n,P_n) = \infty$, we cannot have $K(Q,P) = \infty$ which implies absolute continuity and $K(Q_n,P_n) \leq K(Q,P)$.

Let μ_1 and μ_2 be two bounded measures on $(\Omega,\mathcal{F}_\infty)$. Up to a constant function of the total masses, $K(\mu_1,\mu_2)$ is equal to

$$\mu_1(E)K\left[\frac{\mu_1}{\mu_1(E)}, \frac{\mu_2}{\mu_2(E)}\right]$$

which proves the following theorem, leading back to the case of probabilities.

Theorem 2.5.22. *Let $\mathbb{F} = (\mathcal{F}_n)$ be a filtration on a set Ω. Let μ_1 and μ_2 be two bounded measures on $(\Omega, \mathcal{F}_\infty)$ such that if μ_1^n and μ_2^n are their traces on $\mathcal{F}_n$, $\mu_1^n \ll \mu_2^n$. Then $K(\mu_1^n,\mu_2^n)$ increases to $K(\mu_1,\mu_2)$, whether or not these values are finite.*

2.5.3. Absolute Continuity and Singularity of Poisson Processes

Let us consider two Poisson distributions $p(\lambda_1)$ and $p(\lambda_2)$,

$$\int\sqrt{p(\lambda_1)p(\lambda_2)} = \exp\left[-\frac{1}{2}(\lambda_1 + \lambda_2)\right]\sum_{k\geq 0}\frac{1}{k!}(\sqrt{\lambda_1\lambda_2})^k$$

$$= \exp\left[-\frac{1}{2}(\lambda_1 + \lambda_2 - 2\sqrt{\lambda_1\lambda_2})\right];$$

$$K(p(\lambda_1),p(\lambda_2)) = \sum_{k=0}^{\infty} e^{-\lambda_1}\frac{\lambda_1^k}{k!}\operatorname{Log}\left\{e^{\lambda_2-\lambda_1}\left[\frac{\lambda_1}{\lambda_2}\right]^k\right\}$$

$$= \lambda_2 - \lambda_1 + \lambda_1 \operatorname{Log}\frac{\lambda_1}{\lambda_2};$$

$$\frac{dp(\lambda_1)}{dp(\lambda_2)}(k) = \exp\left[\lambda_2 - \lambda_1 + k\operatorname{Log}\frac{\lambda_1}{\lambda_2}\right].$$

Let $(E,\mathcal{E})$ be a measurable space, $\mathcal{E}$ being separable. We are given, for every n, a σ-algebra $\mathcal{B}_n$ generated by a finite partition $(A_1, \ldots, A_{u(n)})$; and we assume $\mathcal{B}_n \subset \mathcal{B}_{n+1}$ and $\mathcal{E} = \vee \mathcal{B}_n$. Let μ_1 and μ_2 be two σ-finite measures and, for $i = 1,2$, $\{\Omega,\mathcal{A},P_{\mu_i},(Z(A))_{A\in\mathcal{E}}\}$ the canonical Poisson process with intensity μ_i ([1.2.1]).

Assume first of all that μ_1 and μ_2 are bounded.

Let $\mathcal{F}_n = \sigma\{Z(A_1), \ldots, Z(A_{u(n)})\}$; the random vector $(Z(A_1), \ldots, Z(A_{u(n)}))$ has distribution $\otimes_{j=1}^{u(n)} p(\mu_i(A_j))$ for $i = 1,2$. For the traces $P^n_{\mu_1}$ and $P^n_{\mu_2}$ of P_{μ_1} and P_{μ_2} on $\mathcal{F}_n$, we have

$$\int \sqrt{P^n_{\mu_1} P^n_{\mu_2}} = \exp\left\{-\frac{1}{2}\left[\mu_1(E) + \mu_2(E) - 2\sum_{j=1}^{u(n)} \sqrt{\mu_1(A_j)\mu_2(A_j)}\right]\right\}.$$

If n tends to ∞, we obtain

$$\int \sqrt{P_{\mu_1} P_{\mu_2}} = \exp\left\{-\frac{1}{2}\left[\mu_1(E) + \mu_2(E) - 2\int \sqrt{\mu_1\mu_2}\right]\right\}$$

$$= \exp\left\{-\frac{1}{2}\int(\sqrt{\mu_1} - \sqrt{\mu_2})^2\right\}.$$

The same formula holds asuming only that μ_1 and μ_2 are σ-finite. Consider in fact a sequence (K_p) of measurable sets with finite measure for $\mu_1 + \mu_2$. Let $\mathcal{F}^p_n = \sigma\{Z(A_1 \cap K_p), \ldots, Z(A_{u(n)} \cap K_p)\}$; $\hat{\mathcal{F}}^p = \sigma\{Z(A \cap K_p)\}$; $A \in \mathcal{E}\}$. For the traces $\hat{P}^p_{\mu_1}$ and $\hat{P}^p_{\mu_2}$ of P_{μ_1} and P_{μ_2} on $\hat{\mathcal{F}}^p$, we obtain,

$$\int \sqrt{\hat{P}^p_{\mu_1} \hat{P}^p_{\mu_2}} = \exp\left\{-\frac{1}{2}\int_{K_p} (\sqrt{\mu_1} - \sqrt{\mu_2})^2\right\}.$$

And we pass to the limit,

$$\int \sqrt{P_{\mu_1} P_{\mu_2}} = \exp\left[-\frac{1}{2}\int (\sqrt{\mu_1} - \sqrt{\mu_1})^2\right].$$

Proposition 2.5.23. *For two canonical Poisson processes taking values in a measurable space with σ-finite intensities μ_1 and μ_2 and distributions P_{μ_1} and P_{μ_2}, we have*

$$\int \sqrt{P_{\mu_1} P_{\mu_2}} = \exp\left[-\frac{1}{2}\int(\sqrt{\mu_1} - \sqrt{\mu_1})^2\right].$$

Hence they are singular if, and only if, $\int(\sqrt{\mu_1} - \sqrt{\mu_2})^2$ *is infinite.*

Now let us study Kullback information, assuming μ_1 and μ_2 are bounded. Let us denote by μ_1^n and μ_2^n the traces of μ_1 and μ_2 on $\mathcal{B}_n$.

$$K(P^n_{\mu_1}, P^n_{\mu_2}) = \sum_{j=1}^{u(n)} \left[\mu_2(A_j) - \mu_1(A_j) + \mu_1(A_j)\text{Log}\,\frac{\mu_1(A_j)}{\mu_2(A_j)}\right]$$

$$= K(\mu_1^n, \mu_2^n).$$

We obtain, letting $n \to \infty$

$$K(P_{\mu_1}, P_{\mu_2}) = K(\mu_1, \mu_2).$$

Proposition 2.5.24. *For two canonical Poisson processes taking values in a measurable space, with bounded intensities* μ_1 *and* μ_2 *and distributions* P_{μ_1} *and* P_{μ_2}, *we have*

$$K(P_{\mu_1}, P_{\mu_2}) = K(\mu_1, \mu_2).$$

Finally let us study likelihoods. Let us assume once more that μ_1 and μ_2 are bounded and $\mu_1 << \mu_2$: $\mu_1 = f\mu_2$ with $f \leq 1$. Then $P^n_{\mu_1} = L_n P^n_{\mu_2}$ with

$$L_n = \prod_{j=1}^{u(n)} \exp\left[\mu_2(A_j) - \mu_1(A_j) + Z(A_j)\text{Log}\,\frac{\mu_1(A_j)}{\mu_2(A_j)}\right]$$

Let us show that

$$P_{\mu_1} = \exp\left[\mu_2(E) - \mu_1(E) + \int \text{Log}\, f\, dZ\right] P_{\mu_2}.$$

Let

$$f_n = \sum_{j=1}^{u(n)} \frac{\mu_1(A_j)}{\mu_2(A_j)} 1_{A_j}.$$

From [2.5.1] the sequence (f_n) tends μ_2-a.s. to f; by virtue of [2.5.2],

$$E_{\mu_1}\left[\int \operatorname{Log} f_n dZ\right] = \sum_{j=1}^{u(n)} \mu_1(A_j) \operatorname{Log} \frac{\mu_1(A_j)}{\mu_2(A_j)}$$

tends to $\int \operatorname{Log} f \, d\mu_1$ if $n \to \infty$, (these expressions equal $K(\mu_1^n, \mu_2^n)$ and $K(\mu_1, \mu_2)$, up to $\mu_2(E) - \mu_1(E)$). Moreover f and f_n take their values in $[0,1]$ and $(\operatorname{Log} f_n - \operatorname{Log} f)_+$ is majorized by $-\operatorname{Log} f$ which is μ_1-integrable as a result of the relations:

$$-\int \operatorname{Log} f \, d\mu_1 = -\int f \operatorname{Log} f \, d\mu_2 \leqslant \mu_2(E) - \mu_1(E).$$

Hence,

$$E_{\mu_1}\left[\int (\operatorname{Log} f_n - \operatorname{Log} f)_+ dZ\right] = \int (\operatorname{Log} f_n - \operatorname{Log} f)_+ d\mu_1$$

tends to 0 for $n \to \infty$. Since

$$|\operatorname{Log} f_n - \operatorname{Log} f| = 2(\operatorname{Log} f_n - \operatorname{Log} f)_+ - (\operatorname{Log} f_n - \operatorname{Log} f)$$

we obtain

$$E_{\mu_1}\left[\int |\operatorname{Log} f_n - \operatorname{Log} f| dZ\right] \xrightarrow[n \to \infty]{} 0,$$

and $\int \operatorname{Log} f_n dZ$ tends to $\int \operatorname{Log} f \, dZ$ in L^1. A subsequence of $(\int \operatorname{Log} f_n dZ)$ tends a.s. to $\int \operatorname{Log} f \, dZ$. Since (L_n) converges P_{μ_1}-a.s. its limit is,

$$\exp\left[\mu_2(E) - \mu_1(E) + \int \operatorname{Log} f \, dZ\right].$$

Let μ_1 and μ_2 be arbitrary and bounded and let $f = d\mu_1/d(\mu_1+\mu_2)$. We have

$$\mu_1 = 1_{[f=1]} \mu_1 + \frac{f}{1-f} \mu_2;$$

$$P_{\mu_1} = \exp\left[\mu_2(E) + \int \operatorname{Log} f \, dZ\right] P_{\mu_1+\mu_2};$$

$$P_{\mu_2} = \exp\left[\mu_1(E) + \int \operatorname{Log}(1-f) dZ\right] P_{\mu_1+\mu_2}.$$

Z is P_{μ_1}-a.s. the sum of a finite number of point masses,

$$P_{\mu_1}\left[\int \operatorname{Log}(1-f) dZ = -\infty\right] = P_{\mu_1}[Z\{f = 1\} > 0]$$

$$= 1 - \exp[-\mu_1(f = 1)].$$

If μ_1 is not absolutely continuous with respect to μ_2, P_{μ_1} charges $A^c = \{\int \mathrm{Log}(1 - f)dZ = -\infty\}$ while P_{μ_2} does not charge it. Conversely if A^c is not charged by P_{μ_1}, then for every set B,

$$P_{\mu_2}(B) = 0 \quad \text{implies} \quad P_{\mu_1+\mu_2}(A \cap B) = 0$$

hence $P_{\mu_1}(B) = 0$; then,

$$P_{\mu_1} \ll P_{\mu_2} \text{ and } \frac{dP_{\mu_1}}{dP_{\mu_2}} = \frac{dP_{\mu_1}}{dP_{\mu_1+\mu_2}}\left[\frac{dP_{\mu_2}}{dP_{\mu_1+\mu_2}}\right]^{-1}.$$

Proposition 2.5.25. *If μ_1 and μ_2 are two bounded measures on E, $P_{\mu_1} \ll P_{\mu_2}$ is equivalent to $\mu_1 \ll \mu_2$,*

$$\frac{dP_{\mu_1}}{dP_{\mu_2}} = \exp\left[\mu_2(E) - \mu_1(E) + \int \mathrm{Log} \frac{d\mu_1}{d\mu_2}\, dZ\right].$$

and $K(P_{\mu_1}, P_{\mu_2}) = K(\mu_1, \mu_2)$.

Note. In the general case

$$P_{\mu_1} = 1_A P_{\mu_1} + 1_{A^c} P_{\mu_1}$$

with $1_A P_{\mu_1} \ll P_{\mu_2}$, $1_{A^c} P_{\mu_1} \perp P_{\mu_2}$. The total mass of the absolutely continuous part is,

$$P_{\mu_1}(A) = \exp(-\mu_1(f = 1)) \geqslant \exp(-\mu_1(E)).$$

If μ_1 and μ_2 are bounded, P_{μ_1} and P_{μ_2} are never singular. This result also follows from Proposition 2.5.23, since $\int(\sqrt{\mu_1} - \sqrt{\mu_2})^2$ is then finite. This type of phenomenon cannot occur in the Gaussian case, we shall see this in [2.5.4].

2.5.4. Kullback Information of Gaussian Processes

Consider two centered canonical Gaussian sequences $(\Omega, A, (X_n)_{n\geq 1}, P_i)$, $i = 0,1$; $(\Omega, A) = (\mathbb{R}, \mathcal{B}_{\mathbb{R}})$ and $F_n = \sigma(X_1, ..., X_n)$. Let $P_{i,n}$ be the trace of P_i on F_n, the distribution of $X^{(n)} = (X_1, ..., X_n)$ is,

$$X^{(n)}(P_{i,n}) = N_n(0, \Gamma_i^n).$$

We assume that the determinants $|\Gamma_i^n|$ of the matrices Γ_i^n are nonzero. Then,

$$\text{Log } L_n = \text{Log } \frac{dP_{0,n}}{dP_{1,n}} = \frac{1}{2}\text{Log } \frac{|\Gamma_1^n|}{|\Gamma_0^n|}$$

$$+ \frac{1}{2}({}^tX^{(n)}(\Gamma_1^n)^{-1}X^{(n)} - {}^tX^{(n)}(\Gamma_0^n)^{-1}X^{(n)}).$$

We can always find a basis of $\mathbb{R}^n$ orthonormal for the scalar product $(x,y) \longmapsto {}^tx(\Gamma_0^n)^{-1}y$ and orthogonal for the scalar product $(x,y) \longmapsto {}^tx(\Gamma_1^n)^{-1}y$. Let U be the matrix the column vectors of which are these vectors; ${}^tU(\Gamma_0^n)^{-1}U$ is the identity matrix and ${}^tU(\Gamma_1^n)^{-1}U$ is a diagonal matrix the diagonal terms of which are denoted by $\sigma_{j,n}^2$, $j = 1, ..., n$. From which,

$$\frac{|\Gamma_0^n|}{|\Gamma_1^n|} = \left|U^{-1}\Gamma_0^n({}^tU)^{-1}\ {}^tU(\Gamma_1^n)^{-1}U\right| = \prod_{j=1}^n \sigma_{j,n}^2 .$$

Set $Y^{(n)} = U^{-1}X^{(n)}$, $Y^{(n)} = (Y_{j,n})_{1\leq j\leq n}$. We have,

$${}^tX^{(n)}(\Gamma_0^n)^{-1}X^{(n)} = \sum_{j=1}^n Y_{j,n}^2$$

$${}^tX^{(n)}(\Gamma_1^n)^{-1}X^{(n)} = \sum_{j=1}^n \sigma_{j,n}^2 Y_{j,n}^2 .$$

Thus

$$\text{Log } L_n = \frac{1}{2}\sum_{j=1}^n (-\text{Log } \sigma_{j,n}^2 + (\sigma_{j,n}^2 - 1)Y_{j,n}^2).$$

For P_0, the covariance matrix of $Y^{(n)}$ is

$$({}^tU)^{-1}\Gamma_0^nU^{-1} = (U(\Gamma_0^n)^{-1}\ {}^tU)^{-1} = I;$$

and for P_1 this is $({}^tU^{-1})\Gamma_1^nU^{-1} = (U(\Gamma_1^n)^{-1}\ {}^tU)^{-1}$, a diagonal matrix the terms of which are $1/\sigma_{j,n}^2$, $j = 1, ..., n$. From which the calculation of Kullback information follows,

$$K(P_0^n,P_1^n) = E_0(\text{Log } L_n) = \frac{1}{2}\sum_{j=1}^{n}(-\text{Log } \sigma_{j,n}^2 + \sigma_{j,n}^2 - 1)$$

$$K(P_1^n,P_0^n) = -E_1(\text{Log } L_n) = \frac{1}{2}\sum_{j=1}^{n}\left[\text{Log } \sigma_{j,n}^2 - 1 + \frac{1}{\sigma_{j,n}^2}\right].$$

Let us assume that $\sup\{\sigma_{j,n}^2;\ 1 \leqslant j \leqslant n,\ n \geqslant 1\} = \infty$. Then a certain subsequence $(\sigma_{j(k),n(k)}^2)$ tends to ∞ as $k \to \infty$. For P_1, $(\tilde{Y}_{j(k),n(k)})$ tends to 0 in probability, and a subsequence $(\tilde{Y}_{j(k),n(k)})$ tends to 0 a.s. Since, for P_0, $\tilde{Y}_{j(k),n(k)}$ has distribution $N(0,1)$, P_0 does not charge $\{\lim_k \tilde{Y}_{j(k),n(k)} = 0]$; P_0 and P_1 are perpendicular.

If $\inf\{\sigma_{j,n}^2;\ 1 \leqslant j \leqslant n,\ n \geqslant 1\} = 0$ argue similarly by considering the r.v.'s $(\sigma_{j,n}^2 Y_{j,n})$, which are $N(0,1)$ r.v.'s for P_1, a certain subsequence of which tends to 0, P_0-a.s. Assume thus that there exist an m and an M such that for $1 \leqslant j \leqslant n$ and $n \geqslant 1$

$$0 < m \leqslant \sigma_{j,n}^2 \leqslant M < \infty.$$

Consider the **divergence** of P_0^n and P_1^n, $K(P_0^n,P_1^n) + K(P_1^n,P_0^n)$, denoted by $J(P_0^n,P_1^n)$ or J_n. From Theorem 2.5.22, if the sequence (J_n) has a finite limit, P_0 and P_1 are equivalent. Let us assume that (J_n) tends to infinity. We have, for $n \geqslant 1$,

$$J_n = \frac{1}{2}\sum_{j=1}^{n}\frac{(\sigma_{j,n}^2 - 1)^2}{\sigma_{j,n}^2} \geqslant \frac{1}{2M}\sum_{j=1}^{n}(\sigma_{j,n}^2 - 1)^2.$$

Let

$$B_n = \left\{\text{Log } L_n - K(P_0^n,P_1^n) \geqslant -\frac{1}{2}J_n\right\}$$

$$= \left\{\sum_{j=1}^{n}(\sigma_{j,n}^2 - 1)Y_{j,n}^2 \geqslant \sum_{j=1}^{n}(\sigma_{j,n}^2 - 1) + J_n\right\};$$

$$B_n^c = \left\{-\text{Log } L_n - K(P_1^n,P_0^n) > -\frac{1}{2}J_n\right\}.$$

Thus

$$P_0(B_n) \leqslant 4\,\frac{\sigma_{P_0}^2(\text{Log } L_n)}{J_n^2}, \quad P_1(B_n^c) \leqslant 4\,\frac{\sigma_{P_1}^2(\text{Log } L_n)}{J_n^2},$$

with

$$\sigma^2_{P_0}(\text{Log } L_n) = \sum_{j=1}^{n} (\sigma^2_{j,n} - 1)$$

$$\sigma^2_{P_1}(\text{Log } L_n) = 2 \sum_{j=1}^{n} \frac{(\sigma^2_{j,n} - 1)^2}{\sigma^2_{j,n}}$$

$$\leqslant \frac{2}{m} \sum_{j=1}^{n} (\sigma^2_{j,n} - 1).$$

We have assumed that (J_n) tends towards infinity: $P_0(B_n)$ and $P_1(B_n^c)$ tend to 0 if $n \to \infty$. A subsequence $B_{u(n)}$ can be found such that the series $\Sigma P_0(B_{u(n)})$ and $\Sigma P_1(B^c_{u(n)})$ converge. By the Borel-Cantelli lemma

$$P_0[\overline{\lim}\, B_{u(n)}] = 0, \quad P_1[\overline{\lim}\, B^c_{u(n)}] = 0.$$

Hence P_1 and P_0 are perpendicular.

When $K(P_0,P_1)$ is finite, we have $P_0 << P_1$, hence the sequence (J_n) cannot converge to infinity: $K(P_1,P_0)$ is finite.

Theorem 2.5.26. *If P_0 and P_1 are the distributions of two centered Gaussian sequences, we have only two cases:*

(1) *$K(P_0,P_1)$ and $K(P_1,P_0)$ are finite, P_0 and P_1 are equivalent.*
(2) *$K(P_0,P_1)$ and $K(P_1,P_0)$ are infinite, P_0 and P_1 are perpendicular.*

In the case of stationary sequences this theorem takes the following form:

Theorem 2.5.27. *Consider two centered stationary canonical Gaussian sequences $((\mathbb{R},\mathcal{B}_{\mathbb{R}})^{\mathbf{Z}}, (X_n)_{n\in\mathbf{Z}}, P_i)$, $i = 0,1$. Assume that, for $i = 0,1$, the sequence has spectral density f_i, and that there exist two constants m and M such that $0 < m < f_i < M < \infty$. Then*

$$\lim_{n\to\infty} \frac{1}{n} K(P_0^n,P_1^n) = \frac{1}{4\pi} \int_{-\pi}^{\pi} \left[\frac{f_0}{f_1} - 1 - \text{Log } \frac{f_0}{f_1}\right](\lambda)d\lambda.$$

The right hand term, which we shall denote by $K(f_0,f_1)$ is strictly positive unless f_1 and f_0 are equal a.s. *for Lebesgue measure. Thus P_0 and P_1 are perpendicular or equal.*

If L_n^i is the density of $X^{(n)}(P_i)$, we also have,

$$\frac{1}{n} \operatorname{Log} \frac{L_n^0}{L_n^1} \xrightarrow{L^2(P_0)} K(f_0,f_1).$$

Proof. From [1.4.3] we have here, with the above notations, $\Gamma_i^n = 2\pi T_n(f_i), T_n(f_i)$ being the Toeplitz matrix of f_i:

$$\operatorname{Log} \frac{L_n^0}{L_n^1} = \frac{1}{2} \operatorname{Log} \frac{|T_n(f_1)|}{|T_n(f_0)|}$$

$$+ \frac{1}{4\pi} {}^t X^{(n)}([T_n(f_1)]^{-1} - [T_n(f_0)]^{-1})X^{(n)}.$$

We have shown, in [1.3.4], that $(1/n)\operatorname{Log}|T_n(f_i)|$ tends to

$$\frac{1}{2\pi} \int_{-\pi}^{\pi} \operatorname{Log}[2\pi f_i(\lambda)]d\lambda.$$

In order to study ${}^t X^{(n)}((\Gamma_1^n)^{-1} - (\Gamma_0^n)^{-1})X^{(n)}$, let recall the transformation used at the start of [2.5.4]. This r.v. equals $\Sigma_{j=1}^n(\sigma_{j,n}^2 - 1)Y_{j,n}^2$. Denote the trace by Tr.

We have

$$\sum_{j=1}^{n} \sigma_{j,n}^2 = \operatorname{Tr}[{}^tU(\Gamma_1^{(n)})^{-1}U] = \operatorname{Tr}[U{}^tU(\Gamma_1^{(n)})^{-1}]$$

$$= \operatorname{Tr}[\Gamma_0^{(n)}(\Gamma_1^{(n)})^{-1}].$$

Hence

$$E_0[{}^tX^{(n)}[(\Gamma_1^n)^{-1} - (\Gamma_0^n)^{-1}]X^{(n)}] = \operatorname{Tr}[T_n(f_0)(T_n(f_1))^{-1} - I].$$

The stated result for Kullback information is thus true if we can show that

$$\frac{1}{n} \operatorname{Tr}[T_n(f_1)(T_n(f_0))^{-1}] \quad \text{tends to} \quad \int \frac{f_1}{f_0}(\lambda)d\lambda.$$

The following lemma is required.

Lemma 2.5.28. *If f and g are two even bounded and positive functions,*

$$\frac{1}{n} \operatorname{Tr} |T_n(f)T_n(g) - T_n(fg)| \xrightarrow[n\to\infty]{} 0.$$

Proof of the Lemma. Let (f_p) and (g_p) be the sequence of Fourier coefficients of f and of g. We have

$$(T_n(f))_{i,j} = f_{j-i} \quad \text{and} \quad (fg)_i = \sum_{k=-\infty}^{\infty} f_{i-k} g_k .$$

Hence

$$\begin{aligned}
&\operatorname{Tr} |T_n(f)T_n(g) - T_n(fg)| \\
&= \sum_{k=1}^{n} \left| \sum_{i=1}^{n} (f_{i-k} g_{i-k}) - \sum_{i=-\infty}^{\infty} (f_{i-k} g_{i-k}) \right| \\
&= \sum_{k=1}^{n} \left| \sum_{i \leqslant -k} f_i g_i + \sum_{i > n-k} f_i g_i \right| ;
\end{aligned}$$

$$\frac{1}{n} \operatorname{Tr} |T_n(f)T_n(g) - T_n(fg)| \leqslant \sum_{|i|>n} |f_i g_i| + \frac{2}{n} \sum_{i=1}^{n} i |f_i g_i|.$$

The functions f and g being bounded, these quantities tend to 0 if $n \to \infty$.

Proof of the Theorem. It is necessary to compare $\operatorname{Tr}[T_n(f_0)(T_n(f_1))^{-1}]$ and

$$\int_{-\pi}^{\pi} \frac{f_0}{f_1}(\lambda) d\lambda = \frac{1}{n} \operatorname{Tr}\left[T_n\left(\frac{f_0}{f_1}\right)\right].$$

We have

$$\begin{aligned}
&\frac{1}{n} \left| \operatorname{Tr} T_n(f_0)(T_n(f_1))^{-1} - \operatorname{Tr} T_n\left(\frac{f_0}{f_1}\right) \right| \\
&\leqslant \frac{1}{n} \left| \operatorname{Tr} \left[T_n(f_0)(T_n(f_1))^{-1} - T_n(f_0) T_n\left(\frac{1}{f_1}\right) \right] \right| \\
&\quad + \frac{1}{n} \left| \operatorname{Tr} \left[T_n(f_0) T_n\left(\frac{1}{f_1}\right) - T_n\left(\frac{f_0}{f_1}\right) \right] \right| .
\end{aligned}$$

The second term tends to 0 from the lemma. Let A and B be two $n \times n$ matrices; assume A is symmetric and denote by $\|A\|$ the absolute value of the greatest eigenvalue of A;

$$\|A\| = \sup\{|Ax|;\ x \in \mathbb{R}^n,\ |x| \leqslant 1\}.$$

We make an orthonormal change of basis with matrix U which diagonalizes A; $U^{-1}AU$ has diagonal $(\lambda_1, \ldots, \lambda_n)$. The traces do not alter. Denoting $U^{-1}BU = (b_{ij})_{1 \leqslant i,j \leqslant n}$, we have

$$\mathrm{Tr}(AB) = \sum_{i=1}^{n} \lambda_i b_{ii} .$$

From which $|\mathrm{Tr}(AB)| \leqslant \|A\| \mathrm{Tr}|B|$. Thus,

$$\frac{1}{n} \left| \mathrm{Tr}\left[T_n(f_0)(T_n(f_1))^{-1} - T_n(f_0)T_n\left[\frac{1}{f_1}\right]\right]\right|$$

$$\leqslant \frac{1}{n} \|T_n(f_0)\| \mathrm{Tr} \left| (T_n(f_1))^{-1} - T_n\left[\frac{1}{f_1}\right] \right|.$$

We have seen in [1.3.4] that the eigenvalues of $T_n(f_i)$ are in $[m,M]$. Thus this expression is majorized by

$$\frac{1}{n} M \|T_n(f_1)\|^{-1} \mathrm{Tr} \left| I - T_n(f_1)T_n\left[\frac{1}{f_1}\right] \right|;$$

with

$$\|T_n(f_1)\|^{-1} \leqslant \frac{1}{m} .$$

By applying Lemma 2.5.28 again, we obtain the result relative to Kullback information.

The variance of ${}^tX^{(n)}[(\Gamma_1^n)^{-1} - (\Gamma_0^n)^{-1}]X^{(n)}$ is

$$2 \sum_{j=1}^{n} (\sigma_{j,n}^2 - 1)^2 = 2\mathrm{Tr}[T_n(f_0)(T_n(f_1))^{-1} - 1]^2$$

$$\leqslant 2 \|T_n(f_0)(T_n(f_1))^{-1} - I\| \mathrm{Tr}|T_n(f_0)T_n(f_1)^{-1} - I|.$$

Thus,

$$\frac{2}{n^2} \sum_{j=1}^{n} (\sigma_{j,n}^2 - 1) \xrightarrow[n\to\infty]{} 0$$

and

$$\frac{1}{n}\left[\mathrm{Log}\,\frac{L_n^0}{L_n^1} - K(P_0^n, P_1^n)\right]$$

tends to zero in quadratic mean for P_0. This concludes the proof.

2.6. Square Integrable Martingales

Let $M = (M_n)_{n\geqslant 0}$ be a square integrable martingale defined on $(\Omega,\mathcal{A},P)$ adapted to the filtration $\mathbb{F}$. Then $M^2 = (M_n^2)_{n\geqslant 0}$ is a submartingale. Let $<M> = (<M>_n)_{n\geqslant 0}$ be the compensator of M^2: $M^2 - <M>$ is a martingale and $<M>$ is predictable. In what follows, M_0 and $<M_0>$ are taken to be zero and $<M>$ is called the **increasing process associated with** M.

Finally we denote $<M>_\infty = \lim <M>_n$ (an r.v. which is not necessarily finite) and,

$$M_n^* = \sup_{k\leqslant n} |M_k|, \quad M_\infty^* = \sup_k |M_k|.$$

2.6.1. Wald's Theorem for Square Integrable Martingales

Theorem 2.6.29. *Let M be a square integrable martingale with associated increasing process $<M>$ and $M_0 = <M>_0 = 0$. For any finite stopping time such that $<M>_\nu$ is integrable, we have*

$$E[M_\nu] = 0 \quad \textit{and} \quad E[M_\nu^2] = E[<M>_\nu].$$

Proof. It is true for any bounded ν (Stopping Theorem 2.3.8). So, for any integer n, $E[M_{\nu\wedge n}] = 0$ and $E[M_{\nu\wedge n}^2] = E[<M>_{\nu\wedge n}]$. For $n \to \infty$, $E[<M>_{\nu\wedge n}]$ increases to $E[<M>_\nu]$ (Beppo-Levi theorem). So $(M_{\nu\wedge n})$ is a martingale adapted to $(\mathcal{F}_{\nu\wedge n})$ such that $\sup_n E[M_{\nu\wedge n}^2] < \infty$; it converges in L^2 and a.s. to M_ν and

$$\begin{cases} E[M_\nu] = \lim_{n\to\infty} E[M_{\nu\wedge n}] = 0 \\ E[M_\nu^2] = \lim_{n\to\infty} E[M_{\nu\wedge n}^2] = E[<M>_\nu]. \end{cases}$$

Convergence. Coming back to Wald's theorem 2.3.9, we get the proof of the second half: for $(Y_n)_{n\geqslant 1}$ a sequence of independent square integrable r.v. of mean 0 and variance σ^2, take $M_n = S_n = Y_1 + \ldots + Y_n$, $<M>_n = n\sigma^2$ and ν an integrable stopping time: $E[S_\nu] = \sigma^2 E[\nu]$.

2.6.2. Inequalities and the Law of Large Numbers

Proposition 2.6.29. *Let M be a square integrable martingale with associated increasing process $<M>$.*

(a) *On* $\{<M>_\infty < \infty\}$, *M converges* a.s. *to a finite* r.v. M_∞.
(b) *For every stopping time ν, every $\varepsilon > 0$ and $a > 0$, we have,*

$$P(M^*_\nu \geq \varepsilon) \leq \frac{1}{\varepsilon^2} E(<M>_\nu),$$

$$P(M^*_\nu \geq \varepsilon) \leq \frac{1}{\varepsilon^2} E(<M>_\nu \wedge a) + P(<M>_\nu \geq a).$$

(c) $\{M^*_\infty = \infty\}$ *and* $\{<M>_\infty = \infty\}$ *coincide* a.s.

Proof. (a) Let $\sigma = \inf\{p; <M>_{p+1} > a\}$ with $a > 0$. This is a stopping time since $<M>$ is predictable. $M^\sigma = (M_{n\wedge\sigma})$ is a martingale and $<M^\sigma> = (<M>_{n\wedge\sigma})$ is majorized by a. Thus $\sup_n E(M^2_{n\wedge a})$ is majorized by $a < \infty$, and M^σ converges a.s. to an r.v. M^σ_∞ which is a.s. finite. However on $\{<M>_\infty \leq a\} = \{\sigma = \infty\}$, the sequences M^σ and M coincide. M converges a.s. to a finite r.v. M_∞ on $\cup_a\{<M>_\infty \leq a\} = \{<M_\infty> < \infty\}$.

(b) Let us assume ν to be bounded. Let $\sigma' = \inf\{p; |M_p| \geq \varepsilon\}$. We have,

$$\varepsilon^2 P(M^*_\nu \geq \varepsilon) = \varepsilon^2 P(\sigma' \leq \nu) \leq E(M^2_{\sigma'\wedge\nu})$$

$$= E(<M>_{\sigma'\wedge\nu}) \leq E(<M>_\nu);$$

$$P[\sigma' \leq \nu, <M>_\nu \leq a] = P[\sigma' \leq \nu, \sigma \geq \nu]$$

$$\leq P[\sigma' \leq \nu \wedge \sigma]$$

$$\leq \frac{1}{\varepsilon^2} E[<M>_{\nu\wedge\sigma}] \leq \frac{1}{\varepsilon^2} E[<M>_\nu \wedge a].$$

From which part (b) of the statement follows for ν bounded.

For arbitrary ν, the inequalities are written for $\nu \wedge n$; letting n tend to ∞, the two terms of the inequality increase to the stated inequality.

(c) This is deduced from (b):

$$P(M^*_n = \infty) \leq P(<M>_n \geq a).$$

Thus

$$P(M_n^* = \infty) \leqslant P(\langle M\rangle_\infty = \infty).$$

However, from (a), $(M_\infty^* = \infty)$ contains $(\langle M\rangle_\infty = \infty)$, from which (c) follows.

Corollary 2.6.30. The Law of Large Numbers. *Within the framework of the preceding proposition, let V be a square integrable $\mathbb{F}$-predictable sequence; $(V_n \cdot M_n)$ converges* a.s. *to* 0 *on*

$$\{\Sigma V_n^2(\Delta\langle M\rangle_n) < \infty\} \cap \{(V_n) \text{ decreases to } 0\}.$$

In particular, on $\{\langle M\rangle_\infty = \infty\}$, $(M_n/\langle M\rangle_n)$ tends to 0 a.s.; *likewise for $(f(\langle M\rangle_n)M_n)$ when f is a function from $\mathbb{R}_+$ into $\mathbb{R}_+$ which decreases to* 0 *at infinity and for which $\int_0^\infty f(s)^2 ds$ converges.*

Before proving this result, here is a lemma from analysis.

Lemma 2.6.31. *Let (a_n) be a sequence of positive real numbers and (x_n) a sequence of real numbers.*

(a) **Toeplitz' Lemma.** *If $\Sigma a_n = \infty$ and if $(x_n) \to x$ with $x \in \mathbb{R}$, then*

$$\sum_{k=1}^n a_k x_k \Big/ \sum_{k=1}^n a_k$$

tends to x when $n \to \infty$.

(b) **Kronecker's Lemma.** *If (a_n) increases to ∞ and if the series Σx_n converges, then the sequence of means $(1/a_n)\Sigma_{k=1}^n a_k x_k$ tends to* 0.

The proof of (a) is easy. We deduce (b) from this by considering sequences (b_n) and (s_n) with $b_n = a_n - a_{n-1}$, $s_n = \Sigma_{k=1}^n x_k$. If we denote $s_0 = 0$ and $s = \Sigma_{k=1}^\infty x_k$, we have,

$$\frac{1}{a_n}\sum_{k=1}^n a_k x_k = \frac{1}{a_n}\sum_{k=1}^n a_k(s_k - s_{k-1})$$

$$= s_n - \frac{1}{a_n}\sum_{k=1}^n a_k s_{k-1} \to s - s = 0.$$

Proof of Corollary 2.6.30. Let $V \cdot M = ((V \cdot M)_n)$ be the $\mathbb{F}$-martingale defined by $(V \cdot M)_n = \Sigma_{k=1}^n V_k \Delta M_k$. We have,

$$<V\cdot M>_k = \sum_{k=1}^{n} V_k^2 \Delta <M>_k .$$

On $\{<V\cdot M>_\infty < \infty\}$, $V\cdot M$ converges a.s. to a finite value. The first part of the corollary then follows from Kronecker's lemma, setting $a_n = 1/V_n$. Then we take $V_n = f(<M_n>)$ for decreasing f from $\mathbb{R}_+$ into $\mathbb{R}_+$:

$$\sum_{n\geq 1} f^2(<M>_n)\Delta <M>_n \leq \int_0^{<M>_\infty} f^2(s)ds \leq \int_0^\infty f^2(s)ds.$$

From which the result follows (by taking in particular $f(s) = 1/(1+s)$).

2.6.3. Exponential Submartingales and Large Deviations

Let (Z_n) be a sequence of r.v.'s. We say that it converges to an r.v. Z **at an exponential rate** if, for all $\varepsilon > 0$, there exist two constants a and $b > 0$ such that, for all n,

$$P[|Z_n - Z| \geq \varepsilon] \leq ae^{-bn} .$$

The introduction of constants a and b may be avoided by writing,

$$\overline{\lim} \frac{1}{n} \operatorname{Log} P(|Z_n - Z| \geq \varepsilon) < 0.$$

Then the sequence (Z_n) converges a.s. to Z (cf. [Vol. I, 2.5.3]). If $(X_n)_{n\geq 1}$ is a sequence of independent identically distributed r.v.'s such that $\{t;\ E(e^{tX_1}) < \infty\}$ is a neighborhood of 0, we know that $S_n/n = (X_1 + \ldots + X_n)/n$ tends to $E(X_1)$ at an exponential rate [Vol. I, Theorem 4.4.22]. We shall come back to these results in [5.4.1].

In a certain number of applications which we shall find in the following chapters, we can study without difficulty the sequence $<M>_n/n$ and prove that it converges exponentially to a constant α. Then, if the r.v.'s ΔM_n are bounded, the following theorem allows us to state that M_n/n converges exponentially to 0. In the independent case, we find, for bounded r.v.'s, the theorem given below by considering $M_n = S_n - nE(X_1)$ and $<M>_n = n\sigma^2(X_1)$.

Theorem 2.6.32. Large Deviations. *Let M be a square integrable martingale with associated increasing process $<M>$.*

Assume $M_0 = \Delta M_0 = 0$ *and* $|\Delta M_n| \leqslant c$ *for* $c \geqslant 0$ *and every* n.

(a) *Let* $\lambda > 0$. *By setting*

$$\phi_c(\lambda) = \frac{1}{c^2}[e^{\lambda c} - 1 - \lambda c]$$

for $c > 0$, $\phi_0(\lambda) = \lambda^2/2$, *the sequence* $Z_n = [\exp(\lambda M_n - \phi_c(\lambda)\langle M\rangle_n)]$ *is a positive submartingale.*

(b) *For all* $\varepsilon > 0$, *there exists a* $\mu(\varepsilon) > 0$ *such that, for every* $a > 0$,

$$P\left[\sup_{m\geqslant n} \frac{M_m}{\langle M\rangle_n} \geqslant \varepsilon\right] \leqslant P[\langle M\rangle_n \leqslant na] + \exp[-na\mu(\varepsilon)].$$

Proof. For $|y| \leqslant c$, we can write,

$$e^{\lambda y} - 1 - \lambda y = \lambda^2 y^2 \sum_{n\geqslant 2} \frac{(\lambda y)^{n-2}}{n!}$$

$$\leqslant \lambda^2 y^2 \sum_{n\geqslant 2} \frac{(\lambda c)^{n-2}}{n!} \leqslant \phi_c(\lambda) y^2.$$

(a)
$$\Delta Z_n = Z_{n-1}[e^{-\phi_c(\lambda)\Delta\langle M\rangle_n}(e^{\lambda\Delta M_n} - 1) + e^{-\phi_c(\lambda)\Delta\langle M\rangle_n} - 1]$$

$$E^{n-1}[e^{\lambda\Delta M_n} - 1] = E^{n-1}[e^{\lambda\Delta M_n} - 1 - \lambda\Delta M_n]$$

$$\leqslant \phi_c(\lambda)E^{n-1}(\Delta M_n^2) = \phi_c(\lambda)\Delta\langle M\rangle_n;$$

$$E^{n-1}(\Delta Z_n) \leqslant Z_{n-1}[g(\phi_c(\lambda)\Delta\langle M\rangle_n)]$$

with $g(x) = (1 + x)e^{-x} - 1$, a negative function for $x \geqslant 0$. Thus $E^{n-1}(\Delta Z_n) \leqslant 0$, and (Z_n) is a submartingale.

(b) If λ tends to 0, $\phi_c(\lambda)$ is equivalent to $\lambda^2/2$ thus we can always choose $\lambda > 0$ such that $\lambda_\varepsilon - \phi_c(\lambda) = \mu(\varepsilon)$ is > 0. For such a λ,

$$P\left[\sup_{p\geqslant n} \frac{M_p}{\langle M\rangle_p} \geqslant \varepsilon\right] \leqslant P\left[\sup_{p\geqslant n} Z_p \geqslant \exp \mu(\varepsilon)\langle M\rangle_n\right]$$

$$\leqslant P[\langle M\rangle_n \leqslant na]$$

$$+ P\left[\sup_{p\geqslant n} Z_p \geqslant \exp \mu(\varepsilon)na\right].$$

Since $E(Z_0) = 1$, we deduce from the inequality relative to positive submartingales ([2.3.3])

$$P\left[\sup_{p \geqslant n} \frac{M_p}{<M>_p} \geqslant \varepsilon\right] \leqslant P[<M>_n \leqslant na] + \exp(-na\mu(\varepsilon)).$$

2.7. Almost Sure Asymptotic Properties

2.7.1. Random Series

In this section, a filtration $\mathbb{F} = (F_n)_{n \geqslant 0}$ is given, as well as an $\mathbb{F}$-adapted sequence $(\xi_n)_{n \geqslant 1}$. In many applications, ξ_n is independent of F_{n-1} and conditional expectations are expectations. We study $S_n = \xi_1 + \ldots + \xi_n$. If the r.v.'s ξ_n are square integrable, we define $S_n = \Sigma_{k=1}^n E^{k-1}(\xi_k)$, $S_0 = S_0 = 0$. $(M_n) = (S_n - S_n)$ is a square integrable martingale, and,

$$<M>_n = \sum_{k=1}^{n} (E^{k-1}(\xi_k^2) - [E^{k-1}(\xi_k^2)]) = \sum_{k=1}^{n} V^{k-1}(\xi_k).$$

Borel-Cantelli Theorem 2.7.33. *Assume that the* r.v.'s ξ_n *are positive and majorized by a constant* C *(for example,* $\xi_n = 1_{\Gamma_n}$ *for* $\Gamma_n \in F_n$*). Then the sets* $\{\Sigma\xi_n < \infty\}$ *and* $\{\Sigma E^{n-1}(\xi_n) < \infty\}$ *coincide* a.s.

Proof. On $\{<M>_\infty < \infty\}$, (M_n) converges a.s. and the two series are of the same form. On $\{<M>_\infty = \infty\}$, $M_n/<M>_n$ tends to 0 from Corollary 2.6.30. However, $<M>_n$ is majorized by CS_n. Thus (M_n/S_n) tends to 1 a.s.: the two series are again of the same nature.

The 3 Series Theorem 2.7.34. (a) *Let* $C > 0$ *and* ξ_n^C *be the truncated* r.v.,

$$\xi_n^C = \xi_n 1_{\{|\xi_n| \leqslant C\}} + c1_{\{\xi_n > C\}} - c1_{\{\xi_n < -C\}}\,.$$

The series $\Sigma\xi_n$ *converges* a.s. *on the event "the 3 series* $\Sigma P^{n-1}(|\xi_n| > C)$, $\Sigma E^{n-1}(\xi_n^C)$, $\Sigma V^{n-1}(\xi_n^C)$ *converge."*

(b) *If the* r.v.'s (ξ_n) *are independent the convergence of the series* $\Sigma\xi_n$ *with a nonzero probability is equivalent to its almost sure convergence, and to the convergence, for every* $C > 0$, *of the*

3 *series*

$$\Sigma P(|\xi_n| > C), \quad \Sigma E(\xi_n^C), \quad \Sigma\sigma^2(\xi_n^C).$$

Proof. (a) The convergence of the first series proves that (ξ_n^C) and (ξ_n) only differ for a finite number of terms a.s., thus are of the same form. For the truncated sequence,

$$\{\Sigma[\xi_n^C - E^{n-1}(\xi_n^C)] \text{ converges}\} = \{\Sigma V^{n-1}(\xi_n^C) < \infty\}$$

from Proposition 2.6.29.

(b) The series $\Sigma\xi_n$ converges with probability 0 or 1. This follows from Kolmogorov's 0-1 law (Vol. I, Theorem 4.2.15). If $\Sigma\xi_n$ converges a.s., (ξ_n) tends to 0 a.s., and (ξ_n) and (ξ_n^C) coincide a.s. from a certain point onwards. The convergence of the first series follows from this. We can then assume (ξ_n) to be bounded by C. If this sequence is centered, the convergence of $\Sigma\sigma^2(\xi_n)$ is necessary to the convergence of $\Sigma\xi_n$ following Proposition 2.6.29. If (ξ_n) is not centered, consider (ξ_n'), a sequence independent of (ξ_n) and having the same distribution. Then $\Sigma(\xi_n - \xi_n')$ converges a.s., hence $\Sigma\sigma^2(\xi_n - \xi_n') = 2\Sigma\sigma^2(\xi_n)$ converges. Likewise $\Sigma(\xi_n - E(\xi_n))$ converges and the second series also converges.

2.7.2. Exchangeable and Spreading Invariant Sequences

Definition 2.7.35. (1) A sequence (X_n) of r.v.'s is said to be **spreading invariant** if, for every integer k and every $n_1 < n_2 < \dots < n_k$, the distribution of $(X_1, \dots, X_k)$ is the same as that of $(X_{n_1}, \dots, X_{n_k})$.

(2) The sequence (X_n) is **exchangeable** if, for every integer k and every $n_1, n_2, \dots, n_k$, pairwise distinct, the distribution of $(X_1, \dots, X_k)$ is the same as that of $(X_{n_1}, \dots, X_{n_k})$.

(3) The sequence (X_n) is **conditionally independent** with respect to a σ-algebra $\mathcal{B}$ if, for every sequence $f_1, \dots, f_n$ of bounded r.v.'s defined on $\mathbb{R}$, we have

$$E[f_1(X_1) \dots f_n(X_n) | \mathcal{B}] = E[f_1(X_1) | \mathcal{B}] \dots E[f_n(X_n) | \mathcal{B}].$$

(4) The r.v.'s X_n all have the **same distribution conditional**

on $\mathcal{B}$ if, for every bounded r.v. f on $\mathbb{R}$, $E[f(X_n)|\mathcal{B}]$ does not depend on n.

In what follows, we shall denote: $\mathcal{B}_n = \sigma(X_k;\ k \geq n)$, $\mathcal{F}_n = \sigma(X_k;\ k \leq n)$, $\mathcal{B}_\infty = \cap_{n=1}^{\infty}\mathcal{B}_n$. For the notion of exchangeable r.v.'s, see [Vol. I, E4.1.4 and E6.2.3].

Theorem 2.7.36. *The following properties are equivalent:*

(1) (X_n) *is spreading invariant.*
(2) (X_n) *is exchangeable.*
(3) *There exists a* σ*-algebra* $\mathcal{B}$ *such that conditional on* $\mathcal{B}$, *the* r.v.'s $X_1, \ldots, X_n$ *are independent and identically distributed.*
(4) *The same as in* (3), *by taking for* $\mathcal{B}$ *the tail* σ*-algebra* $\mathcal{B}_\infty$.

Proof. It follows from Corollary 2.4.18 that every integrable r.v. Y which is $\mathcal{B}_n$-measurable is the limit, in L^1, of the sequence $E[Y|\sigma(X_j;\ n \leq j \leq n + \ell)]$, hence of a sequence of r.v.'s of the form $\psi(X_n, \ldots, X_{n+\ell})$ ψ being an r.v. on $\mathbb{R}^{\ell+1}$.

Assume that (1) holds, for $n_1 < \ldots < n_k$ the distributions of $(X_1, \ldots, X_k)$ and of $(X_{n_1}, \ldots, X_{n_k})$ conditional on $\mathcal{B}_\infty$ are equal.

Let us consider in fact k bounded r.v.'s on $\mathbb{R}$, $f_1, \ldots, f_k$, and ψ a bounded r.v. on $\mathbb{R}^{\ell+1}$. For $n > n_k$, we have,

$$E[f_1(X_1) \ldots f_k(X_k)\psi(X_n, X_{n+1}, \ldots, X_{n+\ell})]$$
$$= E[f(X_{n_1}) \ldots f_k(X_{n_k})\psi(X_n, X_{n+1}, \ldots, X_{n+\ell})].$$

Thus

$$E[f_1(X_1) \ldots f_k(X_k)|\mathcal{B}_n] = E[f_1(X_{n_1}) \ldots f_k(X_{n_k})|\mathcal{B}_n].$$

Then, by applying corollary 2.4.18,

$$E[f_1(X_1), \ldots, f_k(X_k)|\mathcal{B}_\infty] = E[f_1(X_{n_1}) \ldots f_k(X_{n_k})|\mathcal{B}_\infty].$$

Let us show by recurrence on k, independence conditional on $\mathcal{B}_\infty$,

$$E[f_1(X_1) \ldots f_k(X_k)|\mathcal{B}_\infty] = E[f_1(X_1)f_2(X_{2+n}) \ldots f_k(X_{k+n})|\mathcal{B}_\infty]$$
$$= E[E(f_1(X_1|\mathcal{B}_n))f_2(X_{2+n}) \ldots f_k(X_{k+n})|\mathcal{B}_\infty].$$

For all $\varepsilon > 0$, we can choose n such that,

$$\|E[f_1(X_1)| \mathcal{B}_n] - E[f_1(X_1)| \mathcal{B}_\infty]\|_1 \leqslant \varepsilon.$$

Then denoting $\|f_i\| = \sup\{|f_i(x)|; x \in \mathbb{R}\}$ for $i = 1, ..., k$,

$$\begin{aligned}
&|E[f_1(X_1) \dots f_k(X_k)|\mathcal{B}_\infty] - E[f_1(X_1)|\mathcal{B}_\infty]E[f_2(X_{2+n}) \dots \\
&\qquad \dots f_k(X_{k+n})|\mathcal{B}_\infty]| \\
&= |E[f_1(X_1) \dots f_k(X_k)|\mathcal{B}_\infty] \\
&\quad - E[f_1(X_1)|\mathcal{B}_\infty]E[f_2(X_2) \dots f_k(X_k)|\mathcal{B}_\infty]| \\
&\leqslant \|f_1\| \, \|f_2\| \dots \|f_k\|.
\end{aligned}$$

This is true for all $\varepsilon > 0$, and

$$E[f_1(X_1)\dots f_k(X_k)|\mathcal{B}_\infty] = E[f_1(X_1)|\mathcal{B}_\infty]E[f_2(X_2)\dots f_k(X_k)|\mathcal{B}_\infty].$$

By recurrence on k, we see that $(X_1, ..., X_k)$ are independent and identically distributed conditional on $\mathcal{B}_\infty$: (1) → (4).

The implications (4) → (3) and (2) → (1) are clear; (3) → (2) is easily proved [Vol. I, Exercise 6.2.3].

2.7.3. The Law of Large Numbers

In order to study the distribution of a sequence (X_n) of measurable functions taking values in $(E,\mathcal{E})$, we can consider the canonical version of it, thus take $(\Omega,\mathcal{A}) = (E,\mathcal{E})$ and for X_n the nth coordinate function. Consider then the group $\mathfrak{S}$ of "finite permutations" of $\mathbb{N}$, bijections from $\mathbb{N}$ into $\mathbb{N}$ which leave invariant all points except a finite number, $\mathfrak{S} = \cup_n \mathfrak{S}_n$, where $\mathfrak{S}_n$ is the set of permutations of $\{1, ..., n\}$.

To $\sigma \in \mathfrak{S}$, we associate the transformation of E , which we shall again denote by σ: $(x_n) \longmapsto (x_{\sigma(n)})$. Let P be a probability on $(\Omega,\mathcal{A})$; to say that the sequence (X_n) is exchangeable on $(\Omega,\mathcal{A},P)$ implies that, for every $\sigma \in \mathfrak{S}$, $\sigma(P) = P$. We then say that P is **exchangeable**.

A tail event A does not depend on the first n coordinates, for arbitrary n: it is invariant under σ for arbitrary $\sigma \in \mathfrak{S}$. We shall denote by S the **σ-algebra of exchangeable events** (which contain $\mathcal{B}_\infty$),

$$S = \{A;\ A \in \mathcal{A},\ \sigma(A) = A \text{ for every } \sigma \in \mathfrak{S}\}.$$

Theorem 2.7.37 (Hewitt and Savage). *If P is an exchangeable probability on $(\Omega,\mathcal{A}) = (E,\mathcal{E})$, the tail σ-algebra $\mathcal{B}_\infty$ and the σ-algebra of exchangeable events $\mathcal{S}$ have the same P-completion in $\mathcal{A}$* (i.e. *for every $S \in \mathcal{S}$, there exists an $A \in \mathcal{B}_\infty$ such that S and A coincide P* a.s.).

Proof. Let S be exchangeable and let $\varepsilon > 0$. There exists an n_0 such that for $n > n_0$, $\|1_S - P(S|\mathcal{F}_n)\|_1 \leq \varepsilon$. Set $P(S|\mathcal{F}_n) = \phi(X_0, \dots, X_n)$. Consider $\sigma \in \mathfrak{S}$ such that $\sigma(i) = i + n$ for $i = 0, \dots, n$, the exchangeability of S and of P implies that, for every bounded r.v. ψ on E^{n+1}, we have

$$\int 1_S\psi(X_0, \dots, X_n)dP = \int 1_S \circ \sigma\psi(X_n, \dots, X_{2n})dP$$
$$= \int 1_S\psi(X_n, \dots, X_{2n})dP$$

$$\int 1_S\, \psi\, (X_0, \dots, X_n)dP = \int \phi(X_0, \dots, X_n)\psi(X_0, \dots, X_n)dP$$
$$= \int \phi(X_n, \dots, X_{2n})\psi(X_n, \dots, X_{2n})dP.$$

Thus

$$\phi(X_n, \dots, X_{2n}) = P(S|X_n, \dots, X_{2n}).$$

Since S and P are exchangeable, we always have also,

$$\|1_S - P(S|X_n, \dots, X_{2n})\|_1 = \|1_S - P(S|\mathcal{F}_n)\|_1 \leq \varepsilon.$$

By taking the expectation conditional on $\mathcal{B}_n$, the norm is diminished and, by denoting $\mathcal{B}_{n,2n} = \sigma(X_n, \dots, X_{2n})$,

$$\|P(S|\mathcal{B}_n) - P(S|\mathcal{B}_{n,2n})\|_1 \leq \varepsilon.$$

From which

$$\|P(S|\mathcal{B}_\infty) - 1_S\|_1$$
$$\leq \|P(S|\mathcal{B}_\infty) - P(S|\mathcal{B}_n\|_1 + \|P(S|\mathcal{B}_n) - P(S|\mathcal{B}_{n,2n})\|_1$$
$$+ \|P(S|\mathcal{B}_{n,2n}) - 1_S\|_1$$

$$\leq \|P(S|\mathcal{B}_\infty) - P(S|\mathcal{B}_n)\|_1 + 2\varepsilon.$$

Since $\|P(S|\mathcal{B}_\infty) - P(S|\mathcal{B}_n)\|_1$ tends to 0, if $n \to \infty$, the term on the right can be taken as small as we wish, and $\|P(S|\mathcal{B}_\infty) - 1_S\|_1 = 0$.

Theorem 2.7.38. The Law of Large Numbers. *Let* (X_n) *be an exchangeable sequence of integrable* r.v.'s. *Then* $(X_1 + ... + X_n)/n$ *converges* a.s. *and in* L^1 *to the expectation of* X_1 *conditional on the tail σ-algebra.*

Corollary 2.7.39. *Let* (X_n) *be a sequence of independent, identically distributed and integrable* r.v.'s. *Then* $(X_1 + ... + X_n)/n$ *converges* a.s. *and in* L^1 *to their mean* $E(X_1)$.

Note. If (X_n) is a sequence of independent r.v.'s with the same distribution, but not integrable, $\lim(|X_1 + ... + X_n|)/n$ equals ∞ a.s. (Vol. I, E4.2.15).

Proof of the Theorem. It can be proved using the canonical version of the sequence (X_n).

Let S_n be the σ-algebra of events $A \in \mathcal{A}$, invariant under every permutation $\sigma \in \mathfrak{S}_n$. We have, as a result of the exchangeability of P,

$$E[X_1|S_n] = ... = E[X_n|S_n] = E\left[\frac{X_1 + ... + X_n}{n} \mid S_n\right].$$

However $(X_1 + ... + X_n)/n$ is S_n-measurable; thus

$$E[X_1|S_n] = \frac{X_1 + ... + X_n}{n}.$$

By Corollary 2.4.18, $(E[X_1|S_n])$ converges in L_1 and a.s. to $E[X_1|S]$ and, by Theorem 2.7.37, $E[X_1|S] = E[X_1|\mathcal{B}_\infty]$. Finally, if the sequence (X_n) is independent, $E[X_1|\mathcal{B}_\infty] = E[X_1]$ by Kolmogorov's theorem (Vol. I, Theorem 4.2.15).

2.8. Central Limit Theorems

2.8.1. Triangular Sequences

In numerous problems, we are led to the addition of $\nu(n)$ very small variables $\xi_{n,k}$ where n is the time of the observation and $(\nu(n))$ a sequence of stopping times tending to infinity, the variables $\xi_{n,k}$ having distributions depending on the time n. The study of the behavior of $S_n = \Sigma_{k=1}^{\nu(n)} \xi_{n,k}$ leads first of all to the following definition and notations.

Triangular Sequence. In what follows, we are given a probability space (Ω, A, P) and, for every n, $\mathbb{F}^n = (F_k^n)$ a filtration of A, $\nu(n)$ a stopping time adapted to $\mathbb{F}^n$ and $\xi^{(n)} = (\xi_{n,k})$ a sequence of random vectors of dimension d adapted to $\mathbb{F}^n$. We take $F_0^n = \{\phi, \Omega\}$. We shall denote by $E^{n,k}$ the expectation (resp. $P^{n,k}$, the probability) conditional on F_k^n, and for X a random vector of dimension d,

$$V^{n,k}(X) = \{E^{n,k}(X_i X_j) - E^{n,k}(X_i)E^{n,k}(X_j)\}$$

is the **conditional covariance.**

Denote by $M^n = (M_p^n)$, the $\mathbb{F}^n$-martingale defined by $M_0^n = 0$ and $M_p^n = \Sigma_{k \leqslant p}[\xi_{n,k} - E^{n,k-1}(\xi_{n,k})]$, for $p \geqslant 1$,

$$<M^n>_p = \sum_{k \leqslant p} V^{n,k-1}(\xi_{n,k})$$

in the case where the $\xi_{n,k}$ are square integrable. We study the triangular sequence $(\xi_{n,k})_{n \geqslant 0}$, $k \leqslant \nu(n)$.

The following lemma is a direct consequence of Proposition 2.6.29b.

Lemma 2.8.40. *Let* $d = 1$. *If the* r.v.'s $(\xi_{n,k})$ *are square integrable and if*

$$\sum_{k \leqslant \nu(n)} V^{n,k-1}(\xi_{n,k}) \xrightarrow{P} 0,$$

then

$$\sup_{p \leqslant \nu(n)} \sum_{k \leqslant p} [\xi_{n,k} - E^{n,k-1}(\xi_{n,k})] \xrightarrow[n \to \infty]{P} 0.$$

In particular, replacing $\xi_{n,k}$ by $1_\Gamma(\xi_{n,k})$ for $\Gamma \in \mathcal{B}_{\mathbb{R}^d}$, we obtain, if

$$\sum_{k \leqslant \nu(n)} E^{n,k-1}(1_\Gamma(\xi_{n,k})) \xrightarrow[n\to\infty]{P} 0,$$

then

$$\sum_{k \leqslant \nu(n)} 1_\Gamma(\xi_{n,k}) \xrightarrow[n\to\infty]{P} 0,$$

which implies

$$P\left[\bigcup_{k \leqslant \nu(n)} \{\xi_{n,k} \in \Gamma\}\right] \xrightarrow[n\to\infty]{} 0.$$

Definition 2.8.41. The triangular sequence is **asymptotically negligible** if, for every $\varepsilon > 0$, we have

$$\sum_{k=1}^{\nu(n)} P^{n,k-1}(|\xi_{n,k}| \geqslant \varepsilon) \xrightarrow[n\to\infty]{P} 0.$$

This condition will be denoted by H1. It implies that

$$P\left[\sum_{k-1}^{\nu(n)} \{|\xi_{n,k}| \geqslant \varepsilon\}\right] = P\left[\sup_{1 \leqslant k \leqslant \nu(n)} |\xi_{n,k}| \geqslant \varepsilon\right]$$

tends to 0 with n. Thus,

$$\sup_{1 \leqslant k \leqslant \nu(n)} |\xi_{n,k}| \xrightarrow[n\to\infty]{P} 0.$$

If $\nu(n)$ is constant and if the $\xi_{n,k}$ are independent, H1 implies

$$\sum_{k=1}^{\nu(n)} P(|\xi_{n,k}| \geqslant \varepsilon) \xrightarrow[n\to\infty]{} 0.$$

2.8.2. Central Limit Theorems

For $\varepsilon > 0$, let us denote by $\xi_{n,k}^\varepsilon$, the r.v. $\xi_{n,k}$ truncated at ε, as in Theorem 2.7.34.

Theorem 2.8.42. *Let $(\xi_{n,k})$ be an asymptotically negligible triangular sequence of random vectors of dimension d, and let Γ be a $d \times d$ positive definite matrix. For an $\varepsilon > 0$, assume*

H2) $$\sum_{k=1}^{\nu(n)} E^{n,k-1}(\xi_{n,k}^\varepsilon) \xrightarrow[n\to\infty]{P} 0$$

H3) $$\Gamma_n^\varepsilon = \sum_{k=1}^{\nu(n)} V^{n,k-1}(\xi_{n,k}^\varepsilon) \xrightarrow[n\to\infty]{P} \Gamma .$$

Then $X_n = \Sigma_{k=1}^{\nu(n)}\xi_{n,k}$ *converges in distribution to* $N_d(0,\Gamma)$ *if* $n \to \infty$. *It is not necessary in this framework for the* r.v.'s $\xi_{n,k}$ *to be square integrable. If they are,* H2 *and* H3 *can be replaced by*

H'2) $$\sum_{k=1}^{\nu(n)} E^{n,k-1}(\xi_{n,k}) \xrightarrow[n\to\infty]{P} 0$$

H'3) $$\Gamma_n = \sum_{k=1}^{\nu(n)} V^{n,k-1}(\xi_{n,k}) \xrightarrow[n\to\infty]{P} \Gamma.$$

Corollary 2.8.43. *Let* $\mathbb{F} = (F_n)_{n\geq 0}$ *be a filtration on* (Ω,A,P) *such that* $F_0 = \{\Omega,\phi\}$, *and let* $(\xi_n)_{n\geq 1}$ *be a sequence of square integrable* r.v.'s. *Assume*

(a) $E^{n-1}(\xi_n) = 0$; *we then say that* (ξ_n) *is a centered sequence;*

(b) $\dfrac{1}{S_n^2}\sum_{k=1}^{n} E^{k-1}(\xi_k^2) \xrightarrow[n\to\infty]{P} 1$, *denoting* $S_n^2 = \sum_{k=1}^{n} E(\xi_k^2)$;

(c) *for all* $\varepsilon > 0$,

$$\frac{1}{S_n^2}\sum_{k=1}^{n} E^{k-1}[\xi_k^2 1_{(|\xi_k|\geq s_n\varepsilon)}] \xrightarrow[n\to\infty]{P} 0$$

(Lindeberg's condition).

Then $(1/S_n)\Sigma_{k=1}^{n}\xi_k$ *tends in distribution to* $N(0,1)$, *if* $n \to \infty$.

Corollary 2.8.44. Lindeberg's Theorem. *In the preceding theorem the* r.v.'s ξ_k *are assumed independent, centered, with variance* σ_k^2. *Set* $S_n^2 = \sigma_1^2 + \ldots + \sigma_n^2$. *Then* $(1/S_n)\Sigma_{k=1}^{n}\xi_k$ *tends in distribution to* $N(0,1)$ *if the Lindeberg condition,*

$$\frac{1}{S_n^2}\sum_{k=1}^{n} E[\xi_k^2 1_{(|\xi_k|\geq s_n\varepsilon)}] \xrightarrow[n\to\infty]{} 0,$$

is satisfied, for all $\varepsilon > 0$.

For independent identically distributed r.v.'s, we thus recover the classical central limit theorem (Vol. I, Theorem 4.4.23). In order to deduce Corollary 2.8.43 of Theorem 2.8.42, it is sufficient to take $F_{n,k} = F_k$; $\nu(n) = n$; $\xi_{n,k} = \xi_k/S_n$. Condition (c) implies that the sequence is asymptotically negligible since,

$$P^{k-1}[|\xi_k| > S_n\ ] \leq \frac{1}{S_n^2\ ^2} E^{k-1}(\xi_k^2 1_{\{|\xi_k|>s_n\varepsilon\}}).$$

Corollary 2.8.44 is a copy of the above in the independent case.

Proof of Theorem 2.8.42. Denote for $\varepsilon > 0$,

$$X_n^\varepsilon = \sum_{1 \leqslant k \leqslant \nu(n)} \xi_{n,k}^\varepsilon, \quad \tilde{\xi}_{n,k}^\varepsilon = \xi_{n,k}^\varepsilon - E^{n,k-1}(\xi_{n,k}^\varepsilon)$$

and

$$M_k^{n,\varepsilon} = \sum_{j=1}^{k} \tilde{\xi}_{k,j}^\varepsilon .$$

(a) *Under hypothesis* H1,

$$\sup_{1 \leqslant k \leqslant \nu(n)} |\xi_{n,k}| \xrightarrow{P} 0 \quad \textit{and} \quad X_n - X_n^\varepsilon \xrightarrow{P} 0.$$

Proof. Let $\varepsilon > 0$ and $\eta > 0$; H1 implies that

$$\sum_{1 \leqslant k \leqslant \nu(n)} P^{n,k-1}(\xi_{n,k}^\eta \neq \xi_{n,k}^\varepsilon) \xrightarrow{P} 0,$$

thus

$$\sum_{1 \leqslant k \leqslant \nu(n)} P^{n,k-1}(\tilde{\xi}_{n,k}^\eta \neq \tilde{\xi}_{n,k}^\varepsilon) \xrightarrow{P} 0.$$

By Lemma 2.8.40, we also have

$$\sum_{1 \leqslant k \leqslant \nu(n)} 1_{(\xi_{n,k}^\varepsilon \neq \xi_{n,k}^\eta)} \xrightarrow{P} 0,$$

thus

$$P\left[\bigcup_{1 \leqslant k \leqslant \nu(n)} (\tilde{\xi}_{n,k}^\varepsilon \neq \tilde{\xi}_{n,k}^\eta)\right] \longrightarrow 0.$$

(b) *Let $u \in \mathbb{R}^d$ and $\varepsilon > 0$. Set*

$$Z_n^\varepsilon = \prod_{k=1}^{n} E^{k,n-1}(\exp i\langle u, \tilde{\xi}_{n,k}^\varepsilon\rangle).$$

We then have,

$$Z_n^\varepsilon \xrightarrow{P} \exp\left[-\frac{1}{2}({}^t u \Gamma u)\right].$$

Proof. For every $x \in \mathbb{R}^d$, we have

$$\phi_u(x) = e^{i<u,x>} - 1 - i<u,x> + \frac{1}{2}<u,x>^2$$
$$= \frac{i}{2}\int_0^{<u,x>}(<u,x> - s)^2 e^{is}ds,$$

$$|\phi_u(x)| \leqslant \frac{|<u,x>|^3}{6}.$$

From which

$$Z_n^\varepsilon = \sum_{k=1}^{\nu(n)}\left[1 - \frac{1}{2}\,{}^t u V^{n,k-1}(\xi_{n,k}^\varepsilon)u + E^{n,k-1}(\phi_u(\tilde{\xi}_{n,k}^\varepsilon))\right]$$

with

$$|\phi_u(\tilde{\xi}_{n,k}^\varepsilon)| \leqslant \frac{|<u,\tilde{\xi}_{n,k}^\varepsilon>|^3}{6},$$

thus

$$\left|E^{n,k-1}\phi_u(\tilde{\xi}_{n,k}^\varepsilon)\right| \leqslant \frac{|u|\,\varepsilon}{6}\,{}^t u V^{n,k-1}(\xi_{n,k}^\varepsilon)u.$$

Let $z \in \mathbb{C}$, $|z| \leqslant \rho < 1$,

$$z - \mathrm{Log}(1 + z) = \int_0^z \frac{t\ dt}{1 + t}$$

and

$$|z - \mathrm{Log}(1 + z)| \leqslant |z|^2/2(1 - \rho).$$

Let us study $\mathrm{Log}\ Z_n^\varepsilon + (1/2)^t u\Gamma_n^\varepsilon u$. Set

$$z_k = -\frac{1}{2}\,{}^t u V^{n,k-1}(\xi_{n,k}^\varepsilon)u + E^{n,k-1}[\phi_u(\tilde{\xi}_{n,k}^\varepsilon)].$$

We have

$$|z_k| \leqslant \frac{1}{2}|u|^2\varepsilon^2 + \frac{|u|^3\varepsilon^3}{6} = \rho(\varepsilon).$$

Let ε be such that $\rho(\varepsilon) < 1$, (u fixed),

$$\left|\sum_{k=1}^n (\mathrm{Log}(1 + z_k) - z_k) + \sum_{k=1}^n z_k + \frac{1}{2}\,{}^t u\Gamma_n^\varepsilon u\right|$$

$$\leqslant \frac{\rho(\varepsilon)}{4(1-\rho(\varepsilon))}\left[1 + \frac{|u|\,\varepsilon}{3}\right]{}^t u\Gamma_n^\varepsilon u + \frac{1}{2}\,{}^t u\Gamma_n^\varepsilon u\,\frac{|u|\,\varepsilon}{6}$$

$$= (1/2)\psi(\varepsilon,u)\ {}^t u\Gamma_n^\varepsilon u = \phi_n(\varepsilon,u).$$

From the inequality $|e^x - 1| \leqslant |x|e^{|x|}$, we deduce

$$\left|Z_n^\varepsilon \exp\left(-\frac{1}{2}\ {}^t u\Gamma_n^\varepsilon u\right) - 1\right| \leqslant \phi_n(\varepsilon,u)\exp\ \phi_n(\varepsilon,u);$$

$$\left|Z_n^\varepsilon - \exp\left[-\frac{1}{2}\ {}^t u\Gamma u\right]\right| \leqslant \left|\exp\left[-\frac{1}{2}\ {}^t u\Gamma_n^\varepsilon u\right] - \exp\left(-\frac{1}{2}\ {}^t u\Gamma u\right)\right|$$

$$+ \frac{1}{2}\ {}^t u\Gamma_n^\varepsilon u\psi(\varepsilon,u)\exp\left(\frac{1}{2}\ [(\psi(\varepsilon,u) - 1)^t u\Gamma_n^\varepsilon u]\right).$$

Note that this bound also holds for Z_n^η, Γ_n^η, $\psi(\eta,u)$ with $\eta \leqslant \varepsilon$. Let $\alpha > 0$; choose $\eta < \varepsilon$ such that $H(\Gamma) = (1/2)({}^t u\Gamma u)\psi(\eta,u)\exp[(1/2)(\psi(\eta,u) - 1)^t u\Gamma u] < \alpha/4$. We then have

$$P\left(\left|Z_n^\varepsilon - \exp\left[-\frac{1}{2}\ {}^t u\Gamma u\right]\right| \geqslant \alpha\right) \leqslant P\left(|Z_n^\varepsilon - Z_n^\eta| \geqslant \frac{\alpha}{4}\right)$$

$$+ P\left(\left|\exp\left[-\frac{1}{2}\ {}^t u\Gamma_n^\eta u\right] - \exp\left[-\frac{1}{2}\ {}^t u\Gamma u\right]\right| \geqslant \frac{\alpha}{4}\right)$$

$$+ P\left(|H(\Gamma_n^\eta) - H(\Gamma)| \geqslant \frac{\alpha}{4}\right) + \frac{\alpha}{4}.$$

We deduce from the fact that $\xi_{n,k}^\varepsilon$ and $\xi_{n,k}^\eta$ are bounded and from (a) that

$$|\Gamma_n^\eta - \Gamma_n^\varepsilon| \xrightarrow{P} 0 \quad \text{and} \quad |Z_n^\varepsilon - Z_n^\eta| \xrightarrow{P} 0;$$

from which

$$Z_n^\varepsilon \xrightarrow{P} \exp\left(-\frac{1}{2}\ {}^t u\Gamma u\right).$$

(c) Set

$$N_p^{n,\varepsilon} = \frac{\exp(i\langle u, M_p^{n,\varepsilon}\rangle)}{\prod_{k=1}^{p} E^{n,k-1}(\exp i\langle u, \tilde{\xi}_{n,k}^\varepsilon\rangle)},$$

for $p \geqslant 1$, and $N_0^{n,\varepsilon} = 1$; $(N_p^{n,\varepsilon})_{p\geqslant 0}$ is an $\mathbb{F}$-martingale (check this). The sequence

$$\left(\left|\prod_{k=1}^{p} E^{n,k-1}(\exp i\langle u,\tilde{\xi}^{\varepsilon}_{n,k}\rangle)\right|\right)_{p\geq 0}$$

is decreasing. Consider a number $c > 0$ and

$$T^n_c = \inf\left\{p;\ \left|\prod_{j=1}^{p+1} E^{n,k-1}(\exp i\langle u,\tilde{\xi}^{\varepsilon}_{n,k}\rangle)\right| \leq c\right\}.$$

T^n_c is an $\mathbb{F}^n$ stopping time and the modulus of the martingale $(N^{n,\varepsilon}_{p\wedge\nu(n)\wedge T^n_c})_{p\geq 0}$ is majorized by $1/c$. Hence this martingale converges in L^1 for $p \to \infty$ and

$$E\left[N^{n,\varepsilon}_{\nu(n)\wedge T^n_c}\right] = E[N^{n,\varepsilon}_0] = 1.$$

Take $c < \exp[-(1/2)({}^t u\Gamma u)]$. Then $P[T^n_c < \nu(n)\ ] \leq P[|Z^{\varepsilon}_n| \leq c]$ tends to 0 if $n \to \infty$. Moreover,

$$\left|E\left[N^{n,\varepsilon}_{\nu(n)\wedge T^n_c} Z^{\varepsilon}_n - \exp\left(-\frac{1}{2}\ {}^t u\Gamma u\right)\right]\right|$$

$$= \left|E\left[N^{n,\varepsilon}_{\nu(n)\wedge T^n_c}\ Z^{\varepsilon}_n\right] - \exp\left(-\frac{1}{2}\ {}^t u\Gamma u\right)E(N_{\nu(n)\wedge T^n_c})\right|$$

$$\leq \left(\frac{1}{c}\right)E\left[\left|Z^{\varepsilon}_n - \exp\left(-\frac{1}{2}\ {}^t u\Gamma u\right)\right|\right].$$

This upper bound tends to 0, since $Z^{\varepsilon}_n - \exp[(-(1/2){}^t u\Gamma u)$ is an r.v. bounded by 2 in modulus which tends to zero in probability. Thus

$$\lim E\left[N^{n,\varepsilon}_{\nu(n)\wedge T^n_c}\ Z^{\varepsilon}_n\right] = \exp\left[-\frac{1}{2}\ {}^t u\Gamma u\right].$$

The inequality

$$\left|e^{i\langle u,X^{\varepsilon}_n\rangle} - N^{n,\varepsilon}_{\nu(n)\wedge T^n_c}\ Z^{\varepsilon}_n\right| \leq 1_{(T^c_n<\nu(n))}$$

then implies that $\lim E[e^{i\langle u,X^{\varepsilon}_n\rangle}] = \exp[-(1/2){}^t u\Gamma u]$. (X^{ε}_n) converges in distribution to $N_d(0,\Gamma)$. Since, according to (a), $X_n - X^{\varepsilon}_n \xrightarrow{P} 0$, we have

$$X_n \xrightarrow{\mathcal{D}} N_d(0,\Gamma).$$

Bibliographic Notes

The main thrust to the theory of martingales has been given by Doob's book; Neveu [4] contains in a very learned form essential results and bibliography. All the recent results will be found in Dellacherie-Meyer [2], Chapter 5.

See also Garsia for numerous inequalities not covered in [2.3.3].

Absolute continuity and likelihood theory are some of the essential ideas in the statistics of processes. The case of Gaussian processes is set out in Grenander, Gikhman-Skorohod (Volume 1), Ibragimov-Rozanov, Neveu [1]. The case of Poisson processes is set out in Krickeberg.

Series of random variables have been studied for a long time. See Levy, Gnedenko-Kolmogorov for the independent case; and Loeve, Neveu [4] for the results of [2.7]. See also Spitzer and Stout for some further results.

Another point of view, that of ergodic theory, leads to laws of large numbers. Consult Neveu [1], Billingsley [3]. In what follows we shall remain faithful to martingale methods.

In the last few years a lot has been written on central limit theorems. See Hall-Heyde for a general account and a bibliography. See also Araujo-Gine.

Chapter 3
ASYMPTOTIC STATISTICS

Objectives

In the statistics of processes, it is often natural to use a dominated model at each instant and the associated likelihood process which is a martingale. When the duration of the observation is finite, it is rare to have simple results - unless we have well defined structures like the exponential model ([Vol. I, 7]). However asymptotic theorems will be able to guide the choice of estimators or tests.

The general principle consists of choosing, at each instant t, an r.v. which is a function of the past observations and of the parameter α, which for distribution P_θ tends to be a minimum for $\alpha = \theta$. This process, called the contrast process, chosen in the simplest possible way, leads to a minimum contrast estimator at each instant and to associated tests. In the case of a model dominated at each instant, it may be the opposite of likelihood and we deal with maximum likelihood estimators. In this section we study the asymptotic properties of these estimators or tests.

Technically, this chapter uses no more than Taylor's formula and some previously shown asymptotic properties. The difficulty lies only in a convenient formulation of the statistical problems.

The methods are illustrated here mainly by the case of independent samples, time series and Poisson processes. In the following chapters they will be adapted to other situations:

the fact that the likelihood process is a martingale will then play a more fundamental role.

3.1. Models Dominated at Each Instant

3.1.1. Dominated Model of a Sample

Let Θ be a set of parameters and, for every $\theta \in \Theta$, let F_θ be a distribution on $(E,\mathcal{E})$. A sequence $(X_n)_{n\geq 1}$ of independent observations is observed, with distribution F_θ: we may take $(\Omega,\mathcal{A},P_\theta) = (E,\mathcal{E},F_\theta)^{\mathbf{N}}$, X_n being the nth coordinate ([0.2.3]). Assume that, for every θ, F_θ is dominated by a distribution F on $(E,\mathcal{E})$. Let f_θ, denoted also by $x \longmapsto f(\theta,x)$, be a **likelihood**, i.e. a version of the density of F_θ with respect to F. Then the distribution of $(X_1, ..., X_n)$ is $F_\theta^{\otimes n} = f_\theta^{\otimes n}F^{\otimes n}$, denoting $f_\theta^{\otimes n}(x_1, ..., x_n) = f_\theta(x_1). ... f_\theta(x_n)$.

Denote by $P = F^{\otimes}$, and E (resp. E_θ) the expectation associated with P (resp. P_θ). For every $n \in \mathbb{N}$ and $\Gamma \in \mathcal{E}^{\otimes n}$, we have,

$$P_\theta[(X_1, ..., X_n) \in \Gamma] = \int_\Gamma f_\theta^{\otimes n} dF^{\otimes n}$$

$$= \int f_\theta^{\otimes n}(X_1, ..., X_n)1_\Gamma(X_1, ..., X_n)dP.$$

If $P_{\theta,n}$ and P_n are the restrictions of P_θ and P to $\mathcal{F}_n = \sigma(X_1, ..., X_n)$, we have,

$$P_{\theta,n} = f_\theta^{\otimes n}(X_1, ..., X_n)P_n \ .$$

Thus, for each n, the model $(\Omega, \mathcal{F}_n,(P_{\theta,n})_{\theta\in\Theta})$ is a dominated model in the sense of [Vol. 1, 7.1]. A likelihood of $P_{\theta,n}$ with respect to P_n is

$$L_n(\theta) = f_\theta^{\otimes n}(X_1, ..., X_n).$$

We shall denote in what follows P_θ and P instead of $P_{\theta,n}$ and P_n.

However the model $(\Omega,\mathcal{A},(P_\theta)_{\theta\in\Theta})$ is not dominated by P except in the uninteresting case where $F_\theta = F$ for all θ. In fact, if there exists a $\Gamma \in \mathcal{E}$ such that $F_\theta(\Gamma)$ and $F(\Gamma)$ differ, P_θ and P are singular because they are concentrated respectively on sets of ω such that $(1/n)\Sigma_{p=1}^{n} 1_\Gamma(X_p(\omega))$ converges to $F_\theta(\Gamma)$ and to $F(\Gamma)$.

3.1.2. Dominated Models in the Statistics of Processes

We study a process the state space $(E,\mathcal{E})$ and time space T of which are given ($T \subset \mathbb{N}$, $T \subset \mathbb{Z}$ or T an interval of $\mathbb{R}_+$). We can always consider a canonical version defined on $(\Omega,\mathcal{A}) = (E,\mathcal{E})^T$, the trajectories being $\omega \longmapsto (X_t(\omega))_{t\in T}$, X_t the coordinate with index t ([0.2]). The observations prior to t are described by $\mathcal{F}_t = \sigma(X_s;\ s \leqslant t)$, the history of the process is given by the "filtration" $\mathbb{F} = (\mathcal{F}_t)_{t\in T}$. The distribution of the process is a probability on $(\Omega,\mathcal{A})$. Assume that it is in a family $(P_\theta)_{\theta\in\Theta}$ of probabilities. The study of the independent case leads to the idea of a dominated model along a filtration $\mathbb{F} = (\mathcal{F}_t)_{t\in T}$, an increasing family of sub σ-algebras of $\mathcal{A}$.

Definition 3.1.1. Let $(\Omega,\mathcal{A})$ be a probability space provided with a filtration $\mathbb{F} = (\mathcal{F}_t)_{t\in T}$. We say that the statistical model $(\Omega,\mathcal{A},(P_\theta)_{\theta\in\Theta})$ is a **dominated model along** $\mathbb{F}$, if there exists a nonzero σ-finite measure μ, such that for every t, the trace of P_θ on $\mathcal{F}_t$ is dominated by the trace of μ on $\mathcal{F}_t$.

In what follows we shall take for μ a probability P (which is always possible [Vol. I, 7.1.3]) and we shall denote by E the expectation with respect to P.

Example 1. *Stationary Gaussian Sequence.* Let (X_n) be a centered stationary Gaussian sequence with spectral density f. From [1.4.3], if there are two constants m and M such that $0 < m \leqslant f \leqslant M < \infty$, then $X^{(n)} = (X_1, ..., X_n)$ has density,

$$x^{(n)} = (x_1, ..., x_n) \longmapsto \frac{(2\pi)^{-n}}{\sqrt{T_n(f)}} \exp\left[-\frac{1}{4\pi}\ {}^t x^{(n)}(T_n(f))^{-1}x^{(n)}\right],$$

denoting by $T_n(f)$ the nth Toeplitz matrix of f.

If the set of parameters Θ is given, and a family $(f_\theta)_{\theta\in\Theta}$ of spectral densities such that, for every θ, there exists constants $m_\theta > 0$ and $M_\theta < \infty$ with $m_\theta < f_\theta < M_\theta$, we have a dominated model along the filtration $(\sigma(X_1, ..., X_n))_{n \geq 1}$. If for example we are dealing with ARMA(p,q) models given by equations of the form

$$X_n + a_1 X_{n-1} + ... + a_p X_{n-p} = b_0 \xi_n + b_1 \xi_{n-1} + ... + b_q \xi_{n-q}$$

we take $\theta = (a_1, ..., a_p, b_0, ..., b_q)$ in a subset Θ of $\mathbb{R}^{p+q+1}$. If Θ is the set of θ such that the polynomial $x \longmapsto 1 + a_1 x + ... + a_p x^p$ has no roots of modulus 1, we have a dominated model (Proposition 1.3.32).

Example 2. *Poisson Processes on* $\mathbb{R}_+$. The Poisson process with parameter θ, $(\Omega, A, (N_t)_{t \geq 0}, P_\theta)$ defined in [0.2.3] and [1.2.1] is such that, for $k_1, ..., k_n$ integers and $0 < t_1 < t_2 < ... < t_n = t$, we have

$$P_\theta[N_{t_1} = k_1, N_{t_2} = k_1 + k_2, ..., N_{t_n} = k_1 + ... + k_n]$$

$$= e^{-\theta t_1} \frac{(\theta t_1)^{k_1}}{k_1!} e^{\theta(t_2 - t_1)} \frac{(\theta(t_2 - t_1))^{k_2}}{k_2!}$$

$$... e^{-\theta(t_n - t_{n-1})} \frac{(\theta(t_n - t_{n-1}))^{k_n}}{k_n!}$$

$$= e^{(1-\theta)t} \theta^{k_1 + ... + k_n} P_1[N_{t_1} = k_1, N_{t_2} = k_1 + k_2,$$

$$..., N_{t_n} = k_1 + ... + k_n].$$

On $F_t = \sigma(N_n; n \leq t)$, we have

$$P_\theta = \theta^{N_t} e^{(1-\theta)t} P_1.$$

3.1.3. Decisions in the Statistics of Processes

A decision can only be taken at each instant by taking into account past observations. This is the essence of the following definition.

Definition 3.1.2. We are given a measurable space (Ω, A) equipped with a filtration $\mathbb{F} = (F_t)_{t \in T}$ with $T \subset \mathbb{Z}$ or $T \subset \mathbb{R}_+$, and a family of probabilities $(P_\theta)_{\theta \in \Theta}$. We try to take a decision (estimation, test, ...) taking values in an action space $(A,\mathcal{A})$: an $\mathbb{F}$**-adapted decision** is a process $d = (d_t)_{t\in T}$ taking values in $(A,\mathcal{A})$ and $\mathbb{F}$-adapted. We thus speak of an $\mathbb{F}$-adapted estimator or an $\mathbb{F}$-adapted test ("$\mathbb{F}$-adapted" will be omitted if the context is clear).

Examples. All the statistics on samples studied in Volume I consist of choosing for each n a decision function of the n-sample $(X_1, ..., X_n)$. For example, if we are dealing with an n-sample from a distribution F_θ on $\mathbb{R}$, F_θ having finite mean $m(\theta)$ and variance $\sigma^2(\theta)$, the asymptotic behaviour of the empirical estimator $\bar{X}_n$ of $m(\theta)$ is given by,

$$\bar{X}_n = \frac{X_1 + ... + X_n}{n} \xrightarrow[n\to\infty]{P_\theta\text{-a.s.}} m(\theta),$$

$$\sqrt{n}\,(\bar{X}_n - m(\theta)) \xrightarrow[n\to\infty]{\mathcal{D}(P_\theta)} N(0,\sigma^2(\theta)).$$

Similarly, we have studied the empirical estimators of the mean, covariance, or the spectral measure of a stationary Gaussian sequence in [1.4].

It is certainly hoped that an $\mathbb{F}$-adapted decision will be better when t is larger... in the "good cases" it converges (in senses to be stated) to the best decision. Some of the "good cases" are going to be specified in what follows.

Definition 3.1.3. Given a statistical model $(\Omega,\mathcal{A},(P_\theta)_{\theta\in\Theta})$ and a filtration $\mathbb{F} = (\mathcal{F}_t)_{t\in T}$ with $T = \mathbb{R}_+$ or $T = \mathbb{N}$, an $\mathbb{F}$-adapted estimator $(g_t)_{t\in T}$ of a function g from Θ into $\mathbb{R}^k$ is said to be **consistent at the point** θ if (g_t) tends to $g(\theta)$ in P_θ probability. It is **consistent** if it is consistent for all $\theta \in \Theta$; if the convergence takes place P_θ-a.s., we say that (g_t) is **strongly consistent.**

3.2. Contrasts

3.2.1. Likelihood in a Model Dominated at Each Instant

Following the definition which has been given in Volume I, we define here a likelihood process.

Definition 3.2.4. If $(\Omega,\mathcal{A},(P_\theta)_{\theta\in\Theta})$ is a dominated model along the filtration $\mathbb{F} = (\mathcal{F}_t)_{t\in T}$, a **likelihood** L is a function from $T \times \Theta \times \Omega$ into $\mathbb{R}$, $(t,\theta,\omega) \longmapsto L_t(\theta,\omega)$, satisfying the following properties:

(a) For every t and every θ, the function $\omega \longmapsto L_t(\theta,\omega)$, denoted $L_t(\theta)$, is $\mathcal{F}_t$-measurable.
(b) For every t, $L_t(\theta)$ is a version of the density of P_θ with respect to P on $(\Omega,\mathcal{F}_t)$.

The process $(L_t(\theta))_{t\in T}$ adapted to $\mathbb{F}$ is called the **likelihood process** (or simply the **likelihood**).

The above definition is not unique. Two likelihood processes are equal up to a modification ([0.2.2]). However we speak of "the" likelihood for a likelihood process, chosen, in general to be as regular as possible.

Proposition 3.2.5. (a) *For each* $\theta \in \Theta$, $(L_t(\theta))_{t\in T}$ *is an* $\mathbb{F}$-*martingale on* $(\Omega,\mathcal{A},P)$.
(b) *For* $T \subset \mathbb{Z}$, *we also have a domination property at each stopping time* ν, *finite relative to* $\mathbb{F}$:

$$P_\theta = L_\nu(\theta)P \quad on \quad (\Omega,\mathcal{F}_\nu).$$

Proof. Let $s < t$ and let $\Gamma \in \mathcal{F}_s$. We have

$$E[1_\Gamma L_t(\theta)] = P_\theta(\Gamma) = E[1_\Gamma L_s(\theta)].$$

From which, $E[L_t(\theta)|\mathcal{F}_s] = L_s(\theta)$. Property (a) follows from this. To prove (b), take $A \in \mathcal{F}_\nu$,

$$P_\theta(A) = \sum_{n\geqslant 0} P_\theta(A \cap (\nu = n)) = \sum_{n\geqslant 0} \int_{A\cap(\nu=n)} L_n(\theta)dP$$

$$= \sum_{n\geqslant 0} \int_{A\cap(\nu=n)} L_\nu(\theta)dP = E[L_\nu(\theta)1_A].$$

Definition 3.2.6. Within the framework of Definition 3.2.4, a **maximum likelihood estimator** is an $\mathbb{F}$-adapted estimator $(\hat\theta_t)_{t\in T}$ which satisfies for all t,

$$L_t(\hat\theta_t) = \sup\{L_t(\theta);\ \theta \in \Theta\}.$$

Example 1. *A Sample From a Dominated Model.* We return to the situation of [3.1.1], assuming all the distributions F_α equivalent.

Given an arbitrary $\theta \in \Theta$, the model is then dominated, and the likelihood with respect to θ at the instant n is the **likelihood ratio**

$$\alpha \longmapsto \frac{L_n(\alpha)}{L_n(\theta)} = \frac{f_\alpha(X_1)}{f_\theta(X_1)} \cdots \frac{f_\alpha(X_n)}{f_\theta(X_n)} .$$

If $\operatorname{Log}(f_\alpha/f_\theta)$ is F_θ-integrable, we have

$$\frac{1}{n} \operatorname{Log} \frac{L_n(\alpha)}{L_n(\theta)} \xrightarrow{P_\theta\text{-a.s.}} \int \left[\operatorname{Log} \frac{f_\alpha}{f_\theta}\right] f_\theta dF = -K(F_\theta, F_\alpha),$$

denoting by K the **Kullback information** of F_θ with respect to F_α ([Vol. I, 6.4]). If the model is **identifiable**, ie. if for $\alpha \neq \theta$, F_α and F_θ different, $K(F_\theta, F_\alpha)$ is only zero for $\alpha = \theta$. For $\alpha \neq \theta$, $L_n(\alpha)/L_n(\theta)$ tends P_θ-a.s. to 0: it is this relation which makes it natural to use a maximum likelihood estimator adapted to $(\sigma(X_1, \dots, X_n))$, in order to estimate θ, when n is large enough.

Particular Cases. (a) $\theta = (a,b)$ with $-\infty < a < b < \infty$ and F_θ is the uniform distribution on $[a,b]$. Then ([Vol. I, 7.3]), $\hat{\theta}_n = (X_{(1)}, X_{(n)})$ with $X_{(1)} = \inf_{1 \leq i \leq n} X_i$, $X_{(n)} = \sup_{1 \leq i \leq n} X_i$. If n tends to ∞, $X_{(1)}$ decreases a.s. to a, $X_{(n)}$ increases a.s. to b ([Vol. I, E.4.2.16]), $(\hat{\theta}_n)$ is a strongly consistent estimator.

(b) $\theta = (m, \sigma^2)$ with $m \in \mathbb{R}$ and $\sigma^2 > 0$ and $F_\theta = N(m, \sigma^2)$. Then $\hat{\theta}_n = (\bar{X}_n, S_n)$ with $\bar{X}_n = (1/n)\Sigma_{i=1}^n X_i$, $S_n^2 = (1/n)\Sigma_{i=1}^n (X_i - \bar{X}_n)^2$ ([Vol. I, 5.1.5]). Here $\bar{X}_n$ is distributed as $N(m, \sigma^2/n)$): $\bar{X}_n$ tends P_θ-a.s. to m, $\sqrt{n}(\bar{X}_n - m)$ is distributed as $N(0, \sigma^2)$, and

$$S_n^2 = \frac{1}{n}\left(\sum_{i=1}^n X_i^2\right) - (\bar{X}_n^2) = \frac{1}{n}\left[\sum_{i=1}^n (X_i - \bar{X}_n)^2\right]$$

tends P_θ-a.s. to σ^2, its distribution for each n being $(\sigma^2/n)\chi^2(n-1)$.

Example 2. *The Poisson Process on $\mathbb{R}_+$.* Recall Example 2 of [3.1.2]. The likelihood for the dominant probability P_1 is,

$$L_t(\theta) = \theta^{N_t} e^{t(1-\theta)}.$$

This is maximized by taking $\hat{\theta}_t = N_t/t$. We then have

([Vol. I, E.4.3.6]),

$$\frac{N_t}{t} \xrightarrow[t\to\infty]{P_\theta\text{-a.s.}} \theta,$$

and $[N_t/t]$ is a strongly consistent estimator;

$$\sqrt{t}\left[\frac{N_t}{t} - \theta\right] \xrightarrow[t\to\infty]{\mathcal{D}(P_\theta)} N(0,\theta).$$

Example 3. *Centered Stationary Gaussian Sequences.* Recall Example 1 of [3.1.2]. Here the likelihood does not in general have a simple form, and it is not easy to study maximum likelihood estimators. As a result of Theorem 2.5.27 however, we always have a situation close to the independent case. Denote by P_α^n the distribution of $(X_1, ..., X_n)$ when the parameter equals α, and $L_n(\alpha)$ its density. We have

$$\frac{1}{n} K(P_\theta^n, P_\alpha^n) \xrightarrow[n\to\infty]{} K(f_\theta, f_\alpha),$$

$$\frac{1}{n} \operatorname{Log} \frac{L_n(\theta)}{L_n(\alpha)} \xrightarrow[n\to\infty]{L^2(P_\theta)} K(f_\theta, f_\alpha),$$

with

$$K(f_\theta, f_\alpha) = \frac{1}{4\pi}\int_{-\pi}^{\pi} \left[\frac{f_\theta}{f_\alpha} - 1 - \operatorname{Log}\frac{f_\theta}{f_\alpha}\right](\lambda)d\lambda.$$

3.2.2. Contrasts

If the maximum likelihood is difficult to study, we can try to estimate a "dissimilarity" between the observation and a parameter, and study which values of the parameter minimize this estimation. First of all here are two examples.

Example 1. The χ^2 Distance. If we are studying observations taking values in a finite space $E = \{1,2, ..., r\}$, a probability on E is denoted by $p = (p_1,p_2, ..., p_r)$, p_j being the weight of j for $1 \leqslant j \leqslant r$. If p charges all the points of E and if q is another probability, the χ^2 distance and the Kullback information of q on p are respectively,

$$\chi^2(p,q) = \sum_{j=1}^{r} \frac{(q_j - p_j)^2}{q_j},$$

$$K(q,p) = \sum_{j=1}^{n} q_j \operatorname{Log} q_j - \sum_{j=1}^{n} q_j \operatorname{Log} p_j .$$

A sample from p is observed: $\mathcal{E}$ is the set of subsets of E, $(\Omega,\mathcal{A},P_p) = (E,\mathcal{E},p)$, X_n is the nth coordinate. We then denote, for $1 \leqslant j \leqslant r$ and $n \in \mathbb{N}$, the number N_n^j of observations prior to n equal to j,

$$N_n^j = \sum_{p=1}^{n} 1_{(X_p=j)}, \quad N_n = (N_n^j)_{1 \leqslant j \leqslant r} .$$

The empirical estimator of p is N_n/n.

We are concerned with estimating p. Assume given $\Theta \subset \mathbb{R}^k$ and an injective function $\alpha \longmapsto p(\alpha)$ from Θ into $\Delta = \{(p_1, ..., p_r); p_j > 0 \text{ for } 1 \leqslant j \leqslant r, \Sigma_{j=1}^r p_j = 1\} \subset \mathbb{R}^r$. In order to estimate θ (or $p(\theta)$) we minimize the dissimilarity between the empirical estimator N_n/n and p, which leads, according to the dissimilarity chosen, for example: **the minimal χ^2 estimator** $\overline{\theta}_n$ defined by

$$\chi^2\left[p(\overline{\theta}_n), \frac{N_n}{n}\right] = \inf\left\{\chi^2\left[p(\alpha), \frac{N_n}{n}\right]; \alpha \in \Theta\right\}.$$

the maximum likelihood estimator $\hat{\theta}_n$, defined by

$$\operatorname{Log} L_n(\hat{\theta}_n) = \ell_n(\hat{\theta}_n) = \sup\{\ell_n(\alpha); \alpha \in \Theta\}$$

with

$$\ell_n(\alpha) = \sum_{j=1}^{r} N_n^j \operatorname{Log} p_j(\alpha).$$

This estimator minimizes the Kullback information,

$$K\left[\frac{N_n}{n}, p(\hat{\theta}_n)\right] = \inf\left\{K\left[\frac{N_n}{n}, p(\alpha)\right]; \alpha \in \Theta\right\}.$$

Particular Case. If $\theta \longmapsto p(\theta)$ is the identity mapping of Δ into itself, p is estimated from amongst all the probabilities on E. Then

$$\overline{\theta}_n = \hat{\theta}_n = N_n/n.$$

We know (Vol. 1, Corollary 5.3.12) the asymptotic behavior of this estimator,

$$\frac{N_n}{n} \xrightarrow{P_p\text{-a.s.}} p; \quad n\chi^2\left[\frac{N_n}{n}, p\right] \xrightarrow{\mathcal{D}(P_p)} \chi^2(r-1);$$

$$\sqrt{n}\left\{\frac{1}{\sqrt{p_j}}\left[\frac{N_n^j}{n} - p_j\right]\right\}_{1\leqslant j\leqslant r} \xrightarrow{\mathcal{D}(P_p)} N_r(0,\Gamma)$$

with

$$\Gamma = I - \sqrt{p}\,{}^t\sqrt{p}\,.$$

Example 2. Stationary Gaussian Sequences. Identifying the distribution of a centered stationary Gaussian sequence is equivalent to identifying its spectral measure. By limiting ourselves to the case studied in [3.1.2] and in [3.2.1], it is natural to measure the dissimilarity between θ and α by $K(f_\theta,f_\alpha)$. Consider the spectrogram, and

$$U_n(\alpha) = \frac{1}{4\pi}\int_{-\pi}^{\pi}\left[\operatorname{Log} f_\alpha(\lambda) + \frac{I_n(\lambda)}{f_\alpha(\lambda)}\right]d\lambda.$$

We know (Theorem 1.4.39) that if, for every $\alpha \in \Theta$, the function f_α is continuous and strictly positive on $[-\pi,\pi]$, we have

$$U_n(\alpha) \xrightarrow{P_\theta\text{-a.s.}} \frac{1}{4\pi}\int_{-\pi}^{\pi}\left[\operatorname{Log} f_\alpha(\lambda) + \frac{f_\theta(\lambda)}{f_\alpha(\lambda)}\right]d\lambda$$

$$U_n(\alpha) - U_n(\theta) \xrightarrow{P_\theta\text{-a.s.}} K(f_\theta,f_\alpha).$$

In order to estimate θ, it is thus natural to minimize $U_n(\alpha)$. We shall take an estimator, if it exists, $(\hat\theta_n)$, such that, for all n,

$$U_n(\hat\theta_n) = \inf\{U_n(\alpha);\ \alpha \in \Theta\}.$$

Here is the general framework which we have reached.

Definition 3.2.7. Contrasts. Consider a statistical model $(\Omega,\mathcal{A},(P_\theta)_{\theta\in\Theta})$. We call the **contrast function** of this model relative to θ a function $\alpha \longmapsto K(\theta,\alpha)$ from Θ into $\mathbb{R}_+$ having a strict minimum for $\alpha = \theta$. If the experiments are described by a filtration $\mathbb{F} = (\mathcal{F}_t)_{t\in T}$ ($T = \mathbb{N}$ or $T = \mathbb{R}_+$), a **contrast** (or **contrast process**) relative to θ and to K is a function U (independent of θ) from $\Theta \times T \times \Omega$ into $\mathbb{R}$, denoted by $(\alpha,t,\omega) \longmapsto U_t(\alpha,\omega)$, satisfying the following two properties:

(a) For every $(\alpha,t) \in \Theta \times T$, the r.v. $U_t(\alpha)\colon \omega \longmapsto U_t(\alpha,\omega)$ is $\mathcal{F}_t$-measurable: $(U_t(\alpha))$ is an $\mathbb{F}$-adapted process.

(b) $(U_t(\alpha))$ tends to $K(\theta,\alpha)$ in P_θ probability, for $t \to \infty$.

Finally, a **minimum contrast estimator** associated with U is an $\mathbb{F}$-adapted estimator $(\hat{\theta}_t)_{t \in T}$ such that, for every t,

$$U_t(\hat{\theta}_t) = \inf\{U_t(\alpha);\ \alpha \in \Theta\}.$$

Examples. (a) For a sample from a dominated model (Example 1 of [3.2.1]) all of the distributions of which are equivalent, we can take $K(\theta,\alpha) = K(F_\theta,F_\alpha)$, the Kullback information of the distributions of the sample r.v.'s. If the model is identifiable, then $U_t(\alpha) = -(1/n)\text{Log } L_n(\alpha)$ defines a contrast. In this case, the maximum likelihood and minimum contrast coincide.

(b) For a Poisson process on $\mathbb{R}_+$ (Example 2 of [3.2.1])

$$\frac{1}{t}\ \text{Log } L_t(\alpha) \xrightarrow{P_\theta\text{-a.s.}} \theta \text{ Log } \alpha + 1 - \alpha.$$

The function $\alpha \longmapsto \theta \text{ Log } \alpha + 1 - \alpha$ is negative, zero only for $\theta = \alpha$. We can take $K(\theta,\alpha) = \alpha - 1 - \theta \text{ Log } \alpha$ and $U_t(\alpha) = -(1/t)\text{Log } L_t(\alpha)$. Here again, the minimum contrast implies the maximum likelihood.

(c) For Example 1 of [3.2.2], we can take for the contrast function $K(\theta,\alpha) = \chi^2(p(\alpha),p(\theta))$ and for contrast $U_n(\alpha) = \chi^2(p(\alpha),N/n)$. A minimum contrast estimator is a minimal χ^2 estimator.

(d) In Example 2 of [3.2.2], the notations used are those of the definition.

(e) Let (X_n) be a sample from a distribution F_θ with mean θ, and with variance σ_θ^2 ($\theta \in \Theta$, a subset of $\mathbb{R}$). Setting

$$U_n(\alpha) = \frac{1}{n} \sum_{i=1}^{n} (X_i - \alpha)^2.$$

We have

$$U_n(\alpha) \xrightarrow{P_\theta\text{-a.s.}} \sigma_\theta^2 + (\theta - \alpha)^2 .$$

We are thus within the preceding framework with $K(\theta,\alpha) = \sigma_\theta^2 + (\theta - \alpha)$. A minimum contrast estimator of θ is obtained here by the **method of least squares.**

3.2.3. Consistency of Minimum Contrast Estimators

Theorem 3.2.8. *Within the framework of Definition* 3.2.7, *the following hypotheses are made:* θ *is the true (unknown) value of the parameter.*

(1) Θ *is a compact set of* $\mathbb{R}^k$.
(2) $\alpha \longmapsto K(\theta,\alpha)$ *and, for all* ω, $\alpha \longmapsto U_t(\alpha,\omega)$ *are continuous.*
(3) *Denoting, for* $\eta > 0$,

$$w(t,\eta) = \sup\{|U_t(\alpha) - U_t(\beta)|;\ |\alpha - \beta| \leq \eta\},$$

there exist two sequences (η_k) *and* (ε_k) *both decreasing to* 0 *such that, for all* k,

$$P[w(t,\eta_k) > \varepsilon_k] \xrightarrow[n\to\infty]{} 0 .$$

Then every minimum contrast estimator $(\hat{\theta}_t)$ *is consistent at* θ.

Note. Let ϕ be a function from $\mathbb{R}_+$ to $\mathbb{R}$ such that $\lim_{\eta\to 0}\phi(\eta) = 0$. Assumption 3 will often be obtained through: for each η,

$$\lim_{t\to\infty} P[w(t,\eta) \geq 2\phi(\eta)] = 0.$$

For instance

$$w(t,\eta) \xrightarrow[t\to\infty]{P_\theta} \phi(\eta), \qquad \text{or} \qquad \overline{\lim_{t\to\infty}}\ w(t,\eta) \leq \phi(\eta)$$

a.s., are both sufficient conditions.

Proof. Let D be a countable subset dense in Θ ; $\inf_{\alpha\in\Theta} U_t(\alpha) = \inf_{\alpha\in\Theta\cap D} U_t(\alpha)$ is an $\mathcal{F}_t$-measurable r.v. Similarly $w(t,\eta) = \sup\{|U_t(\alpha) - U_t(\beta)|;\ (\alpha,\beta) \in D^2,\ |\alpha - \beta| \leq \eta\}$ is an r.v. $K(\theta,\theta)$ can be taken to be zero.

Let B be a non-empty open ball centered on θ; $K(\theta,\cdot)$ is minorized on $\Theta\backslash B$ by a positive number 2ε. Take k such that $\varepsilon_k < \varepsilon$, and a covering of $\Theta\backslash B$ by a finite number of balls B_i, $1 \leq i \leq N$, centered on θ_i, with radii less than η_k. For $\alpha \in B_i$,

$$U_t(\alpha) \geq U_t(\theta_i) - |U_t(\alpha) - U_t(\theta_i)|,$$

$$\inf_{\alpha\in\Theta\setminus B} U_t(\alpha) \geq \inf_{1\leq i\leq N} U_t(\theta_i) - w(t,\eta_k)$$

$$\{\hat{\theta}_t \notin B\} \subset \left\{\inf_{\alpha\in\Theta\setminus B} U_t(\alpha) < U_t(\theta)\right\}.$$

From which,

$$P_\theta[\hat{\theta}_t \notin B] \leq P_\theta[w(t,\eta_k) > \varepsilon_k]$$

$$+ P_\theta\left[\inf_{1\leq i\leq N}(U_t(\theta_i) - U_t(\theta)) \leq \varepsilon\right].$$

However,

$$\inf_{1\leq i\leq N}(U_t(\theta_i) - U_t(\theta)) \xrightarrow[t\to\infty]{P_\theta} \inf_{1\leq i\leq N} K(\theta,\theta_i) \geq 2\varepsilon .$$

Thus

$$\lim_{t\to\infty} P_\theta(\hat{\theta}_t \notin B) = 0.$$

Corollary 3.2.9. Maximum Likelihood Estimators for A Sample. *Consider a sample from a statistical model $(E,\mathcal{E},(F_\theta)_{\theta\in\Theta})$ dominated by μ, which is σ-finite. For all $\alpha \in \Theta$, $F_\alpha = f(\alpha,\cdot)\mu$. Assume Θ is a compact set in $\mathbb{R}^k$ and $f(\alpha,x) > 0$ for all $\alpha \in \Theta$, $x \in E$ (measures F_α are equivalent). Then every maximum likelihood estimator $(\hat{\theta}_t)$ is consistent at θ when the three following conditions are satisfied:*

(1) *For $\alpha \neq 0$, $F_\alpha \neq F_\theta$ (identifiable model at θ)*
(2) *For any $x \in E$, $\alpha \longmapsto f(\alpha,x)$ is continuous*
(3) *There exists an F_θ-integrable* r.v. *h such that*

$$\sup_{\alpha\in\Theta} |\mathrm{Log}\, f(\alpha,\cdot)| \leq h.$$

Proof. Set $U_n(\alpha) = -(1/n)\Sigma_{i=1}^n \mathrm{Log}\, f(\alpha,X_i)$. We have

$$U_n(\alpha) - U_n(\theta) \xrightarrow{P_\theta\text{ p.s.}} K(F_\theta,F_\alpha) = E_\theta\left[\mathrm{Log}\,\frac{f(\theta,X)}{f(\alpha,X)}\right].$$

Under conditions (1), (2), and (3), Kullback's information exists, $\alpha \longmapsto K(F_\theta,F_\alpha)$ is continuous, and $K(F_\theta,F_\alpha)$ is zero if and only if $\theta = \alpha$. For $\eta > 0$, let

$$g_\eta(x) = \sup\{|\mathrm{Log}\, f(\alpha,x) - \mathrm{Log}\, f(\beta,x)|;\ |\alpha - \beta| \leq \eta\}.$$

The function g_η is an r.v. (take α and β in a countable subset of Θ dense in Θ) and $g_\eta \leqslant 2h$. Thus $\lim_{\eta\to 0} E_\theta[g_\eta(X)] = 0$ and Theorem 3.2.8 can be applied because, P_θ a.s.

$$\begin{aligned} w(n,\eta) &= \sup\{|U_n(\alpha) - U_n(\beta)|;\ |\alpha - \beta| \leqslant \eta\} \\ &\leqslant \frac{1}{n} \sum_{i=1}^{n} g_\eta(X_i); \end{aligned}$$

$$\lim w(n,\eta) \leqslant E_\theta(g_\eta(X)).$$

Example. Exponential Model. Let T be an r.v. on $\mathbb{R}^k$, μ be a σ-finite measure on $\mathbb{R}^k$ and Θ a compact set contained in the interior of

$$I = \left\{ \theta;\ \exp[\phi(\theta)] = \int \exp^{<\theta,T(x)>} d\mu(x) < \infty \right\}.$$

By Proposition 3.3.22 of Volume I, ϕ is twice continuously differentiable and strictly convex on $\mathring{I}$ unless $T(\mu)$ is concentrated on an hyperplane of $\mathbb{R}^k$. Assume $T(\mu)$ is not concentrated on such a hyperplane, then the exponential model $\{F_\theta = \exp(-\phi(\theta) + <\theta,T>)\cdot\mu\}_{\theta\in\Theta}$ is identifiable. We have Log $f(\alpha,\cdot) = -\phi(\alpha) + <\alpha,T>$ and

$$\begin{aligned} K(F_\theta,F_\alpha) &= \phi(\alpha) - \phi(\theta) + <\alpha - \theta, E_\theta(T)> \\ &= \phi(\alpha) - \phi(\theta) + <\alpha - \theta, \text{grad}\ \phi(\theta)>. \end{aligned}$$

As ϕ is strictly convex, $K(F_\theta,F_\alpha) > 0$ unless $\alpha = \theta$. Then it is easy to apply Corollary 3.2.9: any sequence of maximum likelihood estimators is consistent.

Corollary 3.2.10. *Let Θ be a compact set in $\mathbb{R}^k$ and let $\theta \longmapsto p(\theta)$ be a continuous injective function from Θ into*

$$\Delta = \left\{(p_1, \ldots, p_r);\ p_j > 0 \text{ for } 1 \leqslant j \leqslant r,\ \sum_{j=1}^{r} p_j = 1\right\}.$$

For an n-sample from the distribution $p(\theta)$ ($\theta \in \Theta$), the maximum likelihood and minimal χ^2 estimators of θ are consistent.

Proof. For the maximum likelihood it is an easy consequence of Corollary 3.2.9. For the minimal χ^2 estimators let

$$\chi^2(p(\alpha),q) = \sum_{j=1}^{r} \frac{(p_j(\alpha) - q_j)^2}{q_j}$$

$$U_n(\alpha) = \chi^2\left[p(\alpha), \frac{N_n}{n}\right] \xrightarrow[\theta \to \infty]{P_\theta \text{ p.s.}} \chi^2(p(\alpha),p(\theta))$$

(with the notations of Example 1 of [3.2.2]). Let V be a closed neighbourhood of $p(\theta)$ in $\mathbb{R}^k$ which does not contain 0 and let V_1 be the compact set $V \cap \Delta$. The function $(\alpha,q) \longmapsto \chi^2(p(\alpha),q)$ is uniformly continuous on $\Theta \times V_1$ and

$$\lim_{n\to\infty} P_\theta\left[\frac{N_n}{n} \in V_1\right] = 1.$$

Let

$$g(\eta) = \sup\{\chi^2(p(\alpha),q) - \chi^2(p(\beta),q);\ |\alpha-\beta| \leqslant \eta,\ q \in V_1\}.$$

Then Theorem 3.2.8 applies since

$$P_\theta\left[w(n,\eta) > g(\eta)\right] \leqslant P_\theta\left[\frac{N_n}{n} \in V_1\right] \xrightarrow[n\to\infty]{} 0$$

and $\lim_{\eta\to 0} g(\eta) = 0$.

Notes. (a) The hypothesis of compactness of Θ imposed above may seem unnatural, it has not been made for samples from the classical distributions $(N(m,\sigma^2))_{m\in\mathbb{R},\sigma^2>0}$ for example. However we may often consider that a first rough approximation allows us to limit the domain of the parameter to a compact set.

(b) It is sometimes easy to verify consistency directly without the aid of Theorem 3.2.8 and without assuming Θ to be compact (cf. [3.1.3] and [3.2.1]).

3.3. Rate of Convergence of an Estimator

3.3.1. How is Rate of Convergence Measured?

Consider an adapted estimator $\hat{\theta} = (\hat{\theta}_t)_{t\in T}$ of $\theta \in \Theta$: $\Theta \subset \mathbb{R}^k$, $T = \mathbb{N}$ or $T = \mathbb{R}_+$. Let $t \longmapsto v(t)$ be a function from T into $\mathbb{R}_+$ which tends to ∞ when $t \to \infty$. If, in a sense to be stated, the sequence $(v(t)(\hat{\theta}_t - \theta))$ remains bounded, we say that $(\hat{\theta}_t)$ converges to θ with a rate of the order of v. The two most common cases where we consider the sequence as bounded if the parameter equals θ are the following:

(1) $\sup_t (v(t))^2 E_\theta[\hat{\theta}_t - \theta]^2 < \infty$

(2) $v(t)(\hat{\theta}_t - \theta) \xrightarrow[t\to\infty]{\mathcal{D}(P_\theta)} G_\theta$,

for a nonzero distribution G_θ on $\mathbb{R}^k$; in case (2), the distribution G_θ allows the calculation of approximate confidence regions. For $k = 1$, $a > 0$ such that $G_\theta([-a,a]) = G_\theta(]-a,a[) > 0$, we have

$$P_\theta(\hat{\theta}_t \in [\theta - av^{-1}(t), \theta + av^{-1}(t)]) \cong G_\theta([-a,a]).$$

Example 1. We observe an n-sample $(X_1, ..., X_n)$ from a square integrable distribution F_θ on $\mathbb{R}$, with mean θ, and nonzero variance σ_θ^2. For the empirical mean, the estimator of θ, we have,

$$v(n) = \sqrt{n}, \quad G_\theta = N(0,\sigma_\theta^2).$$

Example 2. Let (X_n) be a stationary Guassian sequence having a continuous spectral density f_θ. In order to estimate the mean, the empirical mean $\overline{X}_n$ has rate $\sqrt{n}$ and $G_\theta = N(0, 2\pi f_\theta(0))$ (Theorem 1.4.35). To estimate the covariance, the empirical estimators again have rate $\sqrt{n}$, under the hypotheses of Corollary 1.4.42.

Example 3. Let us denote by $E(\lambda)$ an exponential distribution with parameter $\lambda > 0$. An n-sample $(X_1, ..., X_n)$ is observed of a breakdown time, exponential $E(1)$ after time θ: $(X_1 - \theta, ..., X_n - \theta)$ is an n-sample from $E(1)$. the likelihood is,

$$L_n(\theta) = \exp\left[-\sum_{i=1}^n X_i + n\theta\right] 1_{(\inf_{1\leq i\leq n} X_i > \theta)} \cdot$$

The maximum likelihood estimator is $\hat{\theta}_n = \inf_{1\leq i\leq n} X_i$; the distribution of $\hat{\theta}_n - \theta$ is $E(n)$ [Vol. I, E.4.2.3], and $n(\hat{\theta}_n - \theta)$ has, for all θ, the distribution $E(1)$: here $v(n) = n$, $G_\theta = E(1)$.

Note. If we want to estimate a translation parameter θ, θ a real number, for an n-sample from a distribution F_θ with density $f(\cdot - \theta)$, the maximum likelihood estimator converges at a rate $\sqrt{n}$ for $F_0 = N(0,1)$ and at a rate n for $F_0 = E(1)$ (or for $F_0 = \mathcal{U}_{[-1/2,1/2]}$, check this). We shall prove in [3.3.3] that the rate $\sqrt{n}$ is obtained when f is "regular"; if θ marks a discontinuity of f, the rate may be more rapid.

Example 4. A Poisson process with parameter θ is observed. The maximum likelihood estimator of θ is $\hat{\theta}_t = N_t/t$ at time t. Its rate if $\sqrt{t}$ [2.2.1].

3.3.2. Asymptotic Normality of an Estimator

To pass from the estimation of θ to that of a function of θ, the following theorem will be useful.

Theorem 3.3.11. *Let (T_n) be a sequence of random vectors in $\mathbb{R}^k$. Assume*

$$T_n \xrightarrow{P} m \quad \text{and} \quad \sqrt{n}\,(T_n - m) \xrightarrow{\mathcal{D}} N_k(0,\Gamma).$$

Let g be a function from a neighbourhood U of m into $\mathbb{R}^p$, twice differentiable. Assume that the second order partial derivatives are bounded on U, and denote by J_g the Jacobian matrix of g,

$$\sqrt{n}\,(g(T_n) - g(m)) \xrightarrow{\mathcal{D}} N_p(0, J_g(m)\Gamma\, {}^tJ_g(m)).$$

This limit may be zero.

Example. This lemma applies to the empirical mean $\overline{X}_n$ of an n-sample from a Bernoulli distribution with parameter $\theta \in]0,1[$,

$$\overline{X}_n \xrightarrow{P_\theta\text{-a.s.}} \theta, \quad \sqrt{n}(\overline{X}_n - \theta) \xrightarrow{\mathcal{D}(P_\theta)} N(0,\theta(1-\theta)).$$

The variance $\theta(1-\theta)$ can be estimated by $\overline{X}_n(1-\overline{X}_n)$, and we have,

$$\overline{X}_n(1-\overline{X}_n) \xrightarrow{P_\theta\text{-a.s.}} \theta(1-\theta)$$

$$\sqrt{n}(\overline{X}_n(1-\overline{X}_n) - \theta(1-\theta)) \xrightarrow{\mathcal{D}(P_\theta)} N(0,\theta(1-\theta)(1-2\theta)^2).$$

For $\theta = 1/2$,

$$\sqrt{n}(\bar{X}_n(1 - \bar{X}_n) - \theta(1 - \theta)) \xrightarrow{P_\theta} 0.$$

Proof of the Theorem. Let us assume first of all that g takes values in $\mathbb{R}$; on U, Taylor's formula is written

$$g(x) - g(m) = <x - m, \text{grad } g(m)> + \frac{1}{2}\, {}^t(u - m)D^{(2)}g(m^*)(u - m),$$

where m^* is a point of the segment $[m,x]$ joining m to x and $D^{(2)}g$ the second differential of g.

Since (T_n) tends in probability to m, we have,

$$1_{(T_n \in U)} \xrightarrow{P} 1.$$

If T_n is in U, we can find an $m_n^* \in [m,T_n]$ such that

$$\sqrt{n}(g(T_n) - g(m)) = <\sqrt{n}(T_n - m), grad\ g(m)> + \frac{1}{2}\sqrt{n}\ {}^t(T_n - m)D^{(2)}g(m_n^*)(T_n - m).$$

For $T_n = (T_n^i)_{1 \leq i \leq k}$, $(X^i)_{1 \leq i \leq k}$ with distribution $N_k(0,\Gamma)$, $m = (m^i)_{1 \leq i \leq k}$,

$$n(T_n^i - m^i)(T_n^j - m^j) \xrightarrow{\mathcal{D}} X^i X^j$$

and

$$\sqrt{n}(T_n^i - m^i)(T_n^j - m^j) \xrightarrow{P} 0.$$

U can be taken to be a ball centered on m; then

$$\alpha_n = \sup_{x \in U} \sqrt{n}\ {}^t(T_n - m)D^{(2)}(g(x))(T_n - m)$$

is an r.v., and the sequence of these r.v.'s tends in probability to 0. The point m_n^* is in the ball U. However,

$$\left| 1_{(T_n \in U)}\sqrt{n}(g(T_n - g(m)) - 1_{(T_n \in U)}<\sqrt{n}(T_n - m),\text{grad } g(m)> \right| \leq \frac{1}{2}\alpha_n .$$

Thus

$$\sqrt{n}(g(T_n) - g(m)) - \langle\sqrt{n}(T_n - m), \text{grad } g(m)\rangle \xrightarrow{P} 0.$$

From which, as a result of Proposition 0.3.11,

$$\sqrt{n}(g(T_n) - g(m)) \xrightarrow{\mathcal{D}} N(0, {}^t\text{grad } g(m)\ \Gamma\ \text{grad } g(m)).$$

For g taking values in $\mathbb{R}^p$, the result is obtained by considering $\langle v,g\rangle$ for arbitrary $v \in \mathbb{R}^p$.

Estimators of an Exponential Model. On $(E,\mathcal{E})$, we are given a σ-finite measure μ, and a random vector T of dimension k. Θ is the convex set of $\mathbb{R}^k$ formed from the θ such that $e^{\langle\theta,T\rangle}$ is μ-integrable. Set

$$\exp \phi(\theta) = \int e^{\langle\theta,T\rangle} d\mu$$

and

$$F_\theta = \exp(-\phi(\theta) + \langle\theta,T\rangle)\mu.$$

For θ in the interior of Θ, $\int T\ dF_\theta = \text{grad } \phi(\theta)$. The covariance matrix of T is

$$D^{(2)}\phi(\theta) = \left\{\frac{\partial^2\phi}{\partial\theta_i\partial\theta_j}(\theta),\ 1 \leqslant i,j \leqslant k\right\}.$$

This is Fisher's information, denoted $I(\theta)$, of the model [Vol. I, 7.2.3].

For a sample from F_θ, $(\Omega,\mathcal{A},P_\theta) = (E,\mathcal{E},F_\theta)^{\mathbb{N}}$, and we have,

$$L_n(\theta) = \exp(-n\phi(\theta) + \langle\theta,T(X_1) + \ldots + T(X_n)\rangle).$$

Setting $T_n = (1/n)(T(X_1) + \ldots + T(X_n))$, T_n is the unbiased estimator with minimum variance of grad $\phi(\theta)$ ([Vol I, 7.1]). It converges P_θ-a.s. to grad $\phi(\theta)$ and,

$$\sqrt{n}\ (T_n - \text{grad } \phi(\theta)) \xrightarrow{\mathcal{D}(P_\theta)} N(0,I(\theta)).$$

If there exists an r.v. $\hat{\theta}$ on $\mathbb{R}^k$ taking values in Θ such that grad $\phi(\hat{\theta}(t)) = t$, then ([Vol I, 7.3]) $\hat{\theta}(T_n) = \hat{\theta}_n$ is a maximum likelihood estimator of θ. If $I(\theta)$ is invertible, there exists an open neighbourhood U of θ and an open neighbourhood V of grad $\phi(\theta)$ such that ϕ is a diffeomorphism from U into V. Let ψ be the inverse function of grad ϕ, from V into U. Denoting by J the Jacobian matrix,

$$J\psi(\text{grad } \phi(\theta)) = (J \text{ grad } \phi(\theta))^{-1} = I^{-1}(\theta).$$

The terms of $I^{-1}(\theta)$ are all continuously differentiable in the neighbourhood of θ. Thus,

$$\begin{aligned}\sqrt{n}(\hat{\theta}_n - \theta) &\xrightarrow{\mathcal{D}(P_\theta)} N(0, I^{-1}(\theta) I(\theta) I^{-1}(\theta)) \\ &= N(0, I^{-1}(\theta)).\end{aligned}$$

Proposition 3.3.12. *For an exponential model, let θ be a point in the interior of Θ for which the Fisher information matrix $I(\theta) = D^{(2)}\phi(\theta)$ is invertible. A sample from this model is observed. If $(\hat{\theta}_n)$ is a sequence of maximum likelihood estimators, we have,*

$$\hat{\theta}_n \xrightarrow{P_\theta\text{-a.s.}} \theta, \quad \sqrt{n}(\hat{\theta}_n - \theta) \xrightarrow{\mathcal{D}(P_\theta)} N(0, I^{-1}(\theta)).$$

3.3.3. Regular Models

We have obtained in [Vol. I, 7.2.3], for the "regular" case, a lower bound to the rate of convergence by the Cramer-Rao bound. Let us recall this result and the associated definitions.

Definition 3.3.13. The model of Definition 3.1.1, where Θ is a subset of $\mathbb{R}^k$ and a neighbourhood θ, is said to be **regular at the point** θ, if a likelihood (L_t) can be chosen satisfying, for all t, the following hypotheses.

H1) On a neighbourhood V of θ, for every ω, $\alpha \longmapsto L_t(\alpha,\omega)$ is twice continuously differentiable.

H2) grad Log $L_t(\theta)$ is a centered r.v. and square integrable for P_θ. For $1 \leq i,j \leq k$ and $\theta = (\theta_1, \dots, \theta_k)$,

$$E_\theta\left[\frac{\partial}{\partial\theta_i} \text{Log } L_t(\theta) \frac{\partial}{\partial\theta_j} \text{Log } L_t(\theta)\right] = -E_\theta\left[\frac{\partial^2}{\partial\theta_i\partial\theta_j} \text{Log } L_t(\theta)\right].$$

This quantity is denoted by $I_t^{ij}(\theta)$. The matrix $I_t(\theta) = \{I_t^{ij}(\theta)\}_{1 \leq i,j \leq k}$ is called the **Fisher information matrix** on θ **at time** t.

H3) $I_t(\theta)$ is invertible.

Note. In spite of their rather unattractive appearance these hypotheses are not very strong: for H2, it is only necessary to be able to interchange differentiations and integrations with respect to P_θ [Vol. 1, 7.2.3].

Denote by D_i the ith partial derivative and $\ell_t(\theta) = \text{Log } L_t(\theta)$. The following theorem has been established [Vol. I, Theorem 7.2.16].

Theorem 3.3.14. Cramer-Rao Inequality. *Let a model be regular in* θ, *and let* $Y = (Y_t)_{t \in T}$ *be a second order process adapted to* $\mathbb{F}$. *Assume*

$$\text{grad } E[L_t(\theta)Y_t] = E[\text{grad } L_t(\theta) \cdot Y_t].$$

Then

$$E_\theta[Y_t - E_\theta(Y_t)]^2 \geq {}^t\text{grad } E_\theta(Y) I_t^{-1}(\theta) \text{grad } E_\theta(Y_t).$$

Consequences. Let ϕ be a differentiable function from Θ into $\mathbb{R}$; in estimating $\phi(\theta)$, the bias of Y_t is $b_t(\theta) = E_\theta(Y_t) - \phi(\theta)$. The quadratic error of Y_t is,

$$E_\theta[Y_t - \phi(\theta)]^2$$
$$\geq b_t^2(\theta) + {}^t[\text{grad}(b_t + \phi)(\theta)] I_t^{-1}(\theta)[\text{grad}(b_t + \phi)(\theta)].$$

An unbiased estimator Y_t is **efficient at time** t if,

$$E_\theta[Y_t - \phi(\theta)]^2 = {}^t\text{grad } \phi(\theta) I_t^{-1}(\theta) \text{ grad } \phi(\theta).$$

Sample From a Regular Model. Let us consider a model regular in θ, $(E, \mathcal{E}, (F_\alpha)_{\alpha \in \Theta})$, dominated by F, $(f_\alpha)_{\alpha \in \Theta}$ its likelihood and $I(\theta)$ its Fisher information at θ; denote $f_\alpha: x \longmapsto f(\alpha, x)$.

A canonical sample from this regular model is observed; hence we denote $(\Omega, \mathcal{A}, P_\theta) = (E, \mathcal{E}, F_\theta)^{\mathbb{N}}$ and X_p the pth coordinate. Let

$$\ell_n(\theta) = \text{Log } L_n(\theta) = \sum_{p=1}^{n} \text{Log } f(\theta,X_p).$$

Let

$$Y_n^i = D_i \ell_n(\theta) = \sum_{p=1}^{n} \frac{D_i f(\theta,X_p)}{f(\theta,X_p)},$$

$Y_n = (Y_n^i)_{1\leqslant i\leqslant k} = \text{grad}(\ell_n(\theta))$. On $(\Omega, \mathcal{A}, P_\theta)$, the random vectors

$$\left[\frac{D_i f(\theta,X_p))}{f(\theta,X_p)}\right]_{1\leqslant i\leqslant k}$$

are, for $p \geqslant 1$, centered and independent, and their covariance matrix is $I(\theta)$. We then deduce from the central limit theorem in $\mathbb{R}^k$ ([Vol. 1, 5.3]),

$$\frac{1}{\sqrt{n}} Y_n(\theta) \xrightarrow{\mathcal{D}(P_\theta)} N_k(0,I(\theta)).$$

Now let us assume that we have a maximum likelihood estimator $(\hat{\theta}_n)$ which converges in P_θ-probability to θ. Let V be a convex neighbourhood of θ in which $\alpha \longmapsto f(\alpha,x)$ is twice continuously differentiable for arbitrary x.

If $\hat{\theta}_n$ is in V, we have grad $\ell_n(\hat{\theta}_n) = 0$. Then,

$$\begin{aligned} 0 &= D_i \ell_n(\hat{\theta}_n) \\ &= D_i \ell_n(\theta) + \sum_{j=1}^{k} (\hat{\theta}_n^j - \theta^j) \int_0^1 D_i D_j \ell_n(\theta + t(\hat{\theta}_n - \theta))dt; \end{aligned}$$

$$0 = \frac{1}{\sqrt{n}} Y_n^i + \sum_{j=1}^{k} \sqrt{n}(\hat{\theta}_n^j - \theta^j) \frac{1}{n} \sum_{p=1}^{n} \int_0^1 \psi_{ij}(\theta + t(\hat{\theta}_n - \theta),X_p)dt,$$

with

$$\psi_{ij}(\alpha,x) = D_i D_j \text{ Log } f(\alpha,x);$$

$$E_\theta[\psi_{ij}(\theta,X_1)] = -I^{ij}(\theta).$$

By the law of large numbers,

$$\frac{1}{n} \sum_{p=1}^{n} \psi_{ij}(\theta,X_p) \xrightarrow{P_\theta\text{-a.s.}} -I^{ij}(\theta).$$

If we can be assured of the convergence in P_θ-probability to 0 of

$$\frac{1}{n} 1_{(\hat{\theta}_n \in V)} \sum_{p=1}^{n} \int_0^1 [\psi_{ij}(\theta,X_p) - \psi_{ij}(\theta + t(\hat{\theta}_n - \theta),X_p]dt,$$

then we have on (Ω,A,P_θ),

$$1_{(\hat{\theta}_n \in V)} \left[\frac{1}{\sqrt{n}} Y_n(\theta) + \sqrt{n}(\hat{\theta}_n - \theta)\left\{\frac{1}{n} \sum_{p=1}^{n} \psi_{ij}(\theta,X_p)\right\}\right] \xrightarrow{P_\theta} 0$$

Since the matrix $I(\theta)$ is invertible and $1_{(\hat{\theta}_n \in V)} \xrightarrow{P_\theta} 1$,

$$1_{(\hat{\theta}_n \in V)} \frac{I(\theta)^{-1}}{\sqrt{n}} Y_n(\theta) \xrightarrow{\mathcal{D}(P_\theta)} N(0,I(\theta)^{-1}I(\theta)I(\theta)^{-1})$$
$$= N(0,I(\theta)^{-1});$$

$$1_{(\hat{\theta}_n \in V)} I(\theta)^{-1}\left\{\frac{1}{n} \sum_{p=1}^{n} \psi_{ij}(\theta,X_p)\right\}$$

tends in P_θ-probability to the identity matrix.

Following Proposition 0.3.11, this is equivalent to

$$1_{(\hat{\theta}_n \in V)}\sqrt{n}(\hat{\theta}_n - \theta) \xrightarrow{\mathcal{D}(P_\theta)} N(0,I(\theta)^{-1}).$$

As $P(\hat{\theta}_n \in V) \to 1$, it follows from Proposition 0.3.14,

$$\sqrt{n}\,(\hat{\theta}_n - \theta) \xrightarrow{\mathcal{D}(P_\theta)} N(0,I(\theta)^{-1}).$$

We obtain the following theorem, proved for the case of exponential models in Proposition 3.3.12.

Theorem 3.3.15. *Let a model $(E,\mathcal{E},(F_\alpha)_{\alpha\in\Theta})$ be regular in θ, with Fisher information $I(\theta)$ and likelihood f. Assume that there exists a neighbourhood V of θ and an F_θ-integrable* r.v. *h on $(E,\mathcal{E})$, such that, for every $x \in E$, $\alpha \in V$, $1 \leq i,j \leq k$,*

$$|D_iD_j\text{Log } f(\alpha,x)| \leq h(x).$$

A sample is then observed from this regular model. If $(\hat{\theta}_n)$ is a maximum likelihood estimator which converges in P_θ-probability to θ, we have,

$$\sqrt{n}(\hat{\theta}_n - \theta) \xrightarrow[n\to\infty]{\mathcal{D}(P_\theta)} N(0,I^{-1}(\theta));$$

$$I(\theta)\sqrt{n}(\hat{\theta}_n - \theta) - \frac{1}{\sqrt{n}} \text{ grad } \ell_n(\theta) \xrightarrow[n\to\infty]{P_\theta} 0.$$

Proof. It remains to examine the term

$$\frac{1}{n} \sum_{p=1}^{n} \int_0^1 [\psi_{ij}(\theta,X_p) - \psi_{ij}(\theta + t(\hat{\theta}_n - \theta),X_p)]dt.$$

Let B be a closed ball $B \subset V$ with nonzero radius centered at θ. Set

$$\sigma_{ij}(B,x) = \sup_{\alpha\in B} |\psi_{ij}(\theta,x) - \psi_{ij}(\alpha,x)|.$$

Then $x \longmapsto \sigma_{ij}(B,x)$ is an r.v. on $(E,\mathcal{E})$ since the sup can be taken over the points with rational coordinates, and $\sigma_{ij}(b,\cdot) \leqslant 2h$. If the radius of B decreases to 0, $\sigma_{ij}(B,\cdot)$ tends to 0. By Lebesgue's theorem, for all $\varepsilon > 0$, we can choose B such that

$$\int \sigma_{ij}(B,x)dF_\theta(x) \leqslant \varepsilon.$$

If $\hat{\theta}_n \in B$ we have

$$\left|\frac{1}{n} \sum_{p=1}^{n} \int_0^1 [\psi_{ij}(\theta,X_p) - \psi_{ij}(\theta + t(\hat{\theta}_n - \theta),X_p)]dt\right|$$

$$\leqslant \frac{1}{n} \sum_{p=1}^{n} \sigma_{ij}(B,X_p).$$

Thus

$$P_\theta\left[\left|\frac{1}{n} \sum_{p=1}^{n} \int_0^1 [\psi_{ij}(\theta,X_p) - \psi_{ij}(\theta+t(\hat{\theta}_n-\theta),X_p)]dt\right| \geqslant 2\varepsilon\right]$$

$$\leqslant P_\theta[\hat{\theta}_n \notin B] + P_\theta\left[\frac{1}{n} \sum_{p=1}^{n} \sigma_{ij}(B,X_p) \geqslant 2\varepsilon\right]$$

tends to zero for $n \to \infty$.

Asymptotic Efficiency. Let a model be regular in θ with Fisher information $I(\theta)$ at θ. The Fisher information at time n of a sample from this model is $nI(\theta)$. Let ϕ be a function from a neighbourhood of θ into $\mathbb{R}$, twice differentiable in the neighbourhood of θ, and having bounded derivatives. If Theorem 3.3.15 applies, we have, by Theorem 3.3.11,

$$\phi(\hat{\theta}_n) \xrightarrow{P_\theta} \phi(\theta),$$

$$\sqrt{n}(\phi(\hat{\theta}_n) - \phi(\theta)) \xrightarrow{\mathcal{D}(P_\theta)} N(0, {}^t\text{grad } \phi(\theta)I^{-1}(\theta)\text{grad } \phi(\theta)$$

Moreover the Cramer-Rao inequality implies that, for every adapted estimator (Y_n) of $\phi(\theta)$, we have,

$$E_\theta[\sqrt{n}(Y_n - \phi(\theta))]^2 \geqslant {}^t\text{grad } \phi(\theta)I^{-1}(\theta)\text{grad } \phi(\theta).$$

Hence, $\phi(\hat{\theta}_n)$ is not in general efficient at each instant, however, for n large, its distribution is close to $N(\phi(\theta), {}^t\text{grad } \phi(\theta)I_n^{-1}(\theta)\text{grad } \phi(\theta))$. It is thus "asymptotically efficient." Note that in the above we have not imposed convergence of the variances

$$E_\theta[n(\phi(\hat{\theta}_n) - \phi(\theta))]^2 \to {}^t\text{grad } \phi(\theta)I^{-1}(\theta)\text{grad } \phi(\theta),$$

or even the existence of the variance of $\phi(\hat{\theta}_n)$. The following definition applies to the case of samples from the regular model and from the usual processes. There are more general definitions of "asymptotic efficiency" than that which follows.

Definition 3.3.16. Consider the regular model of Definition 3.3.13. Assume that there exists an invertible matrix $\bar{I}(\theta)$ such that,

$$\lim_{t\to\infty} \frac{I_t(\theta)}{t} = \bar{I}(\theta).$$

Then an estimator $(\hat{\theta}_t)_{t\in T}$ of θ is said to be **asymptotically efficient** at θ if,

$$\sqrt{t}(\hat{\theta}_t - \theta) \xrightarrow{\mathcal{D}(P_\theta)} N_k(0,\bar{I}^{-1}(\theta)).$$

Consequence. Let ϕ be a function of class C^2 in the neighbourhood of θ, taking real values,

$$\sqrt{n}(\phi(\hat{\theta}_t) - \phi(\theta)) \xrightarrow{\mathcal{D}(P_\theta)} N_k(0, {}^t\text{grad } \phi(\theta)\bar{I}^{-1}(\theta)\text{grad } \phi(\theta)).$$

The estimator $(\phi(\hat{\theta}_t))$ of $\phi(\theta)$ is asymptotically efficient.

Examples. (a) Under the hypotheses of Proposition 3.3.10 or of Theorem 3.3.13, the maximum likelihood estimator is asymptotically efficient.

(b) Let us consider in particular a **translation model** $(\mathbb{R}, \mathcal{B}_{\mathbb{R}}, F_\theta)$, where F_θ is the distribution with density $x \longmapsto f(x-\theta)$. We say that it is a **regular translation model** under the following hypotheses:

(1) f is strictly positive on $\mathbb{R}$ and of class C^2; f and f' tend to 0 at infinity.
(2) The functions f'' and $(f')^2/f$ are integrable with respect to Lebesgue measure.

Hypothesis (1) implies:

$$\int f'(x)dx = \lim_{\substack{a\to-\infty \\ b\to\infty}} \int_a^b f'(x)dx$$

$$= \lim_{b\to\infty} f(b) - \lim_{a\to-\infty} f(a) = 0;$$

and likewise

$$\int f''(x)dx = 0.$$

Let us now consider $(\Omega, \mathcal{A}, (P_\theta)_{\theta\in\mathbb{R}}, (X_n))$, a sample from this regular translation model:

$$P_\theta = \prod_{i=1}^{n} \frac{f(X_i - \theta)}{f(X_i)} P_\theta = L_n(\theta)P_0$$

on $\sigma(X_1, \ldots, X_n)$;

$$\frac{d}{d\theta} \operatorname{Log} L_n(\theta) = -\sum_{i=1}^{n} \frac{f'(X_i - \theta)}{f(X_i - \theta)},$$

$$\frac{d^2}{d\theta^2} \operatorname{Log} L_n(\theta) = -\sum_{i=1}^{n} \frac{f''}{f}(X_i - \theta) + \sum_{i=1}^{n} \left(\frac{f'}{f}(X_i - \theta)\right)^2.$$

On $(\Omega, \mathcal{A}, P_\theta)$, $(X_n - \theta)$ is a sample from the distribution with density f:

$$E_\theta\left[\frac{f'}{f}(X_i - \theta)\right] = \int f'(x)dx = 0,$$

$$E_\theta\left[\frac{f'}{f}(X_i - \theta)\right]^2 = \int \frac{f'^2}{f}(x)dx,$$

the expression being denoted by $I(f)$.

$$E_\theta\left[\frac{f''}{f}(X_i - \theta)\right] = \int f''(x)dx = 0.$$

Hence we are dealing with a regular model in the sense of Definition 3.3.13. The Fisher information at time n is, for all θ, $nI(f)$.

From Corollary 3.2.10, when Log f is uniformly continuous on $\mathbb{R}$ every sequence of maximum likelihood estimators is consistent if Θ is compact.

If the function $((f''/f) - (f'^2/f^2))$ is uniformly continuous on $\mathbb{R}$, Theorem 3.3.15 applies, and if $(\hat{\theta}_n)$ is a sequence of maximum likelihood estimators which tends to θ, P_θ-a.s., then,

$$\sqrt{n}(\hat{\theta}_n - \theta) \xrightarrow{\mathcal{D}(P_\theta)} N(0,(1/I(f))).$$

(c) For a Poisson process on $\mathbb{R}_+$, $(\Omega,A,(P_\theta)_{\theta>0},(N_t)_{t\geqslant 0})$, the likelihood at time t is $\exp\{N_t \mathrm{Log}\,\theta + (1 - \theta)t\}$. From which

$$I_t(\theta) = E_\theta\left[\frac{N_t}{\theta} - t\right]^2 = \frac{t}{\theta}$$

and $\overline{I}(\theta) = 1/\theta$; $E_\theta[(N_t/t) - \theta]^2 = \theta/t$ and N_t/t is efficient at each instant;

$$\sqrt{t}\left[\frac{N_t}{t} - \theta\right] \xrightarrow{\mathcal{D}(P_\theta)} N(0, \theta))$$

and (N_t/t) is asymptotically efficient.

3.3.4. Limit Distributions of Minimum Contrast Estimators

Let us use the framework of Definition 3.2.7. In order to sketch out the proof of Theorem 3.3.15, we can make the following hypotheses (in [3.3.3] $-nU_n = \ell_n$):

K1) $\Theta \subset \mathbb{R}^k$, θ is an interior point of Θ and in a neighbourhood V of θ, the function $\alpha \longmapsto U_t(\alpha,\omega)$ is twice continuously differentiable for all ω.

K2) For the process $(\mathrm{grad}\, U_t(\theta))$, we have a central limit theorem

$$\sqrt{t}\,\mathrm{grad}\, U_t(\theta) \xrightarrow{\mathcal{D}(P_\theta)} N_k(0,\Gamma_U(\theta)).$$

K3) There exists an invertible symmetric matrix $I_U(\theta)$ such that

$$\{D_iD_jU_t(\theta)\}_{1\leqslant i,j\leqslant k} \xrightarrow[t\to\infty]{P_\theta} I_U(\theta).$$

K4) $$\sup_{|\alpha|\leqslant r} |D_iD_jU_t(\theta+\alpha) - D_iD_jU_t(\theta)| \xrightarrow[n\to\infty]{\theta} 0,$$

for $1 \leqslant i,j \leqslant k$ and an $r > 0$.

Then, if $(\hat{\theta}_t)$ is a minimum contrast estimator consistent at θ,

$$\sqrt{t}\,(\hat{\theta}_t - \theta) \xrightarrow{\mathcal{D}(P_\theta)} N(0, I_U^{-1}(\theta)\Gamma_U(\theta)I_U^{-1}(\theta)).$$

This basic result is above all a method of proof which will be used in what follows for various processes. We shall be satisfied with checking hypotheses K1-4 and shall thus obtain the asymptotic rate of convergence of $(\hat{\theta}_t)$. In order to establish this result, let us proceed as in [3.3.3]. Take r such that $\{\alpha;\ |\alpha - \theta| \leqslant r\} \subset V$; for $|\hat{\theta}_t - \theta| \leqslant r$:

$$\begin{aligned} 0 &= \sqrt{t}\,D_iU_t(\hat{\theta}_t) \\ &= \sqrt{t}\,D_iU_t(\theta) + \sum_{j=1}^{k} \sqrt{t}(\hat{\theta}_t^j - \theta^j)\int_0^1 D_iD_jU_t(\theta + s(\hat{\theta}_t - \theta))ds. \end{aligned}$$

If

$$1_{(|\hat{\theta}_t-\theta|\leqslant r)} \int_0^1 [D_iD_jU_t(\theta + s(\hat{\theta}_t - \theta)ds - D_iD_jU_t(\theta)]ds \xrightarrow{P_\theta} 0.$$

which holds under hypothesis K4, then the conditions K1, K2, K3 ensure that $\sqrt{t}(\hat{\theta}_t - \theta)$ converges in distribution like $I_U^{-1}(\theta)\sqrt{t}(D_iU_t(\theta))_{1\leqslant j\leqslant k}$ to

$$_k(0,\ I_U^{-1}(\theta)\Gamma_U(\theta)I_U^{-1}(\theta)).$$

Estimation for Stationary Gaussian Sequences. Recall Example 2 of [3.2.2.] with the contrast function,

$$U_n(\alpha) = \frac{1}{2\pi}\int_{-\pi}^{\pi}\left[\text{Log } f(\alpha,\lambda) + \frac{I_n(\lambda)}{f(\alpha,\lambda)}\right]d\lambda.$$

Take $\Theta \subset \mathbb{R}^k$, θ in the interior of Θ, and V a neighbourhood of θ. Assume that for any $\alpha \in V$, $m \leqslant f_\alpha \leqslant M$, with constants m and M such that $m > 0$. Let us assume that the partial derivatives in α of f of order 1 and 2 exist for $\alpha \in V$ and are continuous on $V \times \mathbb{R}$, with period 2π. We can then differentiate under the integral sign,

$$-D_iU_n(\alpha) = \frac{1}{2\pi}\int_{-\pi}^{\pi}\left[I_n(\lambda)\,\frac{D_if(\alpha,\lambda)}{f^2(\alpha,\lambda)} - \frac{D_if(\alpha,\lambda)}{f(\alpha,\lambda)}\right]d\lambda.$$

From Theorem 1.4.41,

$$\sqrt{n}\ \text{grad}\ U_n(\theta) \xrightarrow{\mathcal{D}(P_\theta)} N_k(0,\Gamma(\theta))$$

with

$$\Gamma(\theta) = \left\{\frac{1}{4\pi}\int_{-\pi}^{\pi}\frac{D_if(\theta,\lambda)D_jf(\theta,\lambda)}{f^2(\theta,\lambda)}d\lambda\right\}_{1\leqslant i,j\leqslant k}.$$

Moreover,

$$D_iD_jU_n(\theta) = \frac{1}{4\pi}\int_{-\pi}^{\pi}[I_n(\lambda)D_iD_j(1/f)(\theta,\lambda) + D_iD_j\text{Log}\ f(\theta,\lambda)]d\lambda.$$

From Theorem 1.4.39,

$$D_iD_jU_n(\theta) \xrightarrow{P_\theta\text{-a.s.}} \frac{1}{4\pi}\int_{-\pi}^{\pi}\left[f(\theta,\lambda)D_iD_j\left(\frac{1}{f}\right)(\theta,\lambda) + D_iD_j\text{Log}\ f(\theta,\lambda)\right]d\lambda.$$

from which we deduce

$$\{D_iD_jU_n(\theta)\}_{1\leqslant i,j\leqslant k} \xrightarrow{P_\theta\text{-a.s.}} \Gamma(\theta).$$

Finally, for B a closed ball, centered on θ, $B \subset V$, and h continuous on $B \times [-\pi,\pi]$, let us denote $h^B(\lambda) = \sup\{|h(\alpha,\lambda) - h(\theta,\lambda)|;\ \alpha \in B\}$. We have

$$\left[D_iD_j\frac{1}{f}\right]^B = \left[\frac{D_iD_jf}{f^2} - 2\,\frac{D_ifD_jf}{f^3}\right]^B$$

$$\leqslant \frac{1}{m^3}\,|fD_iD_jf - 2D_ifD_jf|^B,$$

$$(D_iD_j\text{Log}\ f)^B = \left[\frac{D_iD_jf}{f} - \frac{D_ifD_jf}{f^2}\right]^B$$

$$\leqslant \frac{1}{m^2}|fD_iD_jf - D_ifD_jf|^B.$$

For all $\varepsilon > 0$, we can find B, a ball of radius r, such that these expressions are majorized by ε. Then,

$$\sup_{|\alpha|\leqslant r} |D_iD_jU_n(\theta + \alpha) - D_iD_jU_n(\theta)| \leqslant \frac{\varepsilon}{4\pi}\int_{-\pi}^{\pi} (I_n(\lambda) + 1)d\lambda.$$

However,

$$\frac{\varepsilon}{4\pi}\int_{-\pi}^{\pi} (I_n(\lambda) + 1)d\lambda \xrightarrow{P_\theta\text{-a.s.}} \frac{\varepsilon}{4\pi}\int_{-\pi}^{\pi} (f(\lambda) + 1)d\lambda.$$

This being true for arbitrary $\varepsilon > 0$, we obtain K4. Hypotheses K1, K2, K3, K4 are satisfied. Here, as in the case of samples, we have $I_U = \Gamma_U$.

Taking Θ compact and f continuous and nonzero on $\Theta \times [-\pi,\pi]$, Theorem 3.2.8 applies. From which the following theorem results.

Theorem 3.3.17. *Let $\Theta \subset \mathbb{R}^k$. A continuous function f from $\Theta \times R$ into $]0,\infty[$ of period 2π is given. For θ an interior point of Θ and V a neighbourhood of θ, assume that f has continuous partial derivatives with respect to the first variable of order 1 and 2 on $V \times \mathbb{R}$. Let $(\hat{\theta}_n)$ be a minimum contrast estimator for the contrast,*

$$\frac{1}{4\pi}\int_{-\pi}^{\pi} \left[\text{Log } f(\alpha,\lambda) + \frac{I_n(\lambda)}{f(\alpha,\lambda)}\right]d\lambda.$$

When the estimator is consistent (which is always the case if Θ is compact), we have,

$$\sqrt{n}\,(\hat{\theta}_n - \theta) \xrightarrow{\mathcal{D}(P_\theta)} N(0,\Gamma^{-1}(\theta))$$

with

$$\Gamma(\theta) = \left\{\frac{1}{4\pi}\int_{-\pi}^{\pi} \frac{D_if(\theta,\lambda)D_jf(\theta,\lambda)}{f^2(\theta,\lambda)}\, d\lambda\right\}_{1\leqslant i,j\leqslant k}.$$

Example. Let us take $\Theta = [-a,a]$ with $0 < a < 1$, a Gaussian white noise (ε_n), and the sequence (X_n) defined by $X_n = \varepsilon_n + \theta\varepsilon_{n-1}$. Its spectral density is

$$\lambda \longmapsto f(\theta,\lambda) = \frac{1}{2\pi}\,|1 + \theta^2 + 2\theta \cos \lambda|,$$

and its satisfies the conditions of the theorem. We calculate,

$$\Gamma(\theta) = \frac{1}{\pi}\int_{-\pi}^{\pi} \frac{(\theta + \cos\lambda)^2}{(1 + \theta^2 + 2\theta\cos\lambda)^2}\, d\lambda = \frac{1}{1 - \theta^2}.$$

3.3.5. Newton's Method

Recall the basic idea of Newton's method for searching for a root of a function f of class C^1. Let x_0 be a point close to the root x sought after: we can write $0 = f(x) \simeq f(x_0) + (x - x_0)f'(x_0)$, where $x \simeq x_0 - (f(x_0)/f'(x_0))$. Let us now consider a contrast U (Definition 3.2.7). Take Θ as an open interval of $\mathbb{R}$, $T = \mathbb{N}$ and $\alpha \longmapsto U_n(\alpha,\omega)$ of class C^2. Denote by U_n' and U_n'' its partial derivatives. A minimum contrast estimator $(\hat{\theta}_n)$ necessarily satisfies the relation $U_n'(\hat{\theta}_n) = 0$.

This relation does not allow in general a simple calculation of $\hat{\theta}_n$. On the other hand we very often have available simple adapted estimators $(\bar{\theta}_n)$ which are consistent, for example some empirical estimators.

Newton's method suggests that

$$\hat{\hat{\theta}}_n = \bar{\theta}_n - \frac{U_n'(\bar{\theta}_n)}{U_n''(\bar{\theta}_n)}$$

is an estimator closer to $\hat{\theta}_n$ than $\bar{\theta}_n$. This estimator $(\hat{\hat{\theta}}_n)$ is easy to calculate. It has the same asymptotic risk as $(\hat{\theta}_n)$ under the hypotheses of the following theorem (see [0.3.2] for the notion of "bounded in probability").

Theorem 3.3.18. *Let us assume that the hypotheses* K1,2,3,4 *of* [3.3.4] *are satisfied,* Θ *being an interval of* $\mathbb{R}$ *and* θ *a point interior to* Θ. *Let* $(\hat{\theta}_n)$ *be a sequence of minimum contrast estimators of* θ *and* $(\bar{\theta}_n)$ *another adapted sequence of estimators. Assume that these two sequences are consistent and* $(\sqrt{n}(\bar{\theta}_n - \theta))$ *is bounded in probability. Set*

$$\hat{\hat{\theta}}_n = \bar{\theta}_n - \frac{U'(\bar{\theta}_n)}{U''(\bar{\theta}_n)}.$$

Then

$$\sqrt{n}(\hat{\hat{\theta}}_n - \hat{\theta}_n) \xrightarrow{P_\theta} 0,$$

thus we again have

$$\sqrt{n}(\hat{\hat{\theta}}_n - \theta) \xrightarrow{\mathcal{D}(P_\theta)} N(0, I_U^{-1}(\theta)\Gamma_U(\theta)I_U^{-1}(\theta)).$$

Example 1. Recall the example of the translation model ([3.3.3]). A maximum likelihood estimator $\hat{\theta}_n$ is the solution of the equation

$$-U_n'(\hat{\theta}_n) = \sum_{i=1}^{n} \frac{f'}{f}(X_i - \hat{\theta}_n) = 0.$$

The solution of this equation is not in general very easy (except in simple cases such as those where $F_\theta = N(\theta,1)$.

Now let us assume $\sigma^2 = \int x^2 f(x)dx < \infty$ and $\int x\, f(x)dx = 0$. Then F_θ has mean θ and variance σ^2. A simple estimator of θ is the empirical mean $\overline{X}_n = (X_1 + \dots + X_n)/n$. We have

$$\overline{X}_n \xrightarrow{P_\theta\text{-a.s.}} \theta;$$

and $\sqrt{n}(\overline{X}_n - \theta)$ has, under P_θ, a distribution independent of θ, which tends to $N(0,\sigma^2)$. Under the conditions seen in Example [3.3.3] which assure the asymptotic efficiency of $(\hat{\theta}_n)$, the estimator $(\hat{\hat{\theta}}_n)$ is asymptotically efficient, by setting

$$\hat{\hat{\theta}}_n = \overline{X}_n + \frac{\sum_{i=1}^{n} \frac{f'}{f}(X_i - \overline{X}_n)}{\sum_{i=1}^{n} \left[\left(\frac{f'}{f}\right)^2 - \frac{f''}{f}\right](X_i - \overline{X}_n)}.$$

Example 2. Recall the Gaussian sequence with spectral density

$$f(\theta,\lambda) = \frac{1}{2\pi}\,[1 + \theta^2 + 2\theta \cos \lambda]$$

considered at the end of [3.3.4]. The empirical estimator of θ is

$$\overline{\theta}_n = \frac{1}{n} \sum_{p=1}^{n} X_p X_{p-1}.$$

This is a consistent estimator and (Corollary 1.4.42)

$$\sqrt{n}(\overline{\theta}_n - \theta) \xrightarrow{\mathcal{D}(P_\theta)} N\left(0, \frac{2}{\pi} \int_0^{\pi} \cos^2\lambda[1 + 2\theta\cos \lambda + \theta^2]^2 d\lambda\right).$$

Now

$$\frac{2}{\pi}\int_0^\pi \cos^2\lambda[1 + 2\theta\cos\lambda + \theta^2]^2 d\lambda = (1 + \theta^2)^2 + 3\theta^2 > \frac{1}{\Gamma(\theta)} = 1 - \theta^2.$$

Hence the improved sequence $(\hat{\hat{\theta}}_n)$ converges more quickly than the empirical estimator $(\overline{\theta}_n)$.

Proof of Theorem 3.3.18. Let $\delta > 0$, $I_\delta =]\theta - \delta, \theta + \delta[$ and let $A_n(\delta) = \{\hat{\theta}_n \in I_\delta, \overline{\theta}_n \in I_\delta\}$. The hypothesis of convergence of $\hat{\theta}_n$ and $\overline{\theta}_n$ implies $\lim_{n\to\infty} P[A_n(\delta)] = 1$. Let us take δ small enough in order that, for all (ω,n), $\alpha \longmapsto U_n(\alpha)$ is of class C^2 on I_δ.

Denote $R_n(\alpha) = \int_0^1 U_n''(\theta + s(\alpha - \theta))ds$. On $A_n(\delta)$, we have

$$U_n'(\hat{\theta}_n) = 0 = U_n'(\theta) + (\hat{\theta}_n - \theta)R_n(\hat{\theta}_n),$$

$$U_n'(\overline{\theta}_n) = U_n'(\theta) + (\overline{\theta}_n - \theta)R_n(\overline{\theta}_n);$$

$$\hat{\theta}_n - \overline{\theta}_n + \frac{U_n'(\overline{\theta}_n)}{U_n''(\overline{\theta}_n)} = U_n'(\overline{\theta}_n)\left[\frac{1}{U_n''(\theta)} - \frac{1}{R_n(\overline{\theta}_n)}\right] + U_n'(\theta)\left[\frac{1}{R_n(\overline{\theta}_n)} - \frac{1}{R_n(\hat{\theta}_n)}\right]$$

$$= U_n'(\theta)\left[\frac{1}{U_n''(\overline{\theta}_n)} - \frac{1}{R_n(\hat{\theta}_n)}\right] + (\overline{\theta}_n - \theta)\left[\frac{R_n(\overline{\theta}_n)}{U_n''(\overline{\theta}_n)} - 1\right].$$

However, under hypothesis K2,

$$\sqrt{n}\, U_n'(\theta) \xrightarrow{\mathcal{D}(P_\theta)} N(0,\Gamma_U(\theta)).$$

The sequences

$$1_{A_n(\delta)}\left[\frac{1}{U_n''(\overline{\theta}_n)} - \frac{1}{R_n(\hat{\theta}_n)}\right]$$

and

$$1_{A_n(\delta)}\left[\frac{R_n(\overline{\theta}_n)}{U_n''(\overline{\theta}_n)} - 1\right]$$

tend to 0 in probability; since the sequences $\sqrt{n}\, U_n'(\theta)$ and $\sqrt{n}(\overline{\theta}_n - \theta)$ are bounded in probability, we obtain the result.

3.4. Asymptotic Properties of Tests

3.4.1. Separation of Two Hypotheses

A sample $\{\Omega, \mathcal{A}, (P_\theta)_{\theta \in \Theta}, (X_n)\}$ from an exponential model $(E, \mathcal{E}, (F_\theta)_{\theta \in \Theta})$ is observed, with $\Theta \subset \mathbb{R}$ and F_θ with density $\exp[-\phi(\theta) + \theta T]$, T being an r.v. on $(E, \mathcal{E})$.

Let $\theta_0 < \theta_n$; in order to test "$\theta \leqslant \theta_0$" against "$\theta > \theta_n$", we have seen ([Vol. I, 8.3.2]) that a uniformly most powerful test is obtained at time n by taking the rejection region

$$R_n = \{T(X_1) + \ldots + T(X_n) > C_n\}.$$

Let us assume that $T(X_1)$ has, on $(\Omega, \mathcal{A}, P_\theta)$, mean $m(\theta)$ and variance $\sigma^2(\theta)$.

The level of this test is $P_\theta[T(X_1) + \ldots + T(X_n) > C_n]$. By the central limit theorem, we obtain a test with level close to α by taking,

$$P_{\theta_0}\left[\frac{T(X_1) + \ldots + T(X_n) - nm(\theta_0)}{\sqrt{n}\ \sigma(\theta_0)} > \frac{C_n - nm(\theta_0)}{\sqrt{n}\ \sigma(\theta_0)}\right] \simeq \alpha.$$

Thus, denoting by Φ_α the point of $\mathbb{R}$ such that $N(0,1)([\Phi_\alpha, \infty[) = \alpha$, we shall take $C_n = nm(\theta_0) + \Phi_\alpha \sqrt{n}\ \sigma(\theta_0)$. The power of this test is

$$p_n = P_{\theta_n}\left[\frac{T(X_1) + \ldots + T(X_n) - nm(\theta_n)}{\sqrt{n}\ \sigma(\theta_n)} > \frac{C_n - nm(\theta_n)}{\sqrt{n}\ \sigma(\theta_n)}\right]$$

$$\simeq 1 - \Phi\left[\frac{n(m(\theta_0) - m(\theta_n))}{\sqrt{n}\ \sigma(\theta_n)} + \Phi_\alpha \frac{\sigma(\theta_0)}{\sigma(\theta_n)}\right].$$

However, if θ_0 is an interior point of Θ, we have $m(\theta_0) = \phi'(\theta_0)$, $\sigma^2(\theta_0) = \phi''(\theta_0)$. If (θ_n) is a sequence minorized by θ_0 tending to θ_0, we have

$$p_n \simeq 1 - \Phi[\Phi_\alpha - \sqrt{n}(\theta_n - \theta)\sigma(\theta_0)].$$

If $\sqrt{n}(\theta_n - \theta_0)\sigma(\theta_0)$ tends to γ, $0 < \gamma < \infty$, the power tends to $N(0,1)[\Phi_\alpha - \gamma, \infty[= \beta$, $\beta > \alpha$, if $(\sqrt{n}(\theta_n - \theta))$ tends to ∞, the power tends to 1, the hypotheses separating very quickly using the tests. On the other hand if $(\sqrt{n}(\theta_n - \theta))$ tends to 0, the power tends to α, and the test does not separate the hypotheses very well.

The above can also be stated as: for large n, the hypotheses "$\theta \leqslant \theta_0$" and "$\theta \geqslant \theta_n$" are correctly "separated" for $\sqrt{n}(\theta_n - \theta_0) \simeq \gamma$ and very well separated if $\sqrt{n}(\theta_n - \theta)$ is large. Whereas, for $\sqrt{n}(\theta_n - \theta_0)$ small they are not well separated.

The study of the power of tests of an hypothesis "$\theta \in \Theta_0$" against "$\theta \in \Theta_1$" is difficult. Sometimes it is made easier, when, as in the case of the one-sided test studied above, a $\theta_0 \in \Theta_0$ and a $\theta_1 \in \Theta_1$ are the least well separated... in a heuristic sense. For the asymptotic study of tests, it is natural to fix an hypothesis "$\theta \in \Theta_0$" and to consider at what rate the alternative hypothesis "$\theta \in \Theta_n$" can approximate the null hypothesis in such a way that the sequence (p_n) of powers of the tests used of fixed level α converges to a number β, $\alpha < \beta \leqslant \infty$.

3.4.2. Likelihood Ratio Test

Recall the scheme described in [3.3.4] with the hypotheses K1-4. If $(\hat{\theta}_t)$ is a minimum contrast estimator consistent at θ, the Taylor expansion of order 2 of U_t at $\hat{\theta}_t$ may be written; for $r > 0$ small enough, when $|\hat{\theta}_t - \theta| \leqslant r$,

$$[U_t(\theta) - U_t(\hat{\theta}_t)] = \frac{1}{2} \sum_{i=1}^{k} \sum_{j=1}^{k} (\theta^i - \hat{\theta}_t^i)(\theta^j - \hat{\theta}_t^j) D_i D_j U_t(\theta)$$

$$+ \frac{1}{2} \sum_{i=1}^{k} \sum_{j=1}^{k} (\theta^i - \hat{\theta}_t^i)(\theta^j - \hat{\theta}_t^j) \int_0^1 [D_i D_j U_t(\theta + x(\hat{\theta}_t - \theta)) - D_i D_j U_t(\theta)] dx.$$

We have,

$$\sqrt{t}(\theta - \hat{\theta}_t) \xrightarrow{\mathcal{D}(P_\theta)} N(0, I_u^{-1}(\theta)\Gamma_U(\theta)I_u^{-1}(\theta)),$$

$$(D_i D_j U_t(\theta))_{1 \leqslant i,j \leqslant k} \xrightarrow{P_\theta} \Gamma(\theta),$$

$$1_{(|\hat{\theta}_t-\theta|\leq r)} \int_0^1 [D_iD_jU_t(\theta + x(\hat{\theta}_t - \theta)) - D_iD_jU_t(\theta)]dx \xrightarrow{P_\theta} 0.$$

In the two cases which we have developed, n-samples from a regular model and stationary Gaussian sequences (Theorems 3.3.15 and 3.3.17), we have $\Gamma_U(\theta) = I_U(\theta)$.

However if Y is a Gaussian vector distributed as $N_k(0,\Gamma)$, Γ being invertible, ${}^tY\Gamma^{-1}Y$ is distributed as $\chi^2(k)$ (Vol. 1, E5.2.3). Hence

$$1_{(|\hat{\theta}_t-\theta|\leq r)}2t(U_t(\theta) - U_t(\hat{\theta}_t)) \xrightarrow[t\to\infty]{\mathcal{D}(P_\theta)} \chi^2(k),$$

and by Proposition 0.3.14,

$$2t(U_t(\theta) - U_t(\hat{\theta}_t)) \xrightarrow[t\to\infty]{\mathcal{D}(P_\theta)} \chi^2(k).$$

Theorem 3.4.19. (a) *For an n-sample from a regular model* $(E, \mathcal{E}, (F_\alpha)_{\alpha\in\Theta})$, *under the hypotheses of Theorem* 3.3.15, *we have, denoting by* $f(\alpha,\cdot)$ *the density of* F_α *and* $\ell_n(\alpha) = \Sigma_{i=1}^n \text{Log } f(\alpha,X_i) = -nU_n(\alpha)$,

$$\ell_n(\hat{\theta}_n) - \ell_n(\theta) \xrightarrow{\mathcal{D}(P_\theta)} \frac{1}{2}\chi^2(k).$$

(b) *For a stationary Gaussian sequence, we have under the hypotheses of Theorem* 3.3.17,

$$n[U_n(\theta) - U_n(\hat{\theta}_n)] \xrightarrow{\mathcal{D}(P_\theta)} \frac{1}{2}\chi^2(k).$$

This theorem is the basis of **likelihood tests** (or **contrast tests**) of level asymptotically equal to α. In order to test H_0: "$\theta = \theta_0$" against H_1: "$\theta \in \Theta_1$," the following rejection regions can be used, $\{\ell_n(\hat{\theta}_n) - \ell_n(\theta_0) \geq (1/2)\chi^2_{k,\alpha}\}$ in the independent case

$$\left\{\int_{-\pi}^{\pi} d\lambda \left[I_n(\lambda)\left[\frac{1}{f(\theta_0,\lambda)} - \frac{1}{f(\hat{\theta}_n,\lambda)}\right] + \text{Log}\frac{f(\theta_0,\lambda)}{f(\hat{\theta}_n,\lambda)}\right] \geq \frac{2\pi}{n}\chi^2_{k,\alpha}\right\}$$

in the Gaussian case.

3.4.3. Chi-Squared Test

In this section we study samples from distributions defined on a finite space (cf. [3.2.2], Example 1).

Theorem 3.4.20. *Let $\Theta \subset \mathbb{R}$ and let $\theta \longmapsto p(\theta) = (p_j(\theta))_{1 \leqslant j \leqslant r}$ be a function from Θ into the set of probabilities on $\{1, 2, ..., r\}$. Assume that θ is an interior point of Θ, and that, for $1 \leqslant j \leqslant r$, the functions p_j are of class C^2, and nonzero at θ. Then the Fisher information matrix is $I(\theta) = B(\theta)^t B(\theta)$, where $B(\theta)$ is the following $k \times r$ matrix,*

$$B(\theta) = \left[\frac{1}{\sqrt{p_j(\theta)}} D_i p_j(\theta)\right]_{1 \leqslant i \leqslant k, 1 \leqslant j \leqslant r}.$$

Assume that $I(\theta)$ is invertible, i.e. the vectors $\{D_i p_j(\theta)\}_{1 \leqslant j \leqslant r} = v_i$ in $\mathbb{R}^r$ are, for $i = 1, ..., k$, linearly independent.

For $\alpha \in \Theta$, the canonical sample of $p(\alpha)$ with distribution P_α is observed, and a sequence $(\hat{\theta}_n)$ of maximum likelihood estimators of the parameter is given, consistent at θ. Denote by N_i^n the number of observations equal to i amongst the first n, and $N^n = (N_i^n)_{1 \leqslant i \leqslant r}$. Then

$$n\chi^2\left(\frac{N^n}{n}, p(\hat{\theta}_n)\right) \xrightarrow[n \to \infty]{\mathcal{D}(P_\theta)} \chi^2(r - 1 - k).$$

Proof. We calculate Fisher's information,

$$I^{ij}(\theta) = \sum_{u=1}^{r} p_u(\theta) \frac{D_i p_u(\theta) D_j p_u(\theta)}{[p_u(\theta)]^2}; \quad I(\theta) = B(\theta)^t B(\theta).$$

Let us denote

$$Z_n(\theta) = \left[\frac{N_i^n - np^i(\theta)}{\sqrt{np_i(\theta)}}\right]_{1 \leqslant i \leqslant r}.$$

We have

$$\frac{1}{\sqrt{n}} \operatorname{grad} \ell_n(\theta) = \left[\frac{1}{\sqrt{n}} \sum_{u=1}^{r} N_u^n \frac{D_i p_u(\theta)}{p_u(\theta)}\right]_{1 \leqslant i \leqslant k}.$$

Noting that $B(\theta)\sqrt{p(\theta)}$ is zero, we have,

$$\frac{1}{\sqrt{n}} \operatorname{grad} \ell_n(\theta) = B(\theta) Z_n(\theta).$$

However from Theorem 3.3.15,

$$I^{-1}(\theta)\,\frac{1}{\sqrt{n}}\operatorname{grad} \ell_n(\theta) - \sqrt{n}(\hat{\theta}_n - \theta) \xrightarrow{P_\theta} 0.$$

Moreover, $Z_n(\theta) \xrightarrow{\mathcal{D}(P_\theta)} N_r(0,\Gamma(\theta))$ with $\Gamma(\theta) = I_r - \sqrt{p(\theta)}\,{}^t\sqrt{p(\theta)}$, ($I_r$ the $r \times r$ identity matrix). We deduce,

$$(Z_n(\theta),\sqrt{n}(\hat{\theta}_n - \theta)) \xrightarrow{\mathcal{D}(P_\theta)} N_{r+k}(0,C(\theta)).$$

$C(\theta)$ is calculated by noting that $B(\theta)\sqrt{p(\theta)} = 0$ implies $\Gamma(\theta){}^tB(\theta) = {}^tB(\theta)$ and $B(\theta)\Gamma(\theta) = B(\theta)$,

$$C(\theta) = \begin{bmatrix} I_r \\ I^{-1}(\theta)B(\theta) \end{bmatrix} \Gamma(\theta)[I_r \quad {}^tB(\theta)I^{-1}(\theta)]$$

$$= \begin{bmatrix} \Gamma(\theta) & {}^tB(\theta)I^{-1}(\theta) \\ I^{-1}(\theta)B(\theta) & I^{-1}(\theta) \end{bmatrix}.$$

We then study

$$\frac{N_i^n - np_i(\hat{\theta}_n)}{\sqrt{np_i(\theta)}} = Z_n(\theta) + \sqrt{n}\left[\sqrt{p_i(\theta)} - \frac{p_i(\hat{\theta}_n)}{\sqrt{p_i(\theta)}}\right]_{1 \leq i \leq r}$$

by applying Theorem 3.1.1 to the function

$$g: (x_1, \ldots, x_r, x_{r+1}, \ldots, x_{r+k})$$

$$\longmapsto \left[x_i + \sqrt{n}p_i(\theta) - \sqrt{n}\,\frac{p_i\left(\frac{1}{\sqrt{n}}(x_{r+1}, \ldots, x_{r+k})+\theta\right)}{\sqrt{p_i(\theta)}}\right]_{1 \leq r \leq r}.$$

For $(x_{r+1}, \ldots, x_{r+k}) = 0$, the Jacobian matrix $D(\theta)$ of g is the $r \times (r + k)$ matrix, $D(\theta) = [I_r \; -{}^tB(\theta)]$:

$$DC{}^tD = [I_r \quad -{}^tB]\begin{bmatrix} \Gamma & {}^tBI^{-1} \\ I^{-1}B & I^{-1} \end{bmatrix}\begin{bmatrix} I_r \\ -B \end{bmatrix}$$

$$= [\Gamma - {}^tBI^{-1}B \quad {}^tBI^{-1} - {}^tBI^{-1}]\begin{bmatrix} I_r \\ -B \end{bmatrix}$$

$$= \Gamma - {}^tBI^{-1}B.$$

Since $(p(\hat{\theta}_n)/p(\theta))$ tends to 1 in P_θ-probability, we obtain

$$Z_n(\hat{\theta}_n) \xrightarrow{\mathcal{D}(P_\theta)} N_r(0, I_r - \sqrt{p(\theta)}\ {}^t\sqrt{p(\theta)} - {}^tB(\theta)I^{-1}(\theta)B(\theta))$$

Let $E(\theta) = {}^tB(\theta)I^{-1}(\theta)B(\theta)$. This is a matrix of rank k, and $E(\theta)E(\theta) = E(\theta)$. It is a positive definite symmetric matrix: it is diagonalizable and k of its eigenvalues equal one, the other $r - k$ equal 0; $\sqrt{p(\theta)}$ is in the kernel of $E(\theta)$. Let $(V_1, ..., V_k)$ be an orthonormal basis of the subspace of eigenvectors of 1 and let V_{k+1} be the vector $(1/\sqrt{r})(1, ..., 1)$, orthogonal to V_1, ..., V_k.

Let $D(\theta) = I_r - \sqrt{p(\theta)}\ {}^t\sqrt{p(\theta)} - {}^tB(\theta)I^{-1}(\theta)B(\theta)$; $V_1, ..., V_{k+1}$ are in the kernel of $D(\theta)$ and $V_{k+2}, ..., V_r$ are the eigenvectors of $D(\theta)$ associated with 1. Hence from Proposition 1.1.1, if Y is distributed as $N_r(0,D(\theta))$, $Y = \Sigma_{i=k+2}^r \langle Y,V_i\rangle V_i$ and the r.v.'s $\langle Y,V_i\rangle$ are normal and independent:

$$N_{r-k}(0,I_r - \sqrt{p(\theta)}\ {}^t\sqrt{p(\theta)} - {}^tB(\theta)I^{-1}(\theta)B(\theta))$$

is the distribution of the projection Y of X on the subspace generated by $V_{k+2}, ..., V_r$: $\|Y\|^2$ follows the distribution $\chi^2(r - k - 1)$.

Examples. The above theorem is widely used in statistics. Refer back to Volume I ([Vol. I, 5.4]) for a discussion and numerical examples.

(a) **Independence of Two Characteristics.** Two characteristics are observed, one, X, taking values in $\{1, ..., r\}$ and with distribution p, and the other, Y, taking values in $\{1, ..., s\}$ and with distribution q. The pair (X,Y) takes values in $\{1, ..., r\} \times \{1, ..., s\}$, a space of rs elements. We assume X and Y are independent, without making any other hypothesis on p and q. From the relation $p_1 + ... + p_r = 1 = q_1 + ... + q_s$ we see that we can take for Θ the subspace of $\mathbb{R}^{r+s-2}$ of vectors

$$(p_i,q_j)_{\substack{1\leq i\leq r-1 \\ 1\leq j\leq s-1}} \cdot$$

with positive components and such that

$$\sum_{i=1}^{r-1} p_i \leqslant 1, \quad \sum_{j=1}^{s-1} q_j \leqslant 1.$$

We observe $(X_k,Y_k)_{1\leqslant k\leqslant n}$, an n-sample from $p \otimes q$. If $\theta = ((p_i)_{1\leqslant i\leqslant r-1}, (q_j)_{1\leqslant j\leqslant s-1})$ is in the interior of Θ, i.e. if p and q charge all the points, the likelihood is

$$L_n(\theta) = \prod_{i=1}^{r} \prod_{j=1}^{s} (p_i q_j)^{N_{ij}} = \prod_{i=1}^{r} p_i^{N_{i\cdot}} \prod_{j=1}^{s} q_j^{N_{\cdot j}}$$

with $N_{i\cdot} = \Sigma_{j=1}^{s} N_{ij}$, $N_{\cdot j} = \Sigma_{i=1}^{r} N_{ij}$;

$$L_n(\theta) = \prod_{i=1}^{r-1} p_i^{N_{i\cdot}} \prod_{j=1}^{s-1} q_j^{N_{\cdot j}} \left[1 - \sum_{i=1}^{r-1} p_i\right]^{N_{r\cdot}} \left[1 - \sum_{j=1}^{s-1} q_j\right]^{N_{\cdot s}},$$

$$\frac{\partial \operatorname{Log} L_n(\theta)}{\partial p_i} = \frac{N_{i\cdot}}{p_i} - \frac{N_{r\cdot}}{p_r},$$

$$\frac{\partial \operatorname{Log} L_n(\theta)}{\partial q_j} = \frac{N_{\cdot j}}{q_j} - \frac{N_{\cdot s}}{q_s}.$$

From which the maximum likelihood estimator follows,

$$\hat{p}_i = \frac{N_{i\cdot}}{n}, \quad \hat{q}_j = \frac{N_{\cdot j}}{n}.$$

This estimator is a.s. convergent to θ. Fisher's information $I(\theta)$ is invertible (check this). Hence $n\chi^2((N_n/n),\hat{\theta}_n)$ converges to

$$\chi^2[rs - 1 - (r-1) - (s-1)] = \chi^2[(r-1)(s-1)].$$

Proposition 3.4.1. *If $(X_i,Y_i)_{1\leqslant i,j\leqslant n}$ is an n-sample from a pair of independent discrete distributions p and q, p charging r points and q charging s points, the chi-square for independence*

$$\chi_n^2 = n \sum_{i=1}^{r} \sum_{j=1}^{s} \frac{(N_{ij}^n - N_{i\cdot}^n N_{\cdot j}^n /n)^2}{N_{i\cdot}^n N_{\cdot j}^n}$$

converges in distribution to $\chi^2((r-1)(s-1))$.

(b) **Symmetry of Two Characteristics.** With the notations of the preceding example and $r = s$, we test the following hypothesis: "for every (i,j), $1 \leqslant i,j \leqslant s$, the probabilities $P(X = i, Y = j) = p_{ij}$ and $P(X = j, Y = i) = p_{ji}$ are equal." We can then take for Θ the following subset of $\mathbb{R}_+^{(s(s+1)/2)-1}$.

$$\Theta = \left\{(p_{ij})_{1 \leqslant i<j \leqslant s}, (p_{ii})_{1 \leqslant i<s}; \right.$$

$$\left. 2 \sum_{1 \leqslant i<j \leqslant s} p_{ij} + \sum_{1 \leqslant i<s} p_{ii} \leqslant 1 \right\}.$$

The likelihood at the point $\theta = \{(p_{ij})_{1 \leqslant i<j \leqslant s}, (p_{ii})_{1 \leqslant i<s}\}$ equals,

$$L_n(\theta) = \prod_{1 \leqslant i<j \leqslant s} p_{ij}^{N_{ij}^n + N_{ji}^n} \prod_{1 \leqslant i \leqslant s} p_{ii}^{N_{ii}^n}$$

$$= \prod_{1 \leqslant i<j \leqslant s} p_{ij}^{N_{ij}^n + N_{ji}^n} \prod_{1 \leqslant i<s} p_{ii}^{N_{ii}^n} \left[1 - 2 \sum_{1 \leqslant i<j \leqslant s} p_{ij} - \sum_{1 \leqslant i<s} p_{ii}\right]^{N_{ss}^n}.$$

If the probability (p_{ij}) charges all the points, then θ is an interior point of Θ, and for $i \neq j$,

$$\frac{\partial \operatorname{Log} L_n(\theta)}{\partial p_{ij}} = \frac{N_{ij} + N_{ji}}{p_{ij}} - 2\frac{N_{ss}}{p_{ss}};$$

$$\frac{\partial \operatorname{Log} L_n(\theta)}{\partial p_{ii}} = \frac{N_{ii}}{p_{ii}} - \frac{N_{ss}}{p_{ss}}.$$

From which the maximum likelihood estimator $\hat{p}^n = (\hat{p}_{ij}^n)$ follows with

$$\hat{p}_{ij}^n = \frac{1}{2n} (N_{ij}^n + N_{ji}^n).$$

The Fisher information is seen to be invertible. The r.v.

$$n\chi^2 \left[\frac{N_n}{n}, \hat{p}^n\right] = \sum_{1 \leqslant i,j \leqslant s} \frac{(N_{ij}^n - N_{ji}^n)^2}{N_{ij}^n}$$

tends in distribution to $\chi^2(s^2 - 1 - (s(s+1))/2 + 1) = \chi^2(s(s-1)/2)$ when n tends to infinity.

Bibliographic Notes

Asymptotic statistics are covered in numerous works. Elementary calculations and results on maximum likelihood, likelihood tests and χ^2 tests are found in Kendall and Stuart and Lehmann.

The most important mathematical development are presented in the definitive, but difficult, work of LeCam.

Roussas gives a partial, more elementary, presentation of important ideas of LeCam and Hajek which we have not been able to elucidate in this chapter. In Hajek-Sidak a number of these ideas will be found, in connection with the particular case of asymptotic theorems on rank tests (tests introduced in Volume I, Chapter 4). Ibraguimov-Khasminski study in a thorough manner likelihood processes and limiting properties of standard estimators, including non-regular models and without a compactness hypothesis on the parameter.

For time series, standard results and indications of their applications are given in Grenander-Rosenblatt and in Grenander.

The multidimensional case is covered in Hannan and in Anderson, where guidance on the very important practical problem (not tackled here) of the estimation of the spectral density is also found. The spectral point of view is developed in Koopmans and in Brillinger who, in the multidimensional case, covers data analysis problems, analogous to P.C.A. introduced in Volume I, Chapter 1.

Important statistical problems such as the robustness of the methods used (validity of test procedures and estimation when the correct model is not in the class studied) lead to asymptotic expansions. Asterisque [1] is an introduction to this subject.

Chapter 4
MARKOV CHAINS

Objectives

Numerous random phenomena have the following property: knowledge of the state of the phenomenon at a given time carries as much information on the future as knowledge of the complete past. These are Markov processes, called Markov chains if the time is discrete.

For certain chains which return infinitely often and quickly enough to a point, we obtain a law of large numbers and a central limit theorem. The statistics of these processes may be dealt with according to the schemes defined in Chapter 3.

The autoregressive model of order 1 is studied directly and branching processes are dealt with, because of the practical interest in these processes.

4.1. Introduction and First Tools

4.1.1. Definitions and Construction

We have already defined Markov chains in [Vol. I, 6.3], where various examples will be found.

We are given a **state space** $(E,\mathcal{E})$, an arbitrary measurable space, a transition π from E into E and an initial distribution ν. The σ-algebra $\mathcal{E}$ is often taken as understood. A process

$(\Omega,A,P,(X_n)_{n\geq 0})$ taking values in $(E,\mathcal{E})$ is a **homogeneous Markov chain with transition** π and **initial distribution** ν if the distribution of X_0 is ν and if the distribution of X_{n+1} conditional on $(X_0, ..., X_n)$ is $\pi(X_n, \cdot)$, for every $n \geq 0$.

A canonical version of this process can be constructed by taking $(\Omega,A) = (E,\mathcal{E})^{\mathbb{N}}$, X_n the nth coordinate, and the distribution function of $(X_0, ..., X_n)$, $\nu \otimes \pi^{\otimes n}$, defined by setting, for $(\Gamma_i)_{0\leq i\leq n} \in \mathcal{E}^{\otimes n}$:

$$\nu \otimes \pi^{\otimes n}(\Gamma_0 \times ... \times \Gamma_n) = \int_{\Gamma_0} \nu(dx_0)\int_{\Gamma_1} \pi(x_0,dx_1) \ldots \int_{\Gamma_n} \pi(x_{n-1},dx_n).$$

There exists a probability P_ν on $(E,\mathcal{E})$, such that $(\Omega,A,P_\nu,(X_n)_{n\geq 0})$ is a Markov chain with transition π and initial distribution ν. Ionescu-Tulcea's theorem which ensures this existence generalizes Theorem 0.2.5 (see Neveu [2], p. 161). The advantage of this canonical version is that the trajectories $(\Omega,A,(X_n)_{n\geq 0})$ do not depend on ν or on π. This facilitates the simultaneous study of families of processes corresponding to families of pairs (ν,π).

Let us fix the transition probability π. Setting $P_x = P_\nu$ if ν is Dirac measure at x, we obtain a transition $(x,A) \longmapsto P_x(A)$ from E into Ω and, for every distribution ν, we have: $P_\nu = \int \nu(dx)P_x$ (it is sufficient to check this for finite distribution functions). In what follows, we almost always use the canonical version, and we associate with a transition π, a family of Markov chains $\{\Omega,A,(P_x)_{x\in E},(X_n)\}$: for this family, we speak of the Markov chain with transition π. We use the filtration $\mathbb{F} = (\mathcal{F}_n)_{n\geq 0}$ with $\mathcal{F}_n = \sigma(X_0, ..., X_n)$. The results obtained are translated without difficulty to noncanonical versions.

For $\Gamma \in \mathcal{E}$, we have $P_x(X_n \in \Gamma) = \pi_n(x,\Gamma)$, the transition π_n being defined by: $\pi_0(x,\cdot)$ Dirac measure at x and

$$\pi_n(x,\cdot) = \int \pi_{n-1}(x,dy)\pi(y,\cdot) = \int \pi(x,dy)\pi_{n-1}(y,\cdot).$$

Recall the following notations [Vol. 1, 6.3.1]. If f is an r.v. on E and F an r.v. on E^2, we denote $\pi f(x) = \int \pi(x,dy)f(y)$, $\pi F(x) = \int \pi(x,dyF(x,y)$ when these expressions make sense. If ν is a measure on E, we denote by $\nu\pi$ the measure $\Gamma \longmapsto \int \nu(dx)\pi(x,\Gamma)$ on E, and by $\nu \otimes \pi$ the measure

$$\Gamma \longmapsto \int \nu(dx) \int \pi(x,dy) 1_{\Gamma}(x,y)$$

on E^2. $\pi \otimes \pi$ is the transition from E^2 into E^2 defined by

$$\pi \otimes \pi((x,x'),\Gamma) = \int \pi(x,dy) \int \pi(x',dy') 1_{\Gamma}(y,y').$$

Finally we denote

$$U(x,\cdot) = \sum_{n=0}^{\infty} \pi_n(x,\cdot).$$

4.1.2. Markov Property

On $\Omega = E^{\mathbb{N}}$, the **translation operator** θ is defined by $\theta(\omega) = \{X_{n+1}(\omega)\}_{n\geq 0}$. In other words, $X_n \circ \theta = X_{n+1}$ for every $n \geq 0$. Its pth iterate is denoted θ_p: $X_n \circ \theta_p = X_{n+p}$. For an r.v. T taking values in $\mathbb{N}$, θ_T is the operator defined by $X_n \circ \theta_T(\omega) = X_{n+T(\omega)}(\omega)$. The following result can then be stated, which will be used repeatedly in what follows.

Theorem 4.1.1. Strong Markov Property. *Let T be an $\mathbb{F}$-stopping time and let Y be a bounded measurable* r.v. *on* (Ω,A).
For every initial distribution ν, we have

$$E_\nu[Y \circ \theta_T 1_{(T<\infty)} | F_T] = 1_{(T<\infty)} E_{X_T}(Y).$$

Proof. Let us take first of all $T = n$ and $Y = 1_{\cap_m (X_m \in \Delta_m)}$ with $(\Delta_m)_{1\leq m\leq p} \in E^p$; let $A = \cap_{q=0}^{n}(X_q \in \Gamma_q)$, with $(\Gamma_q)_{0\leq q\leq n} \in E^{n+1}$. We have

$$E_\nu[Y \circ \theta_n 1_A] = P_\nu[X_0 \in \Gamma_0, \dots, X_n \in \Gamma_n, X_n \in \Delta_0, \dots, X_{n+p} \in \Delta_p]$$

$$= \int \nu(dx_0) 1_{\Gamma_0}(x_0) \int \pi(x_0,dx_1) 1_{\Gamma_1}(x_1) \dots \int \pi(x_{n-1},dx_n) 1_{\Gamma_n \cap \Delta_0}(x_n)$$

$$\times \dots \int \pi(x_{n+p-1},dx_{n+p}) 1_{\Delta_p}(x_{n+p})$$

$$= \int \nu(dx_0) 1_{\Gamma_0}(x_0) \ldots \int \pi(x_{n-1}, dx_n) 1_{\Gamma_n}(x_n)$$
$$\times [1_{\Delta_0}(x_n) \int \pi(x_n, dx_{n+1}) 1_{\Delta_1}(x_{n+1})$$
$$\ldots \int \pi(x_{n+p-1}, dx_{n+p}) 1_{\Delta_p}(x_{n+p})].$$

Now the expression in brackets equals $P_{X_n}(X_0 \in \Delta_0, \ldots, X_p \in \Delta_p)$. Thus

$$E_\nu[Y \circ \theta_n 1_A]$$
$$= E_\nu\left[1_{(X_0 \in \Gamma_0)} \ldots 1_{(X_n \in \Gamma_n)} E_{X_n}\left\{1_{(X_0 \in \Delta_0)} \ldots 1_{(X_p \in \Delta_p)}\right\}\right]$$
$$= E_\nu[1_A E_{X_n}(Y)].$$

Keeping Y fixed, we extend this to every $A \in \mathcal{F}_n$ by 3.1.11 of Volume I: $E_\nu(Y \circ \theta_n | \mathcal{F}_n) = E_{X_n}(Y)$. Then we generalize to

every bounded Y by 3.1.13 of Volume I.

Now take a stopping time T. The r.v. $1_{(T<\infty)} E_{X_T}(Y)$ is $\mathcal{F}_T$-measurable and, for every $A \in \mathcal{F}_T$,

$$\int_A Y \circ \theta_T 1_{[T<\infty]} dP_\nu = \sum_{n \in} \int_{A \cap \{T=n\}} \int Y \circ \theta_n dP_\nu$$

$$= \sum_{n \in} \int_{A \cap \{T=n\}} E_{X_n}(Y) dP_\nu$$

$$= \int_A E_{X_T}(Y) 1_{[T<\infty]} \, dP_\nu.$$

4.1.3. Examples of Markov Chains

In the course of reading the remainder of this chapter it will be useful to refer to the following examples which are intuitively described here. These statements will be checked in the course of the following paragraphs.

Example 1. Markov Chain with Finite State Space, $E = \{1,2, \ldots, s\}$. Here its transition is a matrix $\pi = \{\pi(i,j)\}_{1 \leq i,j \leq s}$ said to be **stochastic.** Its terms are positive and the sum of the components of each row vector is 1. The nth power of π is $\pi^n = \{\pi_n(i,j)\}$. An algebraic study of these chains appears in

[Vol. I, E6.3.2]. We shall recover these results in [4.2] and [4.3]. Here are a few simple cases where we see (and shall shortly prove) what is happening.

(a) $$\pi = \begin{bmatrix} \theta & 1-\theta \\ 1-\theta' & \theta' \end{bmatrix} \text{ with } 0 < \theta, \theta' < 1.$$

Whatever the starting point, we revisit infinitely often 1 and 2.

(b) $$\pi = \begin{bmatrix} 0 & 1 \\ 0 & 1 \end{bmatrix};$$

starting from 1 we go to 2 in one step, being in 2 we stay there (2 is "absorbing").

(c) $$\pi = \begin{bmatrix} 0 & 1/2 & 1/2 \\ 0 & 1 & 0 \\ 0 & 0 & 1 \end{bmatrix};$$

starting from 1 we have one chance in two to go to 2 or 3. Being in 2 or in 3 we stay there.

(d) $$\pi = \begin{bmatrix} 0 & 1 \\ 1 & 0 \end{bmatrix}:$$

the chain describes a cycle $1 \to 2 \to 1 \to 2$ ....

Example 2. $E = \mathbb{N}$ and, for $n \geqslant 0$, $\pi(n,n+1) = \theta_n$, $\pi(n,0) = 1 - \theta_n$ with $0 < \theta_n < 1$. Let $T_0 = \inf\{n;\ X_n = 0, n > 0\}$. For $n \geqslant 1$,

$$P_0[T_0 = n] = P_0[X_1 = 1, X_2 = 2, ..., X_{n-1} = n-1, X_n = 0]$$

$$= \theta_0\theta_1 \dots \theta_{n-2}(1 - \theta_{n-1}).$$

If for example $\theta_n = \theta$: $P_0(T_0 = n) = \theta^{n-1}(1 - \theta)$, $P_0(T_0 < \infty) = 1$ and $E_0(T_0) = 1/(1-\theta)$. Then, starting from 0, we certainly return once... and the chain will revisit 0 (a.s.) an infinite number of times. This is also the case if $\Pi\theta_n = 0$ since $P_0(T_0 < \infty) = 1 - \Pi\theta_n$. If $\Pi\theta_n > 0$, starting from 0, the probability of returning there once is < 1 ..., and the chain will only revisit 0 (a.s.) a finite number of times, it tends a.s. to ∞.

Example 3. A G/G/1 Queue. A queue in front of a very simple ticket office can be described by the sequence (A_n) of the arrival times of the customers and the sequence (B_n) of the times at which the service finishes. The nth customer arrives at time A_n and leaves at time B_n. Let us assume that the sequences $(\Delta A_n = A_n - A_{n-1})_{n\geq 1}$ and $(\Delta B_n = B_n - B_{n-1})_{n\geq 1}$ are each sequences of positive independent r.v.'s with the same distribution, α and β respectively, and that these two sequences are independent of each other.

Let W_n be the waiting time of the nth customer. We have the relation,

$$W_{n+1} = (W_n + \Delta B_n - \Delta A_{n+1})_+.$$

In fact, the nth customer's service stops at time $A_n + W_n + \Delta B_n$. The $(n + 1)$th arrives at time $A_n + \Delta A_{n+1}$; he does not wait if ΔA_{n+1} is greater than $W_n + \Delta B_n$, and if not he waits $W_n + \Delta B_n - \Delta A_{n+1}$. Consider $F_n = \sigma(B_1, \ldots, B_{n-1}, A_1, \ldots, A_n)$; ΔB_n and ΔA_{n+1} are independent of F_n. Let γ be the distribution of $\Delta A_n - \Delta B_n$: $\gamma = \alpha * \overline{\beta}$, where $\overline{\beta}$ is the distribution symmetric to β.

$$P[W_{n+1} = 0 \mid F_n] = \gamma([W_n,\infty])$$

$$P[W_{n+1} > y \mid F_n] = \gamma([-\infty,W_n - y[),$$

for $y \geq 0$. Hence $W = (W_n)_{n\geq 0}$ is a Markov chain on $[0,\infty]$. Its transition π satisfies, for f a positive r.v.,

$$\pi f(x) = f(0)(\gamma[x,\infty]) + \int_{[-\infty,x[} f(x - u)\gamma(du).$$

This chain starting at 0 is studied in [Vol. I, E6.3.7] when the distribution γ has a mean $m = E(\Delta A_n) - E(\Delta B_n)$, the difference between the means of arrival and service times.

Let $T_0 = \inf\{n;\ n \geq 1,\ W_n = 0\}$ be the index of the first customer who does not wait. The following three cases are distinguished, assuming $W_0 = 0$.

(a) For $m > 0$, $W_n \xrightarrow{\text{a.s.}} \infty$ and $P_0(T_0 < \infty) < 1$.

(b) For $m = 0$, $W_n \xrightarrow{\mathcal{D}} \infty$, i.e., for all $\lambda > 0$, $P(W_n \leq \lambda) \longrightarrow 0$. However $P_0(T_0 < \infty) = 1$, and, a.s., an infinity of customers do not wait. The expectation of T_0 is infinite.

(c) For $m < 0$, there exists on $[0,\infty[$ a distribution μ such that $W_n \xrightarrow{\mathcal{D}} \mu$. Here the expectation of T_0 is finite and, a.s., an infinity of customers does not wait.

If the distribution γ is diffuse on $\mathbb{R}$, the probability of reaching a given point $a > 0$ starting from 0 is zero. However, for $m \leq 0$ and for every a, $P_a(T_0 < \infty) = 1$. In fact, setting $X_n = \Delta A_n - \Delta B_{n-1}$, we have, for $n < T_0$, $W_0 = a$, $W_n = a + X_1 + \ldots + X_n$. Now, for $m < 0$, $X_1 + \ldots + X_n \xrightarrow{\text{a.s.}} -\infty$, while for $m = 0$ (Vol. I, E4.4.13), $\inf(X_1 + \ldots + X_n) = -\infty$ a.s. Thus, $T_0 = \inf\{n;\ X_1 + \ldots + X_n \leq -a\}$ is a.s. finite.

4.1.4. Excessive and Invariant Functions

Definition 4.1.2. An **invariant function** (resp. **excessive**) for π is a positive r.v. h on E, satisfying $\pi h = h$ (resp. $\pi h \leq h$).

If h is invariant (resp. excessive), $(h(X_n))$ is an $\mathbb{F}$-martingale (resp. supermartingale) on $(\Omega, \mathcal{A}, P_\nu)$ for arbitrary ν. In fact,

$$E_\nu^n(h(X_{n+1})) = E_\nu^n(h(X_1 \circ \theta_n)) = E_{X_n}(h(X_1))$$
$$= \pi h(X_n) = (\text{resp. } \leq)\ h(X_n).$$

This positive supermartingale $h(X_n)$ thus converges P_ν-a.s. for arbitrary ν.

Particular Case. Let $\Gamma \in \mathcal{E}$. Let us consider T_Γ and R_Γ, the **entrance time** and **recurrence set** on Γ defined by

$$T_\Gamma = \inf\{n;\ n > 0,\ X_n \in \Gamma\}$$

(the infimum of an empty set of $\mathbb{N}$ being $+\infty$)

$$R_\Gamma = \overline{\lim_{n\to\infty}}\{X_n \in \Gamma\} = \left\{\omega;\ \sum_{n\in} 1_{(X_n\in\Gamma)}(\omega) = \infty\right\}.$$

Let $p \in \mathbb{N}$;

$$T_\Gamma \circ \theta_p = \inf\{n;\ n > 0,\ X_{n+p} \in \Gamma\}$$

$$= \inf\{n;\ n > p,\ X_n \in \Gamma\} - p.$$

Hence on $\{T_\Gamma > p\}$, we have: $T_\Gamma = p + T_\Gamma \circ \theta_p$. We shall often use this relation for a stopping time T: $T_\Gamma = T + T_\Gamma \circ \theta_T$ on $\{T_\Gamma > T\}$. The function $x \longmapsto P_x(T_\Gamma < \infty)$ is excessive, and the function $x \longmapsto P_x(R_\Gamma)$ is invariant.

In fact,

$$\begin{aligned}
\pi\{P_.(T_\Gamma < \infty)\} &= E_.[P_{X_1}(T_\Gamma < \infty)] \\
&= E_.\left[E_{X_1}(1_{(T_\Gamma<\infty)})\right] \\
&= E_.\left[E_.(1_{[T_\Gamma<\infty]} \circ \theta_1 \mid F_1)\right] \\
&= P_.[T_\Gamma \circ \theta_1 < \infty] \\
&= P_.(\text{there exists an } n;\ n > 1,\ X_n \in \Gamma) \\
&\leq P_.(T_\Gamma < \infty);
\end{aligned}$$

$$\begin{aligned}
\pi[P_.(R_\Gamma)] &= E_.[P_{X_1}(R_\Gamma)] = E_.\,[1_{R_\Gamma} \circ \theta_1] \\
&= P_.\left[\sum_{n\geq 1} 1_{(X_n \in \Gamma)} = \infty\right] = P_.(R_\Gamma).
\end{aligned}$$

4.2. Recurrent or Transient States

We assume in this section that contains $\{x\}$ for every $x \in E$ (but E does not have to be countable). We denote

$$\pi(y,x) = \pi(y,\{x\}) \quad \text{and} \quad U(y,x) = U(y,\{x\}).$$

4.2.1. Successive Returns to a Point

The pth passage time of the chain into $\Gamma \in \underset{\sim}{E}$ is the stopping time T_Γ^p defined by the recurrence relations: $T_\Gamma^0 = 0$,

$$T_\Gamma^1 = \inf\{n;\ n > 0,\ X_n \in \Gamma\} = T_\Gamma,$$

$$T_\Gamma^{p+1} = \inf\{n;\ n > T_\Gamma^p,\ X_n \in \Gamma\}.$$

On $\{T_\Gamma^p < \infty\}$,

$$T_\Gamma^{p+1} = T_\Gamma^p + \inf\{n;\ n > 0,\ X_n \circ \theta_{T_\Gamma^p} \in \Gamma\}$$
$$= T_\Gamma^p + T_\Gamma \circ \theta_{T_\Gamma^p}.$$

We have

$$R_\Gamma = \overline{\lim}\ \{X_n \in \Gamma\} = \bigcap_p \{T_\Gamma^p < \infty\}.$$

For $x \in E$, let us denote T_x^p for $T_{\{x\}}^p$, R_x for $R_{\{x\}}$. By using the strong Markov property and the fact that, on $\{T_x^p < \infty\}$, $X_{T_x^p}$ equals x, we have

$$P_x[T_x^{p+1} < \infty] = P_x[(T_x^p < \infty)(T_x \circ \theta_{T_x^p} < \infty)]$$
$$= E_x[1_{(T_x^p<\infty)} E_x(1_{(T_x \circ \theta_{T_x^p} < \infty)} \mid \mathcal{F}_{T_x^p})]$$
$$= E_x[1_{(T_x^p<\infty)}] E_x[1_{(T_x<\infty)}];$$
$$P_x(T_x^p < \infty) = (P_x(T_x < \infty))^p;$$
$$P_x(R_x) = \lim_p \downarrow (P_x(T_x < \infty))^p;$$
$$U(x,x) = \sum_{n \geq 0} \pi_n(x,x) = E_x\left[\sum_{n \geq 0} 1_{(X_n = x)}\right]$$
$$= \sum_{p \geq 0} P_x(T_x^p < \infty) = \sum_{p \geq 0} (P_x(T_x < \infty))^p.$$

Theorem 4.2.3. *Let $x \in E$. There are only two possibilities*

(a) $P_x(T_x < \infty) = 1$; *x is recurrent. Then*

$$P_x(R_x) = 1; \quad U(x,x) = \infty.$$

(b) $P_x(T_x < \infty) < 1$; *x is transient. Then*

$$P_x(R_x) = 0; \quad U(x,x) < \infty.$$

Proposition 4.2.4. *Let x be a recurrent point of E. For $\Gamma \in \mathcal{E}$, one of the two following cases holds:*

(a) $P_x(R_\Gamma) = P_x(T_\Gamma < \infty) = 1$,

(b) $P_x(R_\Gamma) = P_x(T_\Gamma < \infty) = 0.$

Let h be an excessive function,

$$P_x\left[\bigcap_{n\in\mathbb{N}} (h(X_n) = h(x))\right] = 1.$$

Proof. Let h be an excessive function. The supermartingale $(h(X_n))$ converges a.s. and equals $h(x)$ infinitely often: $(h(X_n))$ tends to $h(x)$, P_x-a.s.

In particular the bounded supermartingales, $P_{X_n}(R_\Gamma)$ and $P_{X_n}(T_\Gamma < \infty)$, converge P_x-a.s. to $P_x(R_\Gamma)$ and $P_x(T_\Gamma < \infty)$,

$$E_x(P_{X_n}(T_\Gamma < \infty)) = E_x(E_x(1_{(T_\Gamma \circ \theta_n < \infty)} \mid F_n))$$

$$= P_x(T_\Gamma \circ \theta_n < \infty).$$

However

$$R_\Gamma = \bigcap_n \{T_\Gamma \circ \theta_n < \infty\}$$

and

$$P_x(T_\Gamma < \infty) = \lim_n P_x(T_\Gamma \circ \theta_n < \infty) = P_x(R_\Gamma).$$

Finally, let A_p = {there exists n, $X_n \in \Gamma$, $T_x^p \leqslant n < T_x^{p+1}$}, with $T_x^0 = 0$; A_{p-1} is in $F_{T_n^p}$ and,

$$E_x[1_{A_p} \mid F_{T_x^p}] = E_x[1_{A_0} \circ \theta_{T_x^p} \mid F_{T_x^p}] = P_x(A_0).$$

The events (A_p) are independent, and $R_\Gamma = \overline{\lim}\,(A_p)$ has probability 0 or 1 according to whether $\Sigma P_x(A_p)$ converges or not (i.e. according to whether $P_x(A_0)$ is zero or not).

Now let us come back to the excessive function h. For $\delta > 0$ and $\Gamma = \{y;\ |h(y) - h(x)| > \delta\}$, $P_x(R_\Gamma)$ is zero since $(h(X_n))$ converges to $h(x)$. Thus $P_x(T_\Gamma < \infty)$ is zero. From which the latter part of the proposition follows.

Definition 4.2.5. Let $x \in E$, $\Gamma \in E$. We say that x **leads to** Γ, and we denote this by $x \to \Gamma$, if $P_x(T_\Gamma < \infty) > 0$.

Thus if x is recurrent, $x \to \Gamma$ implies $P_x(R_\Gamma) = 1$. The relation $x \to y$ is transitive. Let x,y,z be in E, satisfying $x \to y$ and $y \to z$. Let us prove $x \to z$,

$$P_x(T_z < \infty) \geqslant P_x(T_y < \infty\text{: there exists } n \text{ such that } n > T_y, X_n = z)$$
$$= P_x(T_y < \infty, T_z \circ \theta_{T_y} < \infty)$$
$$= P_x(T_y < \infty)P_y(T_z < \infty) > 0.$$

4.2.2. Communication

Corollary 4.2.6. Communication Between Points. *Let x and y be in E. We say that x and y communicate ($x \leftrightarrow y$) if x leads to y and y leads to x, or if x and y coincide.*

(a) *The relation $\leftrightarrow$ is an equivalence relation on E.*

(b) *If x is recurrent, then $x \to y$ implies $x \leftrightarrow y$, and y is then recurrent. Moreover, $P_x(R_y) = P_y(R_x) = 1$.*

Proof. (a) follows from the transitivity of $\to$.

(b) follows from Proposition 4.2.4: if x is recurrent, $P_x(R_y)$ equals 1 and $(P_{X_n}(R_y))$ is P_x-a.s. always equal to 1, and since y is recurrent, $P_y(R_y)$ equals 1.

Definition 4.2.7. Let us assume that E is countable. Let a be a recurrent state and let $C = \{x; a \to x\}$. All the states of C are recurrent and lead only to states of C. C is the **recurrent class** of a. The state space can be restricted to C; the **chain** is then said to be **recurrent**, all its states are recurrent and communicate with each other.

Proposition 4.2.8. *Let x be a transient point. For every $y \in E$,*

$$U(y,x) = P_y(T_x < \infty)U(x,x).$$

The set F of points which lead to x is the increasing limit of a sequence (Γ_n) of E, such that $U(\cdot,\Gamma_n)$ is a bounded function for each n.

Proof. We have

$$U(y,x) = E_y\left[\sum_{n \geq T_x} 1_{(X_n=x)}\right]$$

$$= E_y\left[1_{(T_x<\infty)} E_{X_{T_x}}\left[\sum_{n\geq 0} 1_{(X_n=x)}\right]\right]$$

$$= P_y(T_x < \infty)U(x,x).$$

The set F is the union of the countable family of sets of the form $\Gamma = \{z;\ \pi_n(z,x) \geq (1/i)\}$ for n and i integers > 0;

$$\pi_{m+n}(y,z) \geq \int \pi_m(y,dz) 1_\Gamma(z)\pi_n(z,x) \geq \frac{1}{i}\,\pi_m(y,\Gamma).$$

Summing over m, we obtain: $U(\cdot,\Gamma) \leq iU(x,x)$.

Definition 4.2.9. A set $\Gamma \in E$ is **absorbing** if no $x \in \Gamma$ leads to Γ^c.

This means that, for $x \in \Gamma$, $\pi(x,\cdot)$ is concentrated on Γ. We may take the restriction of π or of the Markov chain to the state space Γ.

Examples. Let $A \in E$.

(a) The set B of points which do not lead to A is absorbing: in fact, on $\{T_{B^c} < \infty\}$, T_A is majorized by the entry time to A after having been in B^c,

$$T_{B^c} + T_A \circ \theta_{T_{B^c}}.$$

Let $x \in B$:

$$P_x(T_A < \infty) = 0 \geq P_x[T_{B^c} < \infty,\ T_{B^c} + T_A \circ \theta_{T_{B^c}} < \infty]$$

$$E_x[1_{(T_{B^c}<\infty)}\, P_{X_{T_{B^c}}}(T_A < \infty)] = 0.$$

However, on B^c, $P_\cdot(T_A < \infty) > 0$; thus x does not lead to B^c.

(b) The set $D = \{x;\ P_x(R_A) = 1\}$ is absorbing. Let $x \in D$. On $\{T_{D^c} < \infty\}$, returning to A an infinite number of times implies returning to A an infinite number of times after T_{D^c}: on $\{T_{D^c} < \infty\}$,

$$P_x(R_A \mid F_{T_{D^c}}) = P_{X_{T_{D^c}}}(R_A) = 1.$$

However, $P_x[R_A] = 1$ implies

$$P_x[R_A \cap (T_{D^c} < \infty)] = P_x(T_{D^c} < \infty)$$
$$= E_x(1_{(T_{D^c}<\infty)} P_{X_{T_{D^c}}}(R_A)).$$

This holds only if x does not lead to D^c. Note that, if E is countable and if a is a recurrent state, then the recurrent class C of a is contained in $D = \{x;\ P_x(R_A) = 1\}$. Sometimes the inclusion is strict: take for example, $E = \{0,1\}$, $a = 0$, $\pi(0,0) = \pi(1,0) = 1$.

4.2.3. Examples

It is not difficult to return to the examples in 4.1.3 for finite E or $E = \mathbb{N}$. For the G/G/1 queue, the point 0 is recurrent for $m \leq 0$; then every $x \neq 0$ is transient, however, we have $P_x(R_0) = 1$. We shall now study another simple example.

Random Walks. Let F be a distribution on $\mathbb{R}^k$, $(Y_n)_{n \geq 1}$ a sequence of independent r.v.'s with distribution F defined on (Ω, A, P). Set $X_n = Y_1 + \ldots + Y_n$. Then (X_n) is a Markov chain with transition π defined by $\pi(x,\Gamma) = F(\Gamma - x)$ ([Vol. I, 6.3.2]). $(X_n + x)$ is, for every $x \in \mathbb{R}^k$, a version of the Markov chain with initial state x and transition π. Such a chain is called a random walk. We have

$$P_0(X_n \in \cdot) = \pi_n(0,\cdot) = F^{*n}(\cdot) = P_x(X_n \in x + \cdot);$$

$U(0,\Gamma) = U(x,\Gamma + x)$, in particular $U(0,0) = U(x,x)$. If 0 is recurrent, then every point x is recurrent. If F has mean m, by the law of large numbers (X_n/n) tends P_0-a.s. to m when $n \to \infty$. For $m \neq 0$, it is clear that (X_n) does not return infinitely often to any bounded set in $\mathbb{R}^k$,

$$(|X_n| \xrightarrow{P_0\text{-a.s.}} \infty);$$

hence 0 is transient.

Let us study the case where F is concentrated on $\mathbb{Z}$. Let $d \geq 1$ be the smallest integer such that F is concentrated on $d\mathbb{Z}$; d is the **period** of F. Let

$$I = \{x;\ x \in \mathbb{Z},\ \sum_{n>0} F^{*n}(x) > 0\} = \{x;\ 0 \to x\}.$$

For every y, $x \in I$ certainly implies, $y \to x + y$. In particular, if x and y are in I, $x + y$ is in I.

Let d' be the smallest integer such that the restriction of F to $\mathbb{N}$ is concentrated on $d'\mathbb{N}$ (d' is a multiple of d and $d' > 0$). We can find an $a > 0$ in I such that $a + d' = b$ is in I. For all $n > 1$, the set I contains the points na, $(n-1)a + b$, ..., $(n-k)a + kb$, ..., nb at spaces d' apart. For $n > a/d'$ we have $nb > (n+1)a$, and I contains the points of $na + d'\mathbb{N}$, hence the sequence $(nd')_{n \geq n_0}$ for a certain n_0. If F is not concentrated on $\mathbb{N}$, I likewise contains the sequence $(-nd'')_{n \geq n_1}$ for a certain n_1. However, d equals d' or d'', and every multiple of d may be written in the form $m'd' - m''d''$ with $m' \geq n_0$, $m'' \geq n_1$: $I = d\mathbb{Z}$.

Proposition 4.2.10. *Let F be a distribution on $\mathbb{Z}$ with period d, the smallest integer $d \geq 1$ such that F is concentrated on $d\mathbb{Z}$. F is said to be aperiodic for $d = 1$. Let*

$$I = \left\{x;\ x \in \mathbb{Z},\ \sum_{n>0} F^{*n}(x) > 0\right\}.$$

(a) *If F is concentrated on $d\mathbb{N}$, I contains $n_0 d + d\mathbb{N}$ for a certain integer n_0.*
(b) *If F is not concentrated either on $d\mathbb{N}$ or on $-d\mathbb{N}$, $I = d\mathbb{Z}$.*

Proposition 4.2.11. *If F is a distribution on $\mathbb{Z}$ with a mean $m = \sum_{n \in \mathbb{Z}} nF(n)$, then the random walk associated with F is recurrent for $m = 0$, transient for $m \neq 0$). In the recurrent case, the recurrent class of 0 is $d\mathbb{Z}$ if d is the period of F.*

Proof. It remains to study the case $m = 0$. By the weak law of large numbers, for $\varepsilon > 0$, we have

$$\lim_{n \to \infty} P[|X_n| \leq \varepsilon n] = 1.$$

For every x, $U(0,x) = P_0[T_x < \infty]U(x,x) \leq U(0,0)$;

$$U(0,[-\varepsilon n, n]) \leqslant (2\varepsilon n + 1)U(0,0);$$

$$\frac{1}{n}\sum_{p=1}^{n} F^{*p}([-\varepsilon p,\varepsilon p]) = \frac{1}{n}\sum_{p=1}^{n} P(|X_p| \leqslant \varepsilon p)$$

$$\leqslant \frac{1}{n}\sum_{p=1}^{n} P(|X_p| \leqslant \varepsilon n)$$

$$\leqslant \frac{U(0,[-\varepsilon n,\varepsilon n])}{n}.$$

The limit of the left hand term is 1, if $n \to \infty$. Thus

$$\lim_{n\to\infty} \frac{(2\varepsilon n + 1)U(0,0)}{n} = 2\varepsilon U(0,0) \geqslant 1.$$

This being true for all $\varepsilon > 0$, thus $U(0,0) = \infty$.

4.3. The Study of a Markov Chain Having a Recurrent State

4.3.1. Invariant Measure

Definition 4.3.12. Let μ be a measure on E. It is invariant ("under π" is understood) if $\mu\pi = \mu$. It is **excessive** if $\mu\pi \leqslant \mu$.

Theorem 4.3.13. *Let a be a recurrent state.*

(1) *The measure μ on E,*

$$\Gamma \longmapsto \mu(\Gamma) = E_a\left[\sum_{n=1}^{T_a} 1_{(X_n \in \Gamma)}\right],$$

is σ-finite and invariant under π: $\mu(\Gamma)$ is the expectation of the time spent in Γ between two visits to a.

(2) *This measure μ is concentrated on the absorbing set*

$$D = \{x;\ P_x(R_a) = 1\}.$$

Let Γ be an arbitrary element of $\mathcal{E}$.

- *If $\mu(\Gamma) > 0$, then, for every $x \in D$, $P_x(R_\Gamma) = 1$.*
- *If $\mu(\Gamma) = 0$, then, for every $x \in D$, $P_x(R_\Gamma) = 0$ and $P_\cdot(T_\Gamma < \infty) = 0$ μ-a.s. on E.*

(3) *Up to a multiplicative factor, μ is the unique σ-finite*

excessive measure concentrated on the set of points leading to a.

Proof. It is easy to check that for any μ-integrable r.v. f on $(E,\mathcal{E})$,

$$\mu(f) = E_a\left[\sum_{n=1}^{T_a} f(X_n)\right] = E_a\left[\sum_{n=0}^{T_a-1} f(X_n)\right].$$

(a) Let us consider a **point-cemetery** c not belonging to E and "kill" the chain at time T_a. In other words, consider $Y_n = X_n 1_{(n\leq T_a)} + c 1_{(n>T_a)}$. (Y_n) is a Markov chain on $E \cup \{c\}$, with transition $\tilde{\pi}$, $\tilde{\pi}(x,\cdot) = \pi(x,\cdot)$ for $x \neq a$, $\tilde{\pi}(a,\cdot) = \delta_c$.

For this chain a is transient, and every point of E leads to a. From Proposition 4.2.8, E is an increasing limit of sets Γ_p of $\mathcal{E}$ such that $U(\cdot,\Gamma_p) = \Sigma_n \tilde{\pi}(\cdot,\Gamma_p)$ if finite. We have $\mu\{a\} = 1$, and for $\Gamma = \Gamma_p\backslash\{a\}$,

$$\mu(\Gamma) = E_a[1_{(T_\Gamma<T_a)} U(X_{T_\Gamma},\Gamma)] < \infty.$$

Hence μ is σ-finite. The measure μ is invariant. In fact, let Γ have finite μ-measure.

$$\mu\pi(\Gamma) = \int d\mu(y)\pi(y,\Gamma) = E_a\left[\sum_{n=0}^{T_a-1} \pi(X_n,\Gamma)\right]$$

$$= \sum_n E_a[\pi(X_n,\Gamma)\, 1_{(n<T_a)}]$$

$$= \sum_n E_a[1_{(n<T_a)} 1_{(X_{n+1}\in\Gamma)}]$$

$$= \sum_n E_a[1_{(n+1<T_a)} + 1_{(T_a=n+1)}) 1_{(X_{n+1}\in\Gamma)}]$$

$$= P_a[X_{T_a} \in \Gamma] - P_a[X_0 \in \Gamma] + E_a\left[\sum_{n=1}^{T_a} 1_\Gamma(X_n)\right]$$

$$= \mu(\Gamma) + 1_\Gamma(a) - 1_\Gamma(a) = \mu(\Gamma).$$

(b) The point a is in the absorbing set D, it does not lead to D^c and $\mu(D^c) = 0$: μ is concentrated on D. Let B be the set of points x which lead to a.

Let Γ be charged by μ: $P_a(R_\Gamma) = 1$. Thus, for $x \in B$, $P_x(R_\Gamma) > 0$; and for $x \in D$, $P_x(R_\Gamma) = 1$.

Let Γ be μ-negligible. For $x \in D$, the martingale $P_{X_n}(R_\Gamma)$ converges P_x-a.s. to $P_a(R_\Gamma)$, hence to 0, and $P_x(R_\Gamma)$ equals 0.

(c) Let α be a nonzero invariant measure concentrated on B. It dominates μ; in fact $\mu(\Gamma) > 0$ implies that every $x \in B$ leads to Γ, and there exists an n such that

$$\alpha\{x;\ \pi_n(x,\Gamma) > 0\} > 0, \quad \alpha\pi_n(\Gamma) = \alpha(\Gamma) > 0.$$

If α is a nonzero excessive measure concentrated on B, it is invariant. In fact, for every $N \in \mathbb{N}$,

$$\alpha(a) \geq \alpha(a) - \alpha\pi_{N+1}(\alpha) = \sum_{n=0}^{N} \int(\alpha(dz) - \alpha\pi(dz))\pi_n(z,a).$$

Let N tend to ∞,

$$\int(\alpha(dz) - \alpha\pi(dz)) \left[\sum_{n=0}^{\infty} \pi_n(z,a)\right]$$

$$= \int(\alpha - \alpha\pi)(dz)U(z,a) < \infty.$$

However $\alpha - \alpha\pi$ is a positive measure and $U(\cdot,a)$ is identically equal to ∞ on B: $\alpha = \alpha\pi$.

(d) Let A be an absorbing set and α an invariant measure concentrated on B. The two measures $1_A\alpha$ and $1_{A^c}\alpha$ are invariant. In fact,

$$1_A\alpha(\Gamma) = \alpha(\Gamma \cap A) = \int\alpha(dz)\pi(x,\Gamma \cap A)$$

$$\geq \int\alpha(dx)1_A(x)\pi(x,\Gamma \cap A)$$

$$= \int(1_A\alpha)(dx)\pi(x,\Gamma)$$

and $1_A\alpha$ is excessive, hence invariant, and similarly for $\alpha - 1_A\alpha = 1_{A^c}\alpha$. If μ charges A, i.e., $1_{A^c}\alpha$, which cannot dominate μ, is zero: $\alpha(A^c) = 0$.

Now take Γ to be μ-negligible; the absorbing set A of points which do not lead to Γ is charged by μ since it contains a. An invariant measure α concentrated on B hence does not charge A^c. In particular, $\mu(A^c) = 0$ (which finishes the proof of (2)); and since A^c contains Γ, $\alpha(\Gamma) = 0$ and μ dominates α.

From (2) α and μ are equivalent.

(e) Let $\alpha = h\mu$ be invariant and equivalent to μ, and let $c > 0$. the measure $\beta = (h \wedge c)\mu$ is also excessive, hence invariant. In fact

$$\beta\pi(\Gamma) = \int (h \wedge c)(x)\mu(dx)\pi(x,\Gamma) \leqslant \inf\{\mu(h1_\Gamma), c\mu(\Gamma)\}.$$

Take $\Gamma \subset \{h \leqslant c\}$: $\beta\pi(\Gamma) \leqslant \mu(h1_\Gamma) = \beta(C)$; then $\Gamma \subset \{c \leqslant h\}$: $\beta\pi(\Gamma) \leqslant c\mu(\Gamma) = \beta(\Gamma)$. Thus the two measures $\alpha - \beta = (h - c)_+\mu$ and $\beta - \alpha = (c - h)_+\mu$ are invariant and perpendicular, one of them is zero, and h is μ-a.s. equal to the constant $\sup\{c;\ \mu(h \geqslant c) > 0\}$.

Definition 4.3.14. The recurrent state a is said to be **positive recurrent** if there exists an invariant probability μ. This is equivalent to assuming that $E_a(T_a) < \infty$, and to taking

$$\mu(\Gamma) = \frac{1}{E_a(T_a)} E_a\left[\sum_{n=1}^{T_a} 1_{(X_n \in \Gamma)}\right].$$

In the alternative case a is said to be **null recurrent.**

Examples. (a) For the G/G/1 queue of [4.1.3], the point 0 is positive recurrent for $m < 0$ and null recurrent for $m = 0$. The invariant measure is the distribution of W. Null recurrence is thus often the intermediate situation between positive recurrence and transience.

(b) For the random walk of [4.2.3], the measure $\Sigma_{n\in\mathbb{N}}\delta_n$ is always invariant. The recurrent case ($m = 0$) always corresponds to null recurrence.

4.3.2. Law of Large Numbers

In this paragraph a is a recurrent point. We take for the state space the absorbing set $\{x;\ P_x(R_a) = 1\}$ on which μ is concentrated, and we denote this space by E.

Theorem 4.3.15. *Let a be a recurrent point such that, for every $x \in E$, $P_x(R_a) = 1$. Let μ be the nonzero σ-finite measure invariant under π. For f and g in $L^1(E,\mathcal{E},\mu)$ and g positive and charged by μ, we have, for arbitrary $x \in E$,*

$$\frac{\sum_{k=0}^{n} f(X_k)}{\sum_{k=0}^{n} g(X_k)} \xrightarrow[n\to\infty]{P_x\text{-a.s.}} \frac{\int f\, d\mu}{\int g\, d\mu}.$$

Notes. (a) In the proof we shall take the measure μ,

$$\Gamma \longmapsto E_a\left[\sum_{n=1}^{T_a} 1_\Gamma(X_n)\right].$$

Every other σ-finite invariant measure is proportional to it and gives the same result. We can take $g = 1_{\{a\}}$, and prove

$$\frac{\sum_{k=0}^{n} f(X_k)}{\sum_{k=0}^{n} 1_{\{a\}}(X_k)} \xrightarrow[n\to\infty]{P_x\text{-a.s.}} \frac{\int f\, d\mu}{\mu\{a\}} = \int f\, d\mu.$$

(b) For a positive recurrent the theorem is most often applied to the invariant probability μ and to $g = 1$. We then have a law of large numbers,

$$\frac{1}{n}\sum_{k=0}^{n} f(X_k) \xrightarrow[n\to\infty]{P_x\text{-a.s.}} \int f\, d\mu.$$

Proof. Let $f \in L^1(\mu)$. We can define on R_a (hence P_x-a.s.) a sequence of r.v.'s $(Z_p)_{p\geqslant 0}$ by setting

$$Z_p = \sum_{n=T_a^p}^{T_a^{p+1}-1} f(X_n).$$

For $p \geqslant 1$, this is a sequence of independent identically distributed r.v.'s. In fact Z_p is $\mathcal{F}_{T_a^{p+1}}$-measurable and, for every $t \in \mathbb{R}$, by setting $Z = Z_0$, we have, P_x-a.s.:

$$P_x(Z_p \leqslant t \mid \mathcal{F}_{T_a^p}) = P_x(Z \circ \theta_{T_a^p} \leqslant t \mid \mathcal{F}_{T_a^p}) = P_a(Z \leqslant t).$$

Moreover, Z_1 is integrable since

$$E_x(|Z_1|) \leqslant E_a\left[\sum_{n=0}^{T_a-1} |f(X_n)|\right] = \int |f|\, d\mu.$$

From which, by the law of large numbers (Corollary 2.7.39),

$$\frac{1}{n}(Z_1 + \dots + Z_{n-1}) = \frac{1}{n}\sum_{k=T_a}^{T_a^n - 1} f(X_k) \xrightarrow[n\to\infty]{P_x\text{-a.s.}} \int f\, d\mu.$$

Let

$$\nu(n) = \sum_{k=1}^{n} 1_{\{a\}}(X_k): \; T_a^{\nu(n)} \leq n < T_a^{\nu(n)+1}.$$

Hence for $f \geq 0$,

$$\frac{\sum_{k=0}^{T_a^{\nu(n)}} f(X_k)}{\nu(n)} \leq \frac{\sum_{k=0}^{n} f(X_k)}{\sum_{k=0}^{n} 1_{\{a\}}(X_k)} \leq \frac{\sum_{k=0}^{T_a^{\nu(n)+1}} f(X_k)}{\nu(n)}.$$

The two extreme terms tend P_x-a.s. to $\int f\, d\mu$: this is also the case for the term in the middle. It proves the theorem for $f \geq 0$. For any $f \in L^1(\mu)$ it is true for f_+ and f_- thus for $f = f_+ - f_-$.

4.3.3. Central Limit Theorem

Theorem 4.3.16. *Let F be an r.v. on E^2. Assume that the point a is positive recurrent and that, for all $x \in E$, $P_x(R_a) = 1$. Then, for every $x \in E$, we have the following properties with $\mu(E) = 1$.*

(a) *If F is $\mu \otimes \pi$-integrable,*

$$\frac{1}{n}\sum_{k=1}^{n} F(X_{k-1}, X_k) \xrightarrow[n\to\infty]{P_x\text{-a.s.}} \int (\pi F) d\mu = \int F\, d\mu \otimes \pi.$$

Convergence in $L^1(P_x)$ also holds.

(b) *If F is $\mu \otimes \pi$-integrable with $\mu \otimes \pi(F) = 0$ and if*

$$\alpha^2(F) = E_a\left[\sum_{k=1}^{T_a} F(X_{k-1}, X_k)\right]^2 < \infty$$

(for instance if $E_a(T_a^2) < \infty$ and F is bounded),

$$\frac{1}{\sqrt{n}}\sum_{k=1}^{n} F(X_{k-1}, X_k) \xrightarrow{\mathcal{D}(P_x)} N\left(0, \frac{\alpha^2(F)}{E_a(T_a)}\right).$$

(c) *If F^2 is $\mu \otimes \pi$-integrable,*

$$\frac{1}{\sqrt{n}}\sum_{k=1}^{n} (F(X_{k-1}, X_k) - \pi F(X_{k-1})) \xrightarrow[n\to\infty]{\mathcal{D}(P_x)} N(0, \sigma^2(F))$$

with

$$\sigma^2(F) = \int F^2 d\mu \otimes \pi - \int (\pi F)^2 d\mu.$$

Proof. Set

$$Z_p = \sum_{k=T_a^p+1}^{T_a^{p+1}} F(X_{k-1},X_k) \quad \text{for } p \geq 1.$$

(a) For P_x-a.s. convergence, we use a simple adaptation of the proof of Theorem 4.3.15. Following Corollary 2.7.39, convergence in $L^1(P_x)$ also holds.

(b) To prove (b) we need the following lemma.

Lemma. *Let (Y_p) be a sequence of random variables adapted to a filtration $\mathbb{G} = (G_p)$. Assume that the* r.v. Y_p *is square integrable with mean* 0 *and variance σ^2, and Y_p is independent of G_{p-1}. Let $(\nu(n))$ be a sequence of $\mathbb{G}$ stopping times such that* $(\nu(n))/n \xrightarrow{\text{a.s.}} c$, for a constant c. Then

$$\frac{1}{\sqrt{n}} \sum_{p=1}^{\nu(n)} Y_p \xrightarrow{\mathcal{D}} N(0,c\sigma^2).$$

Proof of the Lemma. It is an easy consequence of Theorem 2.8.42 taking the filtration $\mathbb{G}$ for all n and $\xi_{n,p} = (1/\sqrt{n})Y_p$. Coming back to the theorem, take

$$\nu(n) = \sum_{k=1}^{n} 1_{\{a\}}(X_k) \quad \text{and} \quad G_p = F_{T_a^{p+1}}.$$

$\nu(n)$ is a stopping time for the filtration $\mathbb{G} = (G_p)$, $T_a^{\nu(n)} \leq n < T_a^{\nu(n)+1}$ and

$$\frac{\nu(n)}{n} \xrightarrow{P_x\text{-a.s.}} \mu(\{a\}) = \frac{1}{E_a(T_a)}.$$

With the hypothesis of (b) the r.v. $(Z_p)_{p\geq 1}$ are square integrable and centered, Z_p being G_p measurable. Set $\nu^1(n) = \sup(1,\nu(n))$,

$$\frac{1}{\sqrt{n}} \sum_{p=1}^{\nu^1(n)} Z_p = \frac{1}{\sqrt{n}} 1_{(T_a \leq n)} \sum_{k=T_a+1}^{T_a^{\nu(n)+1}} F(X_{k-1},X_k)$$

$$\xrightarrow{\mathcal{D}(P_x)} G \quad \text{with} \quad G = N(0,\alpha^2(F)\mu\{a\}).$$

As

$$\frac{1}{\sqrt{n}} \sum_{k=1}^{T_a} F(X_{k-1}, X_k) \xrightarrow{P_x\text{-a.s.}} 0,$$

we deduce

$$\frac{1}{\sqrt{n}} \sum_{k=1}^{T_a^{\nu(n)+1}} F(X_{k-1}, X_k) \xrightarrow{\mathcal{D}(P_x)} G.$$

Finally, set

$$U_n = \sum_{k=n+1}^{T_a^{\nu(n)+1}} F(X_{k-1}, X_k) 1_{(\nu(n)>0)}:$$

$$E_x(|U_n|) \leqslant \sum_{p=1}^{n} E_x\left[1_{(\nu(n)=p)} \sum_{k=T_a^p+1}^{T_a^{p+1}} |F(X_{k-1}, X_k)|\right]$$

$$\leqslant \sum_{p=1}^{n} E_x\left[1_{(\nu(n)=p)} E_a\left[\sum_{k=1}^{T_a} |F(X_{k-1}, X_k)|\right]\right]$$

$$\leqslant E_a(T_a)\mu \otimes \pi(|F).$$

Hence, for $\varepsilon > 0$,

$$P_x[|U_n| \geqslant \varepsilon\sqrt{n}] \leqslant \frac{1}{\varepsilon\sqrt{n}} E_a(T_a)\mu \otimes \pi(|F|):$$

$$\frac{1}{\sqrt{n}} U_n \xrightarrow{P_x} 0.$$

As

$$\frac{1}{\sqrt{n}} \left|\sum_{k=n+1}^{T_a} F(X_{k-1}, X_k) 1_{(\nu(n)=0)}\right| \leqslant \frac{1}{\sqrt{n}} \sum_{k=1}^{T_a} |F(X_{k-1}, X_k)|$$

tends to zero (P_x-a.s.) we obtain,

$$\frac{1}{\sqrt{n}} \sum_{k=n+1}^{T_a^{\nu(n)+1}} F(X_{k-1}, X_k) \xrightarrow{P_x} 0$$

and

$$\frac{1}{\sqrt{n}} \sum_{k=1}^{n} F(X_{k-1}, X_k) \xrightarrow{\mathcal{D}(P_x)} G.$$

Let us prove (c). For every k,

$$E_x^{k-1}[F(X_{k-1},X_k)] = \pi F(X_{k-1}).$$

Thus

$$\left[\sum_{k=1}^{n} F(X_{k-1},X_k)\right]$$

is, on $(\Omega,\mathcal{A},P_x)$, an $\mathbb{F}$-adapted sequence compensated by $(\sum_{k=1}^{n}\pi F(X_{k-1}))$. If F^2 is $\mu \otimes \pi$-integrable, $M = (M_n)$ with

$$M_n = \sum_{k=1}^{n} (F(X_{k-1},X_k) - \pi F(X_{k-1}))$$

is a square integrable martingale. Its associated increasing process is $\langle M\rangle = (\langle M_n\rangle)$ with

$$\langle M\rangle_n = \sum_{k=1}^{n} [\pi F^2(X_{k-1}) - (\pi F(X_{k-1}))^2].$$

Set $\phi = F - \pi F$; we have

$$\alpha^2(\phi) = E_a[M_{T_a}^2].$$

However

$$E_a[\langle M\rangle_{T_a}] = E_a\left[\sum_{k=1}^{T_a} (\pi F^2(X_{k-1}) - (\pi F(X_{k-1}))^2)\right]$$

$$= E_a(T_a) \int d\mu[\pi F^2 - (\pi F)^2]$$

$$= E_a(T_a)\sigma^2(F).$$

Applying Theorem 2.6.29, we get

$$\alpha^2(\phi) = E_a(T_a)\sigma^2(F)$$

and (b) implies (c).

Notes. (1) It could happen in part (b) that

$$\sum_1^{T_a} F(X_{k-1},X_k) = 0 \quad (P_a\text{-a.s.})$$

and $\alpha^2(F) = 0$. We then have δ_0 for the limit. Let us consider for example the Markov chain on $\mathbb{Z}$, the transition probability of which is defined, for $n > 0$, by $\pi(0,0) = 1/2$ and

$$\pi(0,n) = 2^{-n-1}; \quad \pi(n,-n) = \pi(-n,0) = 1.$$

The invariant measure μ of this chain is defined by,

$$\mu(0) = 1/2, \quad \mu(\pm n) = 2^{-n-2}.$$

The point 0 is recurrent. If f is the identity function $\Sigma_{k=1}^{T_0} f(X_k) = 0$ (a.s.), because either $T_0 = 1$ and $X_1 = 0$, or $T_0 = 3$ and $X_1 = n$, $X_2 = -n$, $X_3 = 0$.

(2) Let F be a r.v. on $E \times E$ such that Poisson's equation $[I - \pi]G = F$ can be solved with a $G \in L^2(\mu \otimes \pi)$. Then

$$\frac{1}{\sqrt{n}} \sum_{k=1}^{n} F(X_{k-1},X_k) \xrightarrow{\mathcal{D}(P_x)} N(0,\sigma^2(G)).$$

It is sometimes easier to solve Poisson's equation and apply this result than to apply part (b) of the theorem.

4.3.4. Cyclic Classes

We assume the point a is positive recurrent, and we always restrict the state space to $E = \{x; P_x(R_a) = 1\}$. Let us denote by F the distribution of T_a on $(\Omega,\mathcal{A},P_a)$ and let μ be the invariant probability; F is bounded.

In what follows we use a method called **coupling** which is often useful. Let us **double** the Markov chain: consider a canonical Markov chain with state space E^2 and transition $\pi \otimes \pi$. We denote it by

$$\{(E^2,\mathcal{E}^{\otimes 2})^{\mathbb{N}}, (X_n,\tilde{X}_n),(\tilde{P}_{x,\tilde{x}})_{(x,\tilde{x})\in E^2}\}.$$

For α and β probabilities on $(E,\mathcal{E})$, the sequences (X_n) and $(\tilde{X}_n)$ are independent for the probability $P_{\alpha\otimes\beta}$ and these are both Markov chains with transition π and initial distributions α and β respectively. As above, let us denote by (T_a^p) the sequence of passage times of (X_n) into a, and by $(\tilde{T}_a^p)$ the sequence of passage times of $(\tilde{X}_n)$ into a. The r.v.'s $(T_a^p - \tilde{T}_a^p) - (T_a^{p-1} - \tilde{T}_a^{p-1}) = Z_p$ are independent for $P_{\alpha\otimes\beta}$ and are distributed as F^s, the symmetrizer of F, $F * F(-\cdot)$. The distribution F^s has zero mean.

Assume that the period of F, hence of F^s, is 1. From the recurrent random walk on $\mathbb{Z}$ associated with F^s, every point leads to 0 (Proposition 4.2.11): irrespective of the distribution $\alpha \otimes \beta$ of $T_a^0 - \tilde{T}_a^0$, the sequence $(T_a^p - \tilde{T}_a^p)$ vanishes infinitely often, and the sequence $(X_n, \tilde{X}_n)_{n \geq 0}$ visits (a,a) infinitely often. Moreover the probability $\mu \otimes \mu$ is invariant under $\pi \otimes \pi$. Hence (a,a) is positive recurrent for the doubled chain, and, for every initial point $(x,\tilde{x})$, $\breve{P}_{x,\tilde{x}}(R_{a,a}) = 1$.

We shall denote

$$S = \inf\{n;\ n \geq 1,\ (X_n, \breve{X}_n) = (a,a)\}.$$

Let us now consider a bounded r.v. g on $(E,\mathcal{E})$ ($\|g\| = \sup\{|g(\cdot)|\}$). We apply the strong Markov property to the stopping time S,

$$\alpha\pi_n(g) = \breve{E}_{\alpha\otimes\beta}(g(X_n)) = \breve{E}_{\alpha\otimes\beta}(g(X_n)1_{(S>n)})$$

$$+ \sum_{k=0}^{n} \breve{E}_{\alpha\otimes\beta}[1_{(S=k)}\breve{E}_{a,a}(g(X_{n-k}))]$$

$$= \breve{E}_{\alpha\otimes\beta}(g(X_n)1_{(S>n)}) + \sum_{k=0}^{n} \breve{P}_{\alpha\otimes\beta}(S=k)\pi_{n-k}g(a)$$

$$\alpha\pi_n(g) - \beta\pi_n(g) = \breve{E}_{\alpha\otimes\beta}[g(X_n)1_{(S>n)}] - \breve{E}_{\alpha\otimes\beta}[g(\breve{X}_n)1_{(S>n)}]$$

$$|\alpha\pi_n(g) - \beta\pi_n(g)| \leq 2\|g\|\breve{P}_{\alpha\otimes\beta}(S > n).$$

The **variation distance** of two bounded measures μ and ν is

$$\|\mu - \nu\| = \sup\{|\mu(g) - \nu(g)|;\ \|g\| \leq 1\}$$

(cf. [Vol. I, 6.4.2]). We thus obtain,

$$\|\alpha\pi_n - \beta\pi_n\| \leq 2\breve{P}_{\alpha\otimes\beta}(S > n).$$

Hence

$$\lim_{n\to\infty} \|\alpha\pi_n - \beta\pi_n\| = 0.$$

Orey's Theorem 4.3.17. *Let a Markov chain have a positive recurrent state to which all states lead* a.s. *Let μ be the invariant probability.*

(a) *There exists an integer d, the period of a, such that $\{n; \pi_n(a,a) > 0\}$ is contained in $d\mathbb{N}$ and, for every n greater than a certain integer n_0, $\pi_{nd}(a,a) > 0$. A partition of the state space E can be obtained into a μ-negligible set H and d disjoint cyclic classes $D_1, \ldots, D_d$ such that, for $0 \leqslant r \leqslant d - 1$, we pass* a.s. *in one step from D_r to D_{r+1} (denote $D_0 = D_d$)* i.e. *$\pi(\cdot, D_{r+1})$ equals 1 on D_r and 0 on $(H \cup D_r)^c$.*

(b) *If $d = 1$ (we then say that a is aperiodic), for every initial distribution ν we have,*

$$\lim_{n\to\infty} \|\nu\pi_n - \mu\| = 0.$$

(c) *In the general case, if $D_d = D_0$ is the cyclic class containing a, for every distribution ν concentrated on D_{d-r} we have,*

$$\lim_{n\to\infty} \|\nu\pi_{nd+r} - \pi_{nd}(a,\cdot)\| = 0.$$

Proof. Let d be the period of F,

$$I = \{n; \pi_n(a,a) > 0\} = \left\{n; \sum_{p=0}^{\infty} F^{*p}(n) > 0\right\}.$$

The first part of (a) is a consequence of Proposition 4.2.10. Let us denote, for $r \geqslant 1$,

$$\Gamma_r = \left\{x; \sum_{n\geqslant 0} \pi_{dn+r}(x,a) > 0\right\}.$$

Let $1 \leqslant r < r' \leqslant d$, then $\mu(\Gamma_r \cap \Gamma_{r'})$ is zero. If not we could find an ℓ and m such that $\Delta = \{x; \pi_{\ell d+r}(x,a)\pi_{md+r'}(x,a) > 0\}$ is charged by μ. Then for a certain n, $\pi_n(a,\Delta) > 0$, $n + \ell d + r$ and $n + md + r'$ would be in I, thus $r' - r$ would be a multiple of d, which is impossible.

Let H be the set of all points which lead to

$$\bigcup_{1\leqslant r<r'\leqslant d} (\Gamma_r \cap \Gamma_{r'}),$$

H^c is absorbing ([4.2.2]) and H is μ-negligible. We can again reduce the state space by removing H. Denoting $D_r = \Gamma_{d-r} \backslash H$, part (a) of the proposition is obtained.

Part (b) has been shown before the statement of the theorem. The general case, (c), is obtained by noticing that, on D_0, a is an aperiodic recurrent point of the Markov chain with transition π_d. Hence, for every initial distribution ν concentrated on D_0,

$$\lim_{n\to\infty} \|\nu\pi_{nd} - \pi_{nd}(a,\cdot)\| = 0.$$

Then, if ν is concentrated on D_{d-r}, $\nu\pi_r$ is concentrated on D_d, and the result (c) is obtained.

Notes. (a) In the case where E is countable, by taking for E the recurrent class of a, the set H which is not charged by μ is empty: E is the union of d cyclic classes.

(b) In the case where E is finite, a purely algebraic proof of the above results can be given [Vol. I, E6.3.2].

(c) Let f be an r.v. on E. We call the **Poisson equation** associated with f and π the equation $g - g\pi = f$. Within the framework of Theorem 4.3.17(b), if f is bounded with zero integral then $\Sigma_{n=0}^{\infty}\pi_n f$ is the solution of the Poisson equation.

4.3.5. Rate of Convergence to the Stationary Distribution

Let a be positive recurrent and let $H = \{x;\, E_x(T_a) < \infty\}$,

$$\infty > E_a(T_a) \geqslant E_a[1_{(T_a > T_{H^c})}(E_{X_{T_{H^c}}}(T_a) + T_{H^c})].$$

From which it follows that $P_a(T_{H^c} < T_a) = 0$ and $\mu(H^c) = 0$. Assume in the above that a is aperiodic. The inequality

$$\|\alpha\pi_n - \beta\pi_n\| \leqslant 2\tilde{P}_{\alpha\otimes\beta}(S > n)$$

implies, by assumption,

$$\sum_{n=0}^{\infty} \|\alpha\pi_n - \beta\pi_n\| \leqslant 2\tilde{E}_{\alpha\otimes\beta}(S).$$

We know that $\tilde{E}_{a,a}(S)$ is finite since (a,a) is positive recurrent. What can be said about $\tilde{E}_{\alpha\otimes\beta}(S)$?

Let S^p be the pth return time of the doubled chain to (a,a): S^p is greater than the first passage time to (a,a) after $p-1$, $S^p \geqslant p-1+S\circ\theta_{p-1}$. From which

$$\widetilde{E}_{a,a}(S \circ \theta_p + p) \leqslant \widetilde{E}_{a,a}(S^{p+1}) = (p+1)\widetilde{E}_{a,a}(S).$$

Since $S \geqslant T_a$, we have $S = T_a + S \circ \theta_{T_a}$, and

$$\widetilde{E}_{\alpha\otimes\delta_a}(S) = \widetilde{E}_{\alpha\otimes\delta_a}(T_a) + \widetilde{E}_{\alpha\otimes\delta_a}(S \circ \theta_{T_a})$$

$$= E_\alpha(T_a) + \sum_n \widetilde{E}_{\alpha\otimes\delta_a}(1_{(T_a=n)}\widetilde{E}_{a,X_n}(S))$$

$$= E_\alpha(T_a) + \sum_n P_\alpha(T_a = n)\widetilde{E}_{a,a}[\widetilde{E}_{a,X_n}(S)]$$

$$= E_a(T_a) + \sum_n \frac{P_\alpha(T_a = n)}{\pi_n(a,a)}\widetilde{E}_{a,a}[1_{(X_n=a)}\widetilde{E}_{a,X_n}(S)]$$

(with the convention that $0/0 = 0$);

$$\widetilde{E}_{\alpha\otimes\delta_a}(S) \leqslant E_\alpha(T_a) + \sum_n \frac{P_\alpha(T_a = n)}{\pi_n(a,a)}\widetilde{E}_{a,a}[\widetilde{E}_{X_n,X_n}(S)]$$

$$\leqslant E_\alpha(T_a) + \sum_n \frac{P_\alpha(T_a = n)}{\pi_n(a,a)}\widetilde{E}_{a,a}(S \circ \theta_n)$$

$$\leqslant E_\alpha(T_a) + \sum_n \frac{P_\alpha(T_a = n)}{\pi_n(a,a)}(n+1)\widetilde{E}_{a,a}(S).$$

However $(\pi_n(a,a))$ tends to $\mu(a)$. Thus, if $E_\alpha(T_a)$ is finite, then $\widetilde{E}_{\alpha\otimes\delta_a}(S)$ is finite.

Take for example $\alpha = \mu$, with

$$\mu(\cdot) = E_a\left[\sum_{n=0}^{T_a-1} 1_{(X_n\in\cdot)}\right];$$

$$E_\mu(T_a) = E_a\left[\sum_{n=0}^{T_a-1} E_{X_n}(T_n)\right] = \sum_{n=0}^{\infty} E_a[E_{X_{T_n}}(T_a)1_{(T_a>n)}],$$

However

$$E_a[E_{X_n}(T_a)1_{(T_a>n)}] = E_a[E_a(T_a \circ \theta_n 1_{(T_a>n)} | \mathcal{F}_n)]$$
$$= E_a[1_{(T_a>n)}(T_a - n)];$$

$$E_\mu(T_a) = E_a\left[\sum_{n=0}^{T_a-1} (T_a - n)\right] = E_a\left[\frac{T_a(T_a+1)}{2}\right].$$

Thus $E_\mu(T_a) < \infty$ is equivalent to $E_a(T_a^2) < \infty$.

Theorem 4.3.18. *Let a be positive recurrent and aperiodic such that every point of E leads* a.s. *to a. Let* μ *be the stationary distribution.*

The series

$$\sum_{n=0}^{\infty} \|\pi_n(x,\cdot) - \pi_n(a,\cdot)\| = K(x)$$

converges if $E_x(T_a) < \infty$; *if* $E_a(T_a^2)$ *is finite, K is* μ*-integrable. Denote by*

$$G(x,\cdot) = \sum_{n=0}^{\infty} (\pi_n(x,\cdot) - \pi_n(a,\cdot))$$

the measure thus defined if $E_x(T_a) < \infty$. *If g is a bounded* r.v. *on* $(E,\mathcal{E})$, *such that* $\mu(g) = 0$, *the function Gg satisfies the relation* $(I - \pi)[Gg] = g$ *on* $\{x;\ E_x(T_a) < \infty\}$ *(Poisson's equation).*

Proof. Let $x \in E$. We have seen that

$$\|\pi_n(x,\cdot) - \pi_n(a,\cdot)\| \leq 2\tilde{P}_{x,a}(S > n).$$

From which

$$K(x) = \sum_{n=0}^{\infty} \|\pi_n(x,\cdot) - \pi_n(a,\cdot)\| \leq 2\tilde{E}_{x,a}(S),$$

and

$$\int K\, d\mu \leq 2\tilde{E}_{\mu\otimes\delta_a}(S)$$

is finite when $E_a(T_a^2) < \infty$. The function Gg is defined on $\{K < \infty\}$, hence on $\{x;\ E_x(T_a) < \infty\}$. On this set,

$$(I - \pi)Gg(x) = \lim_{N\to\infty} (I - \pi) \sum_{n=0}^{N} (\pi_n g(x) - \pi_n g(a))$$
$$= \lim_{N\to\infty} (g(x) - \pi_N g(x)) = g(x).$$

Note. Let us assume that, for a $k \geq 1$, we have $\inf_{x\in E} \pi_k(x,a) = \rho > 0$. Then, for all x, we have,

$$P_x(T_a > nk) \leq (1 - \rho)^n$$
$$E_x[e^{\lambda T_a}] \leq \sum_{n=0}^{\infty} P_x((n-1)k < T_a \leq nk)e^{\lambda nk}$$
$$\leq \sum_{n=0}^{\infty} (1 - \rho)^{n-1} e^{\lambda nk}.$$

For $\lambda > 0$ small enough, $\sup_x E_x(e^{\lambda T_a}) = C < \infty$. We then say that the Markov chain is **recurrent** in **Doeblin's sense.** The doubled chain is also recurrent in Doeblin's sense and, if α and β are two distributions on E, we have,

$$\tilde{E}_{\alpha\otimes\beta}(e^{\lambda S}) = \tilde{C}, \quad \tilde{P}_{\alpha\otimes\beta}(S > n) \leq \tilde{C}\, e^{-\lambda n}.$$

Then

$$\|\alpha\pi_n - \mu\| \leq \tilde{C}\, e^{-\lambda n},$$

$\alpha\pi_n$ converges to μ at an exponential rate. In this particular case, the central limit theorem is certainly valid. A recurrent aperiodic Markov chain on a finite space E satisfies, for every $(x,y) \in E$,

$$\pi_n(x,y) \xrightarrow[n\to\infty]{} \mu(y)$$

and $\mu(y) > 0$. It is always recurrent in the Doeblin sense.

4.4. Statistics of Markov Chains

4.4.1. Markov Chains with Finite State Space

Let $E = \{1, ..., s\}$ be a space with s elements and let π be a transition on E. Let $(\Omega, A, (P_x)_{x \in E}, (X_n)_{n \geq 0})$ be a canonical Markov chain associated with π.

Let us denote by

$$N_n^{ij} = \sum_{p=1}^{n} 1_{\{X_{p-1}=i, X_p=j\}}$$

the number of jumps from i to j up till n and let

$$N_n^{i \cdot} = \sum_{p=0}^{n-1} 1_{\{X_p=i\}}$$

be the number of passages to i up till $n - 1$. Let us assume that the initial distribution ν is known, but that the transition π is unknown. The likelihood at time n is,

$$\begin{aligned} L_n(\pi) &= \pi(X_0, X_1) \ldots \pi(X_{n-1}, X_n) \\ &= \prod_{(i,j) \in E^2} (\pi(i,j))^{N_n^{ij}}; \end{aligned}$$

$$\ell_n(\pi) = \text{Log } L_n(\pi) = \sum_{(i,j) \in E^2} N_n^{ij} \text{ Log } \pi(i,j).$$

The parameter here is $(\pi(i,1), ..., \pi(i,s-1);\ 1 \leq i \leq s)$, an element of $\mathbb{R}^{s(s-1)}$, and

$$\ell_n(\pi) = \sum_{i=1}^{s} \left[\sum_{j=1}^{s-1} N_n^{ij} \text{Log } \pi(i,j) + N_n^{is} \text{Log} \left(1 - \sum_{j=1}^{s-1} \pi(i,j) \right) \right]$$

From which we pull out the maximum likelihood estimator which is the empirical estimator,

$$\hat{\pi}(i,j) = \frac{N_n^{ij}}{N_n^{i \cdot}}.$$

Assume that E is a recurrent class: if μ is the invariant probability, we have, for every $x \in E$, from Theorems 4.3.15 and 4.3.16,

$$\frac{1}{n} N_n^{ij} \xrightarrow{P_x\text{-a.s.}} \mu(i)\pi(i,j), \quad \frac{1}{n} N_n^{i \cdot} \xrightarrow{P_x\text{-a.s.}} \mu(i);$$

$$\frac{1}{\sqrt{n}}(N_n^{ij} - N_n^{i\cdot}\pi(i,j)) \xrightarrow{\mathcal{D}(P_x)} N(0,\mu(i)\pi(i,j)(1-\pi(i,j))).$$

Hence

$$\hat{\pi}_n(i,j) = \frac{N_n^{ij}}{N_n^{i\cdot}} \xrightarrow{P_x\text{-a.s.}} \pi(i,j);$$

$$\sqrt{n\mu(i)}(\hat{\pi}_n(i,j) - \pi(i,j)) \xrightarrow{\mathcal{D}(P_x)} N(0,\pi(i,j)[1-\pi(i,j)]).$$

We can put forward a goodness of fit test. Have we a Markov chain with transition π? The compensator of the sequence (N_n^{ij}) is $(N_n^{i\cdot}\pi(i,j))$; from which follows the idea of using the following chi-square statistic analogous to that for samples

$$\chi_n^2 = \sum_{\{(i,j);\pi(i,j)>0\}} \frac{(N_n^{ij} - \pi(i,j)N_n^{i\cdot})^2}{\pi(i,j)N_n^{i\cdot}}$$

and to take a rejection region of the form $R = \{\chi_n^2 \geqslant C\}$. Let k be the number of pairs (i,j) such that $\pi(i,j) > 0$. Since the chain is recurrent, the following theorem proves that

$$\chi_n^2 \xrightarrow{\mathcal{D}(P_x)} \chi^2(k-s).$$

This is the analogue of Theorem 3.4.20 of [3.4.3], from which a goodness of fit test follows, of level asymptotically equal to α, by taking $R = \{\chi_n^2 \geqslant \chi_{k-s,\alpha}^2\}$. If we are dealing with a Markov chain with transition distinct from π, χ_n^2 tends. a.s. to ∞, and the power of the test tends to 1 for $n \to \infty$.

Theorem 4.4.19. *Let a Markov chain with transition π on a space E with s elements be given, which forms a single recurrent class, and let k be the number of pairs (i,j) for which $\pi(i,j) > 0$. For every $x \in E$, we have,*

$$\chi_n^2 = \sum_{\{(i,j);\pi(i,j)>0\}} \frac{(N_n^{ij} - \pi(i,j)N_n^{i\cdot})^2}{\pi(i,j)N_n^{i\cdot}} \xrightarrow{\mathcal{D}(P_x)} \chi^2(k-s).$$

Proof. Let us denote $\Delta = \{(i,j);\ \pi(i,j) > 0\}$, and let $(u_{ij})_{(i,j)\in\Delta}$ be in $\mathbb{R}^k$. Let us apply Theorem 4.3.16 to

$$\sum_{(i,j)\in\Delta} u_{ij}\left[\frac{1_{(i,j)}}{\sqrt{\mu(i)\pi(i,j)}}\right] = F;$$

$$\pi F = \sum_{(i,j)\in\Delta} u_{ij} \frac{\mathbf{1}_{\{i\}}\sqrt{\pi(i,j)}}{\sqrt{\mu(i)}}$$

$$(\pi F)^2 = \sum_{\{(i,j,\ell);(i,j)\in\Delta,(i,\ell)\in\Delta\}} u_{ij}u_{i\ell}\mathbf{1}_{\{i\}} \frac{\sqrt{\pi(i,j)\pi(i,\ell)}}{\mu(i)}$$

$$\sigma^2(F) = \sum_{\{(i,j)\in\Delta\}} u_{ij}^2$$

$$- \sum_{\{(i,j,\ell);(i,j)\in\Delta,(i,\Lambda)\in\ell\}} u_{ij}u_{i\ell} \sqrt{\pi(i,j)\pi(i,\ell)}.$$

Hence the random vector

$$\frac{1}{\sqrt{n}} \sum_{p=1}^{n} \left\{\frac{\mathbf{1}_{\{i,j\}}(X_{p-1},X_p) - \mathbf{1}_{\{i\}}(X_{p-1})\pi(i,j)}{\sqrt{\mu(i)\pi(i,j)}}\right\}_{(i,j)\in\Delta}$$

$$= \left\{\frac{1}{\sqrt{n}} \frac{N_n^{ij} - N_n^{i\cdot}\pi(i,j)}{\sqrt{\mu(i)\pi(i,j)}}\right\}_{(i,j)\in\Delta}$$

tends in distribution to a Gaussian distribution $N_k(0,\Gamma)$, with

$$\Gamma = I - A \quad \text{where } A = \{a_{(i,j),(i',j')}\}_{(i,j)\in\Delta,(i',j')\in\Delta},$$

$$a_{(i,j)(i',j')} = \begin{cases} 0 & \text{where } i \neq i', \\ \sqrt{\pi(i,j)\pi(i,j')} & \text{for } i = i'. \end{cases}$$

Since $N_n^{i\cdot}/n$ tends a.s. to $\mu(i)$, we also have

$$\left\{\frac{N_n^{ij} - N_n^{i\cdot}\pi(i,j)}{\sqrt{N_n^{i\cdot}\pi(i,j)}}\right\} \xrightarrow{\mathcal{D}(P_x)} N_k(0,\Gamma).$$

Let Z be distributed as $N_k(0,\Gamma)$, $Z = (Z_{ij})_{(i,j)\in\Delta}$. Let $i \in E$ and let $\Delta_i = \{j;\ (i,j) \in \Delta\}$. Assume that Δ_i has p_i elements. $Z^{(i)} = (Z_{ij})_{j\in\Delta_i}$ has distribution

$$N_{p_i}(0, I - \sqrt{\pi(i,\cdot)}\ {}^t\sqrt{\pi(i,\cdot)}),$$

denoting by $\sqrt{\pi(i,\cdot)}$ the vector $\{\sqrt{\pi(i,j)}\}_{j\in\Delta_i}$: a random vector with distribution $N_{p_i}(0,I)$ can be found of which $Z^{(i)}$ is the orthogonal projection on the orthogonal of the unitary vector $\sqrt{\pi(i,\cdot)}$, and $\|Z^{(i)}\|^2$ has distribution $\chi^2(p_i - 1)$. Moreover, the vectors $Z^{(i)}$, for $i = 1, ..., s$ are independent. Thus $\|Z\|^2 = \chi_n^2$ has distribution $\chi^2(\Sigma_{i\in E}(p_i - 1))$, which finishes the proof.

4.4.2. Dominated Models of Markov Chains

Given a state space $(E,\mathcal{E})$, the natural domination hypothesis for the study of Markov chains is the following:

Definition 4.4.20. Dominated Model for Markov Chains. Let Θ be a set of parameters and let π be a transition from $(E,\mathcal{E})$ into $(E,\mathcal{E})$. A family $(\pi_\theta)_{\theta\in\Theta}$ of transitions from $(E,\mathcal{E})$ into $(E,\mathcal{E})$ is said to be dominated by π if, for every θ, there exists an r.v. f_θ on $(E,\mathcal{E})^2$ such that, for arbitrary $x \in E$,

$$\pi_\theta(x,\cdot) = f_\theta(x,\cdot)\pi(x,\cdot).$$

Rather we shall denote: $f_\theta(x,y) = f(\theta,x,y)$ and $\pi_\theta(x,dy) = \pi(\theta,x;dy)$. Now let us consider $(\Omega, \mathcal{A}(P_{\theta,x})_{x\in E},(X_n)_{n\geq 0})$, the canonical Markov chain associated with π_θ. Let $(\Omega,\mathcal{A},(P_x)_{x\in E}, (X_n)_{n\geq 0})$ be the canonical Markov chain associated with π. For every initial distribution ν we have, on $\mathcal{F}_n = \sigma(X_0, ..., X_n)$,

$$P_{\theta,\nu} = \left[\prod_{i=1}^{n} f(\theta,X_{i-1},X_i)\right] P_\nu.$$

The model $(\Omega,\mathcal{A},(P_{\theta,\nu})_{\theta\in\Theta},(X_n)_{n\geq 0})$ is thus a model dominated by P_ν, and

$$L_n(\theta) = \prod_{i=1}^{n} f(\theta,X_{i-1},X_i),$$

$$\ell_n(\theta) = \text{Log } L_n(\theta) = \sum_{i=1}^{n} \text{Log } f(\theta,X_{i-1},X_i).$$

It would also be possible to take a dominated model (ν_θ) for the initial distribution: $\nu_\theta = \phi(\theta,\cdot)\nu$. Then,

$$\ell_n(\theta) = \text{Log } \phi(\theta,X_0) + \sum_{i=1}^{n} \text{Log } f(\theta,X_{i-1},X_i).$$

We shall assume in what follows that ν is known, and the study of asymptotic results when (ν_θ) is a dominated model follows easily by examining the initial term.

When we have laws of large numbers and central limit theorems for the Markov chain, we can easily adapt the asymptotic statistics of samples.

Hypothesis 1. *Assume that the chain associated with π_θ has a positive recurrent point towards which all points lead* a.s. *Let μ_θ be the stationary probability under π_θ. We assume that, for every $\alpha \in \Theta$,* Log $f(\alpha,\cdot)$ *is integrable with respect to $\mu_\theta \otimes \pi_\theta$.*

Consider the Kullback information of $\pi(\alpha,x;\cdot)$ on $\pi(\theta,x;\cdot)$:

$$K(\theta,\alpha;x) = \int \left[\text{Log}\, \frac{f(\theta,x,\cdot)}{f(\alpha,x,\cdot)}\right] \pi(\theta,x;dy).$$

The compensator of the sequence $\ell_n(\theta) - \ell_n(\alpha)$ is $\Sigma_{i=1}^n K(\theta,\alpha;X_{i-1})$. By the law of large numbers, we have, for every $x \in E$,

$$\frac{1}{n}(\ell_n(\theta) - \ell_n(\alpha)) \xrightarrow{P_{\theta,x}\text{-a.s.}} \int d\mu_\theta(x) K(\theta,\alpha;x).$$

This limit term may be denoted $K(\theta,\alpha)$ and considered as a distance between θ and α (the average Kullback information). To say that $K(\theta,\alpha)$ is zero implies that $K(\theta,\alpha;\cdot)$ is zero μ_θ-a.s., hence that, for μ_θ-almost all x,

$$\pi(\theta,x;\cdot) = \pi(\alpha,x;\cdot).$$

Hypothesis 2. *We assume that the model is* **identifiable** *in θ,* i.e. *for all $\alpha \neq \theta$: $\mu_\theta\{x;\ \pi(\theta,x;\cdot) \neq \pi(\alpha,x;\cdot)\} > 0$.*

Under Hypotheses 1 and 2, $(-(\ell_n/n))$ is a contrast. Now, a minimum contrast estimator implies a maximum likelihood estimator. From [3.2.3], we obtain the following result.

Theorem 4.4.21. *Under Hypotheses* 1 *and* 2, *every maximum likelihood estimator $(\hat{\theta}_n)$ is consistent, if Θ is a compact set in $\mathbb{R}^k$ and if:*

(1) *For any $(x,y) \in E^2$, $\alpha \longmapsto f(\alpha,x,y)$ is continuous.*
(2) *There exists a $\mu_\theta \otimes \pi_\theta$-integrable* r.v. *$k$ such that*

$$\sup_{\alpha \in \Theta} |\text{Log}\, f(\alpha,\cdot,\cdot)| \leq k.$$

Note. Consider a general dominated model adapted to $\mathbb{F} = (F_n)_{n\in\mathbf{N}}$. Let (L_n) be the likelihood. If it is integrable [Vol. I, p. 161], $(\ell_n(\theta)) = (\text{Log } L_n(\theta))$ is, under P_θ, an $\mathbb{F}$-submartingale. Its compensator $(\tilde{K}_n(\theta))$ is thus an increasing process which expresses the increase in information in the course of observations. It can be considered as a **stochastic Kullback information**. For the above Markov chain,

$$\widetilde{K}_n(\theta) = \sum_{i=1}^{n} K(\theta;X_{i-1}),$$

denoting

$$K(\theta;x) = K(\pi(\theta,x;\cdot),\ \pi(x;\cdot)).$$

For a sample from a distribution F_θ dominated by F, $\widetilde{K}_n(\theta) = nK(F_\theta,F)$.

4.4.3. Regular Models

Now we try to apply the scheme of [3.3.4].

Hypothesis 3. *For the dominated model of Definition* 4.4.20, *assume* $\Theta \subset \mathbb{R}^k$ *is a neighbourhood of* θ. *The following conditions are added.*

(a) *There exists a neighbourhood* V *of* θ *in which* $\alpha \longmapsto f(\alpha,x,y)$ *is twice continuously differentiable for arbitrary* (x,y): *let* D_i $(1 \leqslant i \leqslant k)$ *be the operators of partial differentiation in* α.

(b) *For each* x, *the model* $\{\pi(\alpha,x;\cdot)\}_{\alpha\in\Theta}$ *is a regular model in* θ, i.e. *for* $1 \leqslant i \leqslant k$,

$$\int \left[\frac{D_i f(\theta,x,y)}{f(\theta,x,y)}\right]^2 \pi(\theta,x;dy) < \infty$$

and

$$\int \frac{D_i f(\theta,x,y)}{f(\theta,x,y)}\,\pi(\theta,x;dy) = 0,$$

$$I_{ij}(\theta,x) = \int \frac{D_i f(\theta,x,y) D_j f(\theta,x,y)}{f^2(\theta,x,y)}\pi(\theta,x;dy)$$

$$= -\int D_i D_j \text{Log } f(\theta,x,y)\pi(\theta,x;dy).$$

The matrix $I(\theta,x) = \{I_{ij}(\theta,x)\}_{1 \leqslant i,j \leqslant k}$ *is the Fisher information matrix at* θ *of the model* $\{\pi(\alpha,x;\cdot)\}_{\alpha\in\Theta}$.

(c) $x \longmapsto I_{ij}(\theta,x)$ *is integrable with respect to* μ_θ. *Denote* $I_{ij}(\theta) = \int d\mu_\theta(x) I_{ij}(\theta,x)$, *and call* $I(\theta) = \{I_{ij}(\theta)\}$ *the* **Fisher information matrix of the process.** *We assume that* $I(\theta)$ *is invertible.*

(d) *Let us denote by* B_r *the ball of* Θ *centered at* θ *with radius* r,

$$\int \mu_\theta(dx)\pi(\theta,x;dy) \sup_{\alpha\in B_r} \left| D_i D_j \text{Log } f(\alpha,x,y) - D_i D_j \text{Log } f(\theta,x,y) \right|$$

tends to zero if r *tends to* 0 (*for* $1 \leqslant i,j \leqslant k$).

Notes. (1) Although not very attractive, these conditions are not too difficult to check, since they deal only with transition probabilities, and not with the entire process.

(2) For the initial distribution μ_θ, under Hypotheses 1 and 3 (a), (b) and (c), we have a regular model at each time n, with Fisher information $nI(\theta)$. The situation is analogous to the independent case. For any $P_{\theta,\nu}$,

$$\sum_{p=1}^{n} \frac{D_i f(\theta,X_{p-1},X_p)}{f(\theta,X_{p-1},X_p)}$$

is a centered martingale. For every initial distribution ν such that,

$$E_{\theta,\nu}\left[\sum_{p=1}^{n} \frac{D_i f(\theta,X_{p-1},X_p)}{f(\theta,X_{p-1},X_p)}\right]^2 = \sum_{p=1}^{n} E_{\theta,\nu}\left[\frac{D_i f(\theta,X_{p-1},X_p)}{f(\theta,X_{p-1},X_p)}\right]^2$$

is finite, we again have a regular model at each time n, with Fisher information

$$\left\{ \sum_{p=1}^{n} E_{\theta,\nu}(I_{ij}(\theta,X_{p-1}))\right\}.$$

In the two cases, $I(\theta)$ is the limit of $(1/n)$ times the Fisher information at time n.

(3) Without the recurrence hypothesis, the **stochastic Fisher information**

$$\tilde{I}_n(\theta) = \sum_{p=1}^{n} I(\theta, X_p)$$

is an increasing sequence in n which is often used for studying the rate of convergence of estimators (see [4.4.5]).

Theorem 4.4.22. *Assume Hypotheses* 1, 2 *and* 3. *Then, for every sequence* $(\hat{\theta}_n)$ *of maximum likelihood estimators consistent at* θ,

(a) $\sqrt{n}(\hat{\theta}_n - \theta) \xrightarrow{\mathcal{D}(P_{\theta,x})} N(0, I^{-1}(\theta))$, *and the sequence* $(\hat{\theta}_n)$ *is asymptotically efficient.*

(b) $\ell_n(\hat{\theta}_n) - \ell_n(\theta) \xrightarrow{\mathcal{D}(P_{\theta,x})} \frac{1}{2} \chi^2(k)$.

Proof. We follow the scheme [3.3.6],

$$0 = \frac{1}{\sqrt{n}} D_i \ell_n(\theta) + \sum_{j=1}^{k} \sqrt{n}(\hat{\theta}_n^j - \theta^j) \frac{1}{n} D_i D_j \ell_n(\theta)$$

$$+ \sum_{j=1}^{n} \sqrt{n}(\hat{\theta}_n^j - \theta^j) \frac{1}{n} \int_0^1 [D_i D_j \ell_n(\theta + t(\hat{\theta}_n - \theta))$$

$$- D_i D_j \ell_n(\theta)] dt.$$

We can apply part (c) of Theorem 4.3.16 to the function $F = D_i f(\theta, \cdot, \cdot)/f(\theta, \cdot, \cdot)$ because $\pi_\theta F = 0$ and $F \in L^2(\mu_\theta \otimes \pi_\theta)$,

$$\left(\frac{1}{\sqrt{n}} D_i \ell_n(\theta)\right) = \left(\frac{1}{\sqrt{n}} \sum_{p=1}^{n} \frac{D_i f(\theta, X_{p-1}, X_p)}{f(\theta, X_{p-1}, X_p)}\right)$$

$$\xrightarrow{\mathcal{D}(P_{\theta,x})} N(0, I(\theta)).$$

From the law of large numbers 4.3.15,

$$\left(\frac{1}{\sqrt{n}} D_i D_j \ell_n(\theta)\right) = \left(\frac{1}{n} \sum_{p=1}^{n} D_i D_j \mathrm{Log}\, f(\theta, X_{p-1}, X_p)\right)$$

$$\xrightarrow{P_{\theta,x}\text{-a.s.}} -I(\theta).$$

With Hypothesis (d) in order to show that the residual term tends to 0, the proof is finished as in [3.3.4].

The proof of (b) is then that of [3.4.2], this allows a test of "$\theta = \theta_0$" against "$\theta \neq \theta_0$," by using the rejection region

$$R = \left\{ \ell_n(\hat{\theta}_n) - \ell_n(\theta_0) > \frac{1}{2}\chi^2_{\alpha,k} \right\}$$

thus have a test of asymptotic level α.

Application to Markov Chains with Finite State Space. Starting from the preceding theorem, it is not too difficult to state and to prove the extension of Theorem 3.4.20 to recurrent chains with finite state space.

Let us assume for example that E has s elements and that $\pi(i,j) > 0$ for r $(i,j) \in E^2$. Let us apply Theorem 4.4.19 to the Markov chain (X_n, X_{n+1}) on $\Delta = \{(i,j),\ \pi(i,j) > 0\}$ with transition $\tilde{\pi}$ defined by $\tilde{\pi}((i,j),(j,k)) = \pi(j,k)$. Let us denote

$$N_{ijk} = \sum_{p=1}^{n-2} 1_{(X_p=i, X_{p+1}=j, X_{p+2}=k)}.$$

Then

$$\sum_{(i,j,k)\in E^3} \frac{(N_{ijk} - N_{ij}\pi(j,k))^2}{N_{ij}\pi(j,k)} \xrightarrow{\mathcal{D}} \chi^2(rs - r),$$

r the cardinality of Δ. If we estimate $\pi(j,k)$ by its maximum likelihood estimator, we obtain

$$Z_n = \sum_{(i,j,k)\in E^3} \frac{(N_{ijk} - N_{ij}N_{jk}/n)^2}{N_{ij}N_{jk}/n} \longrightarrow \chi^2(rs - r - r+s)$$

This may be used to test whether we are really observing a Markov chain. We have a test of asymptotic level α by taking the rejection region $\{Z_n \geq \chi^2_{s(s-1)^2,\alpha}\}$, assuming $\pi(i,j) > 0$ for all (i,j).

4.4.4. Markov Chains AR1

In this section we deal with a family of Markov chains having no recurrent state. Nevertheless, we are going to prove asymptotic theorems analogous to those of [4.3]. It is a glimpse of the possible generalizations of [4.3].

We are given $\theta \in \mathbb{R}$, and on a probability space $(\Omega', \mathcal{A}', P')$ a sequence $(\varepsilon_n)_{n\geq 1}$ of independent centered variables, of variance 1 and with density f strictly positive with respect to Lebesgue measure L. Finally let X_0 be an r.v. independent of $(\varepsilon_n)_{n\geq 1}$ with distribution ν. The sequence $(X_n)_{n\geq 1}$ is defined by recurrence as

$$X_n = \theta X_{n-1} + \varepsilon_n, \quad n \geqslant 1.$$

We are dealing with a Markov chain on $\mathbb{R}$ with transition,

$$\pi_\theta : (x,\Gamma) \longmapsto \pi(\theta,x;\Gamma) = \int f(y - \theta x) 1_\Gamma(y) dy.$$

For $\Gamma \in \mathcal{B}_{\mathbb{R}}$, we have ([Vol. I, 6.1]),

$$\begin{aligned} P'(X_n \in \Gamma | X_0, \ldots, X_{n-1}) &= P'(\theta X_{n-1} + \varepsilon_n \in \Gamma | X_0, \ldots, X_{n-1}) \\ &= \int f(y) 1_\Gamma(y + \theta X_{n-1}) dy. \end{aligned}$$

Studying the Markov chain defined above is equivalent to studying its canonical version defined in $(\Omega, A) = (\mathbb{R}, \mathcal{B}_{\mathbb{R}})^{\mathbb{N}}$. Let $P_{\theta,\nu}$ be the distribution of this chain. We again denote by (X_n) the sequence of coordinates.

Proposition 4.4.23. *Consider the* AR1 *chain defined above.*

(a) *For* $|\theta| < 1$, *the sequence* (X_n) *converges in distribution to a distribution* μ_θ, *invariant under* π_θ. *The distribution* μ_θ *is the distribution of* $\Sigma_{n=1}^{\infty} \theta^{n-1} \varepsilon_n$. *It has zero mean and a variance equal to* $1/(1-\theta^2)$. *If the sequence* (ε_n) *is Gaussian,*

$$\mu_\theta = N\left(0, \frac{1}{1-\theta^2}\right)$$

and

$$\frac{1}{n}\sum_{i=1}^{n} X_i \xrightarrow{\text{a.s.}} 0, \quad \frac{1}{n}\sum_{i=1}^{n} X_i^2 \xrightarrow{\text{a.s.}} \frac{1}{1-\theta^2},$$

$$\frac{1}{n}\sum_{i=1}^{n} X_i X_{i-1} \xrightarrow{\text{a.s.}} \frac{\theta}{1-\theta^2};$$

$$\frac{1}{\sqrt{n}}\sum_{i=1}^{n} X_i \xrightarrow{\mathcal{D}} N\left(0, \frac{1}{(1-\theta)^2}\right);$$

$$\frac{1}{\sqrt{n}}\left(\sum_{i=1}^{n} X_i X_{i-1} - \theta \sum_{i=1}^{n} X_i^2\right) \xrightarrow{\mathcal{D}} N\left(0, \frac{1}{1-\theta^2}\right).$$

(b) *For* $|\theta| > 1$, *there exists a nonzero* r.v. Z_θ *such that*

$$X_n/\theta^n \xrightarrow[\text{a.s.}]{L^2} Z_\theta.$$

Proof. Set $U_n = \theta^{n-1}\varepsilon_1 + \theta^{n-2}\varepsilon_2 + \ldots + \varepsilon_n$. We have $X_n = \theta^n X_0 + U_n$ and U_n has the same distribution as $V_n = \varepsilon_1 + \theta\varepsilon_2 + \ldots + \theta^{n-1}\varepsilon_n$.

(a) For $|\theta| < 1$, $(U_n - X_n) \xrightarrow{\text{a.s.}} 0$ and

$$(V_n) \xrightarrow{\text{a.s.}} V = \sum_{n=1}^{\infty} \theta^{n-1}\varepsilon_n$$

from Theorem 2.6.29. Hence $(X_n) \xrightarrow{\mathcal{D}} V$. The r.v. V is centered, with variance $1/(1-\theta^2)$. If ε_0 is an r.v. independent of (ε_n) and with the same distribution, $\theta V + \varepsilon_0$ has the distribution of V: thus $\mu_\theta \pi_\theta = \mu_\theta$. The measure μ_θ is equivalent to the measure L as is the distribution of the ε_n.

If the r.v.'s ε_n are Gaussian then μ_θ, the limit of Gaussian distributions is Gaussian: $\mu_\theta = N(0,1/(1-\theta^2))$. Let us show the asymptotic properties by assuming first of all $\nu = \mu_\theta$; (X_n) is stationary, with variance $1/(1-\theta^2)$. Hence, from Theorem 1.4.35,

$$\frac{1}{n}\sum_{i=1}^{n} X_i \xrightarrow{\text{a.s.}} 0, \quad \frac{1}{\sqrt{n}}\sum_{i=1}^{n} X_i \xrightarrow{\mathcal{D}} N\left(0, \frac{1}{(1-\theta)^2}\right),$$

$$\frac{1}{n}\sum_{i=1}^{n} X_i^2 \xrightarrow{\text{a.s.}} \frac{1}{1-\theta^2}.$$

Finally, the sequence $(\Sigma_{i=1}^n X_i X_{i-1})$ is compensated by $(\theta\Sigma_{i=1}^n X_{i-1}^2)$, and the increasing process associated with

$$\left[\sum_{i=1}^{n} (X_i X_{i-1} - \theta X_{i-1}^2)\right]$$

is $\Sigma_{i=1}^n X_{i-1}^2$: from the law of large numbers (Corollary 2.6.30)

$$\frac{\sum_{i=1}^{n} X_i X_{i-1} - \theta \sum_{i=1}^{n} X_{i-1}^2}{\sum_{i=1}^{n} X_{i-1}^2} \xrightarrow{\text{a.s.}} 0,$$

$$\frac{1}{n}\sum_{i=1}^{n} X_i X_{i-1} \xrightarrow{\text{a.s.}} \frac{\theta}{1-\theta^2}.$$

As in [1.4.2] we see that the sequence (X_n^4) is stationary and has a continuous spectral density. From which it follows that

$$\frac{1}{n}\sum_{i=1}^{n} X_{i-1}^4 \xrightarrow{\text{a.s.}} \frac{3}{(1-\theta^2)^2}.$$

We obtain the central limit theorem stated for

$$\frac{1}{\sqrt{n}}\sum_{i=1}^{n} (X_i X_{i-1} - \theta X_{i-1}^2)$$

by applying Corollary 2.8.43. In fact, for $\varepsilon > 0$ and $a > 0$,

$$\frac{1}{n}\sum_{i=1}^{n} E_{\theta,\nu}\left[(X_i X_{i-1} - \theta X_{i-1}^2)^2 \mathbf{1}_{(|X_i X_{i-1} - \theta X_{i-1}^2| \geq \varepsilon\sqrt{n})}| X_0, \dots, X_{i-1}\right]$$

$$= \frac{1}{n}\sum_{i=1} X_{i-1}^2 \int x^2 f(x) \mathbf{1}_{(|xX_{i-1}| \geq \varepsilon\sqrt{n})} dx$$

$$\leq \frac{1}{n}\sum_{i=1}^{n} X_{i-1}^2 \left[\mathbf{1}_{(|X_{i-1}| \geq a)} + \int x^2 f(x) \mathbf{1}_{(|x| \geq (\varepsilon\sqrt{n}/a))} dx\right]$$

$$\leq \frac{1}{na^2}\sum_{i=1}^{n} X_{i-1}^4$$

$$+ \left[\int x^2 f(x) \mathbf{1}_{(|x| \geq (\varepsilon\sqrt{n}/a))} dx\right] \frac{1}{n}\sum_{i=1}^{n} X_{i-1}^2.$$

The right hand term tends, as $n \to \infty$, to $3/a^2(1-\theta^2)^2$. However a is taken as arbitrary, and the sequence considered satisfies the Lindeberg condition,

$$\frac{1}{\sqrt{n}}\sum_{i=1}^{n} (X_i X_{i-1} - \theta X_{i-1}^2) \xrightarrow{\mathcal{D}} N\left(0, \frac{1}{1-\theta^2}\right).$$

Similar asymptotic results are obtained for $\nu \neq \mu_\theta$ by noting that $X_n - U_n = \theta^n X_0$, and thus that the obtained limits on (Ω', A', P') do not depend on X_0.

(b) (X_n/θ^n) is a martingale:

$$E'(X_n/\theta^n | X_0, \dots, X_{n-1}) = X_{n-1}/\theta^{n-1}$$

$$E'(X_n^2/\theta^{2n}) = \sigma^2(X_0) + \frac{1 - \dfrac{1}{\theta^{2n}}}{\theta^2 - 1} + (E'(X_0))^2.$$

For $|\theta| > 1$, $E'(X_n^2/\theta^{2n}) \leq E'(X_0^2) + 1/(\theta^2-1)$. From Theorems 2.4.12 and 2.4.13, there exists a centered r.v. Z_θ, with variance $E'(X_0^2) + 1/(\theta^2 - 1)$, such that,

$$X_n/\theta^n \xrightarrow[\text{a.s.}]{L^2} Z_\theta.$$

Thus, for every initial distribution ν,

$$P_{\theta,\nu}(|X_n| \to \infty) \geqslant P_{\theta,\nu}(Z_\theta \neq 0) > 0.$$

Note. For $|\theta| = 1$, the study of the AR1 chain leads to sums of independent identically distributed r.v.'s.

Let us now consider the statistical model $(\Omega, A, (P_{\theta,\nu})_{\theta \in \mathbb{R}}, (X_n)_{n \in \mathbb{N}})$, with $F_n = \sigma(X_0, \ldots, X_n)$. This is a dominated model,

$$P_{\theta,\nu} = \prod_{i=1}^{n} \frac{f(X_i - \theta X_{i-1})}{f(X_i)} P_{0,\nu} \,.$$

The likelihood follows from this,

$$L_n(\theta) = \prod_{i=1}^{n} \frac{f(X_i - \theta X_{i-1})}{f(X_i)} \,.$$

In the Gaussian case, we have,

$$\ell_n(\theta) = \text{Log } L_n(\theta) = \theta \sum_{i=1}^{n} X_i X_{i-1} - \frac{\theta^2}{2} \sum_{i=1}^{n} X_{i-1}^2.$$

The maximum likelihood estimator follows:

$$\hat{\theta}_n = \frac{\sum_{i=1}^{n} X_i X_{i-1}}{\sum_{i=1}^{n} X_{i-1}^2}.$$

For $|\theta| < 1$, we have $\hat{\theta}_n \xrightarrow{P_{\theta,\nu}\text{-a.s.}} \theta$, and

$$\sqrt{n}(\hat{\theta}_n - \theta) \xrightarrow{\mathcal{D}(P_{\theta,\nu})} N(0, 1 - \theta^2).$$

We are dealing with minimum contrast estimators for the contrast (U_n) defined by,

$$U_n(\alpha) = \frac{1}{n}\left[\alpha^2 \sum_{i=1}^{n} X_{i-1}^2 - 2\alpha \sum_{i=1}^{n} X_i X_{i-1}\right],$$

$$U_n(\alpha) \xrightarrow{P_{\theta,\nu}\text{-a.s.}} K(\theta,\alpha) = \frac{\alpha^2 - 2\alpha\theta}{1 - \theta^2}$$

and

$$\inf_{\alpha} K(\theta,\alpha) = K(\theta,\theta).$$

Is the AR1 model regular? Let us assume f to be of class C^2, f and f' tending to 0 at infinity. Moreover, assume f'' and (f'^2/f) are integrable for Lebesgue measure. Then

$$Y_n(\theta) = \frac{d}{d\theta}\ell_n(\theta) = -\sum_{i=1}^{n} \frac{f'(X_i - \theta X_{i-1})X_{i-1}}{f(X_i - \theta X_{i-1})}.$$

This sequence is compensated by,

$$-\sum_{i=1}^{n} X_{i-1}\int \frac{f'(x - \theta X_{i-1})}{f(x - \theta X_{i-1})} f(x - \theta X_{i-1})dx$$

$$= -\sum_{i=1}^{n} X_{i-1}\int f'(x)dx = 0.$$

This is a square integrable martingale and the increasing process which is associated with it is,

$$J_n(\theta) = \sum_{i=1}^{n} \int X_{i-1}^2 \frac{f'^2(x - \theta X_{i-1})}{f(x - \theta X_{i-1})}dx$$

$$= \sum_{i=1}^{n} X_{i-1}^2 I(f).$$

Moreover,

$$Z_n(\theta) = \frac{d^2}{d\theta^2}\ell_n(\theta)$$

$$= \sum_{i=1}^{n} X_{i-1}^2\left[-\frac{f''(X_i - \theta X_{i-1})}{f(X_i - \theta X_{i-1})} + \frac{f'(X_i - \theta X_{i-1})}{f^2(X_i - \theta X_{i-1})}\right].$$

Since $\int f''(x)dx = 0$, the compensator of $(Z_n(\theta))$ is $(J_n(\theta))$. In the Gaussian case, $I(f) = 1$ and

$$J_n(\theta)/n \xrightarrow{\text{a.s.}} 1/(1 - \theta^2).$$

Hence, according to Definition 3.3.16, $I(\theta) = 1/(1 - \theta^2)$ and

$$\sqrt{n}\,(\hat{\theta}_n - \theta) \xrightarrow{\mathcal{D}} N(0,1 - \theta^2):$$

$(\hat{\theta}_n)$ is asymptotically efficient.

Proposition 4.4.24. *For the Gaussian* AR1 *sequence, the maximum likelihood estimator of* θ *is*

$$\hat{\theta}_n = \frac{\sum_{i=1}^{n} X_i X_{i-1}}{\sum_{i=1}^{n} X_{i-1}^2}.$$

The sequence $(\hat{\theta}_n)$ *is consistent and asymptotically efficient,*

$$\hat{\theta}_n \xrightarrow{P_{\theta,\nu}\text{-a.s.}} \theta, \qquad \sqrt{n}(\hat{\theta}_n - \theta) \xrightarrow{\mathcal{D}(P_{\theta,\nu})} N(0, 1 - \theta^2).$$

4.4.5. Branching Processes

A simplified description of the evolution of a population is the following. Admitting the possibility of isolating the successive generations, assume that the number of descendents of each individual has distribution F and that the various individuals of a generation reproduce independently of each other (for a human population, individual may imply individual of a given sex...). Let (X_n) be the sequence of sizes of successive generations. We are dealing with a Markov chain with transition $(i,j) \longmapsto F^{*i}(j)$ on $\mathbb{N}$.

Let us study this chain. Assume that $F(0)$ is nonzero, and that F has finite mean m and variance σ^2. The state 0 is a recurrent class since it leads only to itself. All other points lead to 0 with nonzero probability and are thus transients. For $m \leqslant 1$, they lead to 0 with probability 1; see Vol. I, E6.3.4 for this result and for an elementary proof of certain of the following results. The case $m > 1$ is more interesting. Let us study in this case the canonical chain $\{\Omega, A, P, (X_n)_{n \geqslant 0}\}$ with initial state 1. We have,

$$E^{n-1}(X_n) = mX_{n-1} \quad \text{and} \quad E(X_n) = mE(X_{n-1}) = m^n;$$

$$E^{n-1}(X_n^2) = X_{n-1}\sigma^2 + m^2 X_{n-1}^2$$

and

$$E(X_n^2) = m^{n-1}\sigma^2 + m^2 E(X_{n-1}^2);$$

$$\sigma^2(X_n) = \sigma^2 m^{n-1} + m^2 \sigma^2(X_{n-1}) = \sigma^2 m^{n-1} \frac{m^n - 1}{m - 1}.$$

The sequence (X_n/m^n) is a martingale, and

$$E\left[\frac{X_n^2}{m^{2n}}\right] \leqslant \sigma^2 \frac{1}{m(m-1)};$$

hence there exists a positive r.v. W such that,

$$\frac{X_n}{m^n} \xrightarrow[L^2]{\text{a.s.}} W, \quad E(W) = 1, \quad \sigma^2(W) = \frac{\sigma^2}{m(m-1)}.$$

Set $S_n = X_0 + \ldots + X_n$; from Lemma 2.6.31 we have,

$$\frac{S_n}{m^n} = \frac{1}{m^n} \sum_{k=0}^{n} m^k \frac{X_k}{m^k} \xrightarrow{\text{a.s.}} \frac{m}{m-1} W.$$

The set $\{W = 0\}$ coincides with the extinction set $E = \{\lim_{n\to\infty} X_n = 0\}$ (Vol. I, E6.3.4). We can study what happens on E^c by using the probability P^{E^c} conditional on E^c.

Here is a version of the branching process. Let us consider on a probability space $(\Omega,\mathcal{A},P)$, a sequence (Y_n) of independent r.v.'s with distribution F. We define $U_n = 1 + Y_1 + \ldots + Y_n$ and, by recurrence,

$$X_0 = S_0 = 1, \ldots, X_n = U_{S_{n-1}} - U_{S_{n-2}},$$

$$S_n = S_{n-1} + X_n, \ldots .$$

Then (X_n) is a version of the branching process and $S_n = X_0 + X_1 + \ldots + X_n$. The sequence (S_n/m^n) tends. a.s. to $m/(m-1)W$. We then have,

$$\frac{1}{\sqrt{S_{n-1}}} \sum_{k=1}^{S_{n-1}} (Y_k - m) = \frac{S_n - mS_{n-1}}{\sqrt{S_{n-1}}}.$$

The study of the asymptotic behavior of these r.v.'s necessitates a central limit theorem for sums of a random number of independent r.v.'s. Let us assume part (b) of the following theorem. It will be shown in Chapter 7 (Proposition 7.4.30).

Theorem 4.4.25. *Let $(\Omega,\mathcal{A},P,(X_n)_{n\geqslant 0})$ be a branching process. Assume $X_0 = 1$. The distribution of the number of descendants of each individual is F. F is assumed to be square integrable, with mean m strictly greater than 1 and with variance σ^2.*

(a) *There exists an* r.v. *W with mean 1 and with variance $\sigma^2/m(m-1)$ such that, if $n \to \infty$,*

$$\frac{X_n}{m^n} \xrightarrow[L^2]{\text{a.s.}} W .$$

The extinction set $E = \{\lim X_n = 0\}$ *coincides with* $\{W = 0\}$.

(b) *Set* $S_n = X_0 + \ldots + X_n$. *Then if* $n \to \infty$,

$$\frac{S_n}{m^n} \xrightarrow{\text{a.s.}} \frac{m}{m-1} W;$$

conditional on $W > 0$,

$$\left[W, \frac{S_n - mS_{n-1}}{\sigma\sqrt{S_{n-1}}}\right]$$

tends in distribution to (W,Y), *as* $n \to \infty$, *where* Y *is an* r.v. *with distribution* $N(0,1)$ *independent of* W.

Exponential Family of Branching Processes. Let h be a function from $\mathbb{N}$ into $\mathbb{R}_+$ and let Θ be the open interval which is the interior of

$$\left\{\theta;\ \sum_{j=0}^{\infty} h(j)e^{j\theta} < \infty\right\}.$$

We define, for each $\theta \in \Theta$,

$$\exp \phi(\theta) = \sum_{j=0}^{\infty} h(j)e^{j\theta} \quad \text{and} \quad F_\theta(j) = h(j)\exp[-\phi(\theta)+j\theta].$$

The family of distributions (F_θ) is an exponential family. Let $m(\theta)$ and $\sigma^2(\theta)$ be the mean and variance of F_θ. From [Vol. I, 3.3.2] we have,

$$m(\theta) = \phi'(\theta), \quad \sigma^2(\theta) = \phi''(\theta).$$

We calculate $F_\theta^{*i}(j) = H(i,j)\exp[-i\phi(\theta) + j\theta]$, with

$$H(i,j) = \sum_{j_1+\cdots+j_i=j} h(j_1) \ldots h(j_i).$$

Now let us denote by $(\Omega,A,P_\theta,(X_n)_{n\geq 0})$, the canonical branching process, with initial state 1 and with transition $(i,j) \longmapsto F_\theta^{*i}(j)$. For $\theta \in \Theta$, we are dealing with a model dominated by $\nu^{\mathbb{N}}$, ν being counting measure on $\mathbb{N}$. Setting $S_n = \Sigma_{k=0}^n X_k$, we obtain the log-likelihood at time n,

$$\ell_n(\theta) = \text{Log } L_n(\theta) = -\phi(\theta)S_{n-1} + \theta(S_n - X_0) + \sum_{k=1}^{n} \text{Log } H(X_{k-1}, X_k).$$

From which

$$\frac{d}{d\theta} \ell_n(\theta) = -m(\theta)S_{n-1} + S_n - X_0.$$

Hence the maximum likelihood estimator of $m(\theta)$ is

$$\hat{m}_n = \frac{\sum_{i=1}^{n} X_i}{\sum_{i=0}^{n-1} X_i}.$$

For $m(\theta) > 1$, on $E = \{\lim X_n = 0\}$, we do not have an asymptotic result. However from Theorem 4.4.25,

$$\frac{S_n}{(m(\theta))^n} \xrightarrow{P_\theta\text{-a.s.}} \frac{m(\theta)}{m(\theta) - 1} W_\theta,$$

where W_θ is an r.v. with mean 1 and with variance $(\sigma^2(\theta)/m(\theta)(m(\theta)-1))$. Hence $(\hat{m}_n)$ converges P_θ-a.s. on E^c to $m(\theta)$. The model is regular. The sequence $\ell_n'(\theta)$ is a square integrable martingale, the increasing process of which is

$$\tilde{I}_n(\theta) = \sum_{i=1}^{n} E_\theta[(X_i - m(\theta)X_{i-1})^2 | F_{i-1}]$$

$$= \sigma^2(\theta) \sum_{i=1}^{n} X_{i-1} = \sigma^2(\theta)S_{n-1} = \frac{-d^2}{d\theta^2} \ell_n(\theta).$$

The information at time n is thus,

$$I_n(\theta) = \sigma^2(\theta) \sum_{i=1}^{n} E_\theta(X_{i-1}) = \sigma^2(\theta) \frac{m^n(\theta) - 1}{m(\theta) - 1};$$

$$\frac{\tilde{I}_n(\theta)}{m(\theta)^n} \xrightarrow{\text{a.s.}} \frac{\sigma^2(\theta)}{m(\theta) - 1} W_\theta,$$

thus we cannot speak in this case of a deterministic information limit. However we have,

$$\lim \frac{\tilde{I}_n(\theta)}{m^n(\theta)} = \frac{\sigma^2(\theta)}{m(\theta) - 1} = \lim \frac{I_n(\theta)}{m^n(\theta)}.$$

In order to study the rate of convergence of $(\hat{m}_n)$ to $m(\theta)$, we calculate,

$$\sqrt{S_{n-1}}\,[\hat{m}_n - m(\theta)] = \frac{S_n - m(\theta)S_{n-1} - 1}{\sqrt{S_{n-1}}}.$$

However

$$\frac{S_n}{(m(\theta))^n} \xrightarrow{P_\theta\text{-a.s.}} W_\theta;$$

from Theorem 4.4.25,

$$\left[\frac{S_n}{(m(\theta))^n}, \frac{S_n - m(\theta)S_{n-1}}{\sqrt{S_{n-1}}}\right] \xrightarrow{(P_\theta^{E^c})} \left[\frac{m(\theta)}{m(\theta)-1}W_\theta,\ Y_\theta\right].$$

where Y_θ has distribution $N(0,\sigma^2(\theta))$ and is independent of W_θ. Thus

$$[m(\theta)]^{n/2}(\hat{m}_n - m(\theta)) \xrightarrow{\mathcal{D}(P_\theta^{E^c})} [m(\theta) - 1]^{1/2}Y_\theta W_\theta^{-1/2}.$$

But $\theta = \phi'^{-1}(m(\theta))$ and $\hat{\theta}_n = \phi'^{-1}(\hat{m}_n)$ is the maximum likelihood estimator of θ. By adapting Theorem 3.3.11, p. 131 we obtain,

$$\sigma(\theta)\sqrt{S_{n-1}}\,(\hat{\theta}_n - \theta) \xrightarrow{\mathcal{D}(P_\theta^{E^c})} Y,$$

where $Y = Y_\theta/\sigma(\theta)$ has the distribution $N(0,1)$. Thus, $\tilde{I}_n(\theta)$ describes the rate of convergence of $\hat{\theta}_n$ better than does $I_n(\theta)$,

$$\sqrt{\tilde{I}_n(\theta)}\ (\hat{\theta}_n - \theta) \xrightarrow{\mathcal{D}(P_\theta^{E^c})} Y$$

$$\sqrt{I_n(\theta)}(\hat{\theta}_n - \theta) \xrightarrow{\mathcal{D}(P_\theta^{E^c})} \sqrt{m(\theta) - 1}\ Y(W(\theta))^{-1/2}.$$

Bibliographic Notes

In order to appreciate the usefulness of Markov chains, consult books in which examples are given: Bartlett [2], Bharucha-Reid, Chung, Feller (Volume 1), Freedman [2], and Ewens for genetic applications. In Kemeny-Snell-Knapp a detailed study of chains with a countable state space is found. Spitzer studies random walks and Borovkov G/G/1 queues as well as other models of the same form.

In order to arrive at asymptotic theorems essential to the statistics of Markov chains as quickly as possible, we have limited ourselves to chains having a recurrent point. The more general case of recurrent Markov chains in the sense of Harris is studied in Revuz and in Orey. Our account of [4] uses Pitman's coupling method. There are various Markov chain central limit theorems (Doob, Billingsley [1], Cogburn, and Maigret).

Billingsley [1] treats in detail the statistics of finite Markov chains. Basawa-Rao give an overview of Markov chain statistics and a bibliography. See also Hall-Heyde where the notion of stochastic information is developed. Branching processes are studied in Harris, Gikhman-Skorokhod; and the statistics of these processes in Basawa-Rao.

Chapter 5
STEP BY STEP DECISIONS

Objectives

At what time do we decide to stop playing roulette or to replace the car? When will the statistician stop his sampling? This chapter gives methods of dealing with this type of problem, where the decision taken at each instant can only be based on past experiments. These themes are regrouped under the headings optimal stopping, control or sequential statistics which, with distinct vocabularies, make use of the same ideas.

5.1. Optimal Stopping

Let $(Z_n)_{n \geqslant 0}$ be the sequence of successive fortunes of a gambler (a negative fortune is a debt): Z_0 is his initial fortune, Z_n his fortune after n games. These r.v.'s Z_n are defined on a probability space (Ω, A, P). We assume that they are majorized by an integrable r.v. The gambler must choose when to stop: let ν be the random number of times that he plays. At time n, the events which he knows, form the σ-algebra F_n: for $\mathbb{F} = (F_n)$, we assume that the sequence (Z_n) is $\mathbb{F}$-adapted. The decision of the gambler to stop at time n depends only on the past: $(\nu = n) \in F_n$, and ν is a stopping time. The problem of optimal stopping is the following:

(a) determine $V = \sup\{E(Z_\nu);\ \nu$ a finite stopping time$\}$, **the optimal average gain.**
(b) determine if possible, ν_0 a finite stopping time, called the **optimal stopping time**, such that $V = E(Z_{\nu_0})$.

5.1.1. Essential Upper Bound

The first difficulty is the impossibility of talking about

$$\sup\{Z_\nu;\ \nu \text{ a finite stopping time}\}$$

since this is not in general an r.v.

Let us consider on a probability space $(\Omega, \mathcal{A}, P)$ a family $(Y_i)_{i \in I}$ of r.v.'s majorized by an integrable r.v. Z. The function $\sup_{i \in I} Y_i$ is not in general measurable, if I is not countable. On the other hand, if G is a countable subset of I, $Y_G = \sup_{i \in G} Y_i$ is an r.v. Consider a sequence of countable subsets (G_n) of I such that

$$\sup_n E[Y_{G_n}] = \sup\{E[Y_G];\ G \text{ countable},\ G \subset I\}.$$

If $H = \cup_n G_n$, Y_H is an r.v. which is the smallest possible majorant in the sense of almost sure inequality. In fact, for all $i \in I$, $Y_H = Y_{H \cup \{i\}} \geq Y_i$ (a.s.), and, if X is an r.v. which majorizes all the Y_i almost surely, then $X \geq Y_H$ (a.s.). We denote $Y_H = \operatorname{ess\,sup}_{i \in I} Y_i$, or $P - \operatorname{ess\,sup}_{i \in I} Y_i$ if there is ambiguity about the probability P used.

The study of Y_H is easier when there exists a sequence $(i_k) \subset I$ such that: $Y_H = \lim \uparrow Y_{i_k}$. This is the case if we impose on $(Y_i)_{i \in I}$ the following **lattice property**: for $(i,j) \in I^2$, there exists a $k \in I$ such that Y_k is greater than $\sup(Y_i, Y_j)$.

Proposition 5.1.1. *Let* $(Y_i)_{i \in I}$ *be a lattice of* r.v.'s *the modulus of which is majorized by an integrable* r.v.: *there exists an* a.s. *unique* r.v., $Y = \operatorname{ess-sup}_{i \in I} Y_i$ *satisfying the following properties*

(1) *For every* $i \in I$, $Y \geq Y_i$ a.s.
(2) *If, for every* $i \in I$, $X \geq Y_i$, *then* $X \geq Y$ a.s.
(3) *There exists a sequence* $(i_k) \subset I$ *such that* $Y = \lim \uparrow Y_{i_k}$
(4) $E(Y) = \sup_{i \in I} E(Y_i)$
(5) *For every* σ*-algebra* $\mathcal{B}$ *contained in* $\mathcal{A}$:

$$E(Y|\mathcal{B}) = \text{ess}-\sup_{i\in I} E(Y_i|\mathcal{B}).$$

We sometimes substitute for the lattice property the following **approximate countable lattice property**. For a countable G included in I, and for every $\varepsilon > 0$, there exists an $i \in I$ such that

$$Y_i \geq Y_G - \varepsilon.$$

By using this property for H, we obtain the following proposition.

Proposition 5.1.2. *Let* $(Y_i)_{i\in I}$ *be a family of* r.v.'s *majorized by an increasing* r.v., *which satisfies the approximate countable lattice property. Then the* r.v. $Y = \text{ess} - \sup Y_i$ *satisfies* (1), (2), (4) *and* (5) *and the following property*:

(3') *For all* $\varepsilon > 0$, *there exists an* $i_\varepsilon \in I$ *such that*: $Y_{i_\varepsilon} \geq Y - \varepsilon$.

The proofs of (5) (hence (4) by taking $\mathcal{B}$ trivial) follow from Lebesgue's theorem for Proposition 5.1.1. For Proposition 5.1.2, we note that Y majorizes Y_i, hence $E(Y|\mathcal{B})$ majorizes $E(Y_i|\mathcal{B})$ (a.s.). From (3'), for all $\varepsilon > 0$:

$$\text{ess} - \sup E(Y_i|\mathcal{B}) \geq E(Y_{i_\varepsilon}|\mathcal{B}) \geq E(Y|\mathcal{B}) - \varepsilon.$$

5.1.2. Optimal Stopping

Let us return to the optimal stopping problem posed at the start of [5.1]. If the gambler has already played n times, ν can now only be chosen from the family T_n of finite $\mathbb{F}$-stopping times greater than n. Let us denote by E^n the expectation conditional on F_n. The maximum possible gain after n is then

$$G_n = \text{ess} - \sup_{\nu\in T_n} Z_\nu.$$

The family $(Z_\nu)_{\nu\in T_n}$ satisfies the approximate countable lattice property. Let (ν_k) be a sequence of T_n and let $\varepsilon > 0$. For $U = \sup_k Z_{\nu_k}$, we define $\nu^\varepsilon \in T_n$ in the following way:

$$\nu^\varepsilon = \nu_1 \quad \text{on} \quad A_1 = \{Z_{\nu_1} \geq U - \varepsilon\}$$

$$\nu^{\varepsilon} = \nu_2 \quad \text{on} \quad A_2 = \{Z_{\nu_2} \geqslant U - \varepsilon\} \setminus A_1$$

$$\cdots$$

$$\nu^{\varepsilon} = \nu_k \quad \text{on} \quad A_k = \{Z_{\nu_k} \geqslant U - \varepsilon\} \setminus A_1 \cup A_2 \cup \ldots \cup A_{k-1}.$$

We certainly have: $Z_{\nu^\varepsilon} \geqslant U - \varepsilon$.

G_n is not known, however, at time n, we know $E^n(G_n)$, the **optimal gain conditional on the past**. From Proposition 5.1.2, this r.v. denoted $V(n)$ satisfies,

$$V(n) = E^n(G_n) = \underset{\nu \in \mathcal{T}_n}{\text{ess}-\sup}\, E^n(Z_\nu)$$

and the optimal average gain after n is

$$E(V(n)) = \sup_{\nu \in \mathcal{T}_n} E(Z_\nu).$$

However, for $\nu \in \mathcal{T}_n$:

$$E^n(Z_\nu) = Z_n 1_{(\nu=n)} + E^n(Z_\nu) 1_{(\nu>n)}.$$

Let $\nu' = \sup(\nu, n+1)$,

$$\begin{aligned} E^n(Z_\nu) &= Z_n 1_{(\nu=n)} + E^n(Z_{\nu'}) 1_{(\nu>n)} \\ &\leqslant Z_n 1_{(\nu=n)} + E^n(G_{n+1}) 1_{(\nu>n)} \\ &= Z_n 1_{(\nu=n)} + E^n(V(n+1)) 1_{(\nu>n)} \\ &\leqslant \sup[Z_n, E^n(V(n+1))]. \end{aligned}$$

But n is in $\mathcal{T}_n$, hence Z_n minorizes $V(n)$. We also have,

$$V(n) \geqslant \underset{\nu \in \mathcal{T}_{n+1}}{\text{ess}-\sup}\, E^n(Z_\nu) = E^n(V(n+1)).$$

Thus

$$V(n) = \sup[Z_n, E^n(V(n+1))].$$

The following intuitive result is obtained: at time n, we stop if $V(n) = Z_n$, and we continue if $V(n) > Z_n$... . The preceding relation allows $V(n)$ to be calculated using $V(n+1)$. It is a **backward recurrence** equation which is often difficult to solve. On the other hand, it is easy with the framework of the following paragraph where $V(H)$ is given for an $H > n$.

5.1.3. Optimal Stopping with a Finite Horizon

Let $H \in \mathbb{N}$; we are looking for the optimal stopping time ν^H in the class of those which are $\leqslant H$. We then obtain the backward recurrence relations, where $V^H(n) = \text{ess} - \sup[E^n(Z_\nu); n \leqslant \nu \leqslant H]$,

$$\begin{cases} V^H(H) = Z_H \\ V^H(n) = \sup(Z_n, E^n(V^H(n+1)), \text{ for } 0 \leqslant n < H. \end{cases}$$

They allow $V^H(n)$ to be calculated for all n.

Let $\nu^H = \inf\{n; Z_n \geqslant E^n(V^H(n+1))\}$. The stopping time ν^H is optimal with horizon H, and the optimal average gain is,

$$E(Z_{\nu^H}) = \sup_{0 \leqslant \nu \leqslant H} E(Z_\nu).$$

5.1.4. Optimal Stopping with an Infinite Horizon

When H increases to $+\infty$, ν^H increases to a stopping time ν^∞ which may or may not be finite, and $V^H(n)$ increases to $V^\infty(n)$, the optimal gain conditional on the past for stopping times $\nu \leqslant H$ for arbitrary H, hence for bounded stopping times:

$$\begin{aligned} V^\infty(n) &= \text{ess} - \sup\{E^n(Z_\nu); \nu \in T_n \text{ and } \nu \text{ bounded}\} \\ &= \sup_H V^H(n) \leqslant V(n). \end{aligned}$$

Assume that $\sup_n |Z_n|$ is integrable. Then, for $\nu \in T_n$, we can apply Lebesgue's theorem, and

$$E^n(Z_\nu) = \lim_{H \to \infty} E^n(Z_{\nu \wedge H}) \leqslant V^\infty(n).$$

Hence, $V(n) \leqslant V^\infty(n)$ and $V(n) = V^\infty(n)$. Besides this,

$$\nu^\infty = \inf\{n; Z_n \geqslant E^n(V(n+1))\} = \lim \uparrow \nu^H.$$

Hence we can obtain the value for the problem of optimal

stopping with an infinite horizon as the limiting value for problems with finite horizon. For every $\varepsilon > 0$, we can thus find a bounded stopping time ν^H ε**-optimal**, such that

$$E(Z_{\nu^H}) + \varepsilon \geqslant \sup E(Z_\nu).$$

There is an optimal finite stopping time if, and only if, the stopping time ν^∞ is finite, and ν^∞ is then optimal:

$$E(Z_{\nu^\infty}) = \sup_{\nu \in T_0} E(Z_\nu).$$

Theorem 5.1.3. *Let* (Z_n) *be a sequence of* $\mathbb{F}$*-adapted* r.v.'s *majorized by an integrable* r.v., *let* T_n *be the set of* $\mathbb{F}$*-stopping times greater than n and let*

$$V(n) = \text{ess} - \sup\{E^n(Z_\nu); \nu \in T_n\}.$$

(a) $V(n) = \sup[Z_n, E^n(V(n+1))]$.

(b) *For* $1 \leqslant n \leqslant H < \infty$, *set* $V^H(H) = Z_H$, $V^H(n) = \sup[Z_n, E^n(V^H(n+1))]$ *and* $\nu^H = \inf\{n;\ Z_n = V^H(n)\}$ *then* ν^H *is an optimal stopping time with horizon H,*

$$E(Z_{\nu^H}) = \sup\{E(Z_\nu);\ \nu \in T_0,\ \nu \leqslant H\}.$$

(c) *If* $\sup|Z_n|$ *is integrable,* $V(n) = \lim \uparrow V^H(n)$ *and* $\nu^\infty = \inf\{n;\ Z_n = V(n)\} = \lim \uparrow \nu^H$. *Then if* ν^∞ *is finite, it is optimal,*

$$E(Z_{\nu^\infty}) = \sup\{E(Z_\nu):\ \nu \in T_0\}.$$

We can always find, for every $\varepsilon > 0$, *a bounded stopping time* ν_ε, *said to be* ε*-optimal, such that* $E(Z_{\nu_\varepsilon}) \geqslant \sup E(Z_\nu) - \varepsilon$.

5.1.5. The Marriage Problem

We are dealing with a problem with horizon H: choosing the best from amongst H objects.

Assume that we observe H objects in succession: they are of differing quality and we enumerate them in decreasing order 1, ..., H of their quality. This problem is often called the "marriage problem," where the aim is to choose a spouse. In fact we observe a permutation $\{\sigma(1), \ldots, \sigma(H)\}$ of $\{1, \ldots, H\}$.

The event space is Ω, the set of permutations σ of $\{1, ..., H\}$, and the probability P is uniform on Ω: $P(\sigma) = 1/H!$ for every σ. As soon as an object is rejected we can no longer choose it. The problem is to optimize the probability of stopping on object 1.

At the moment of the nth try, we observe $Y_n(\sigma)$, the **relative rank** of the nth object drawn, amongst the n objects already observed: $P(Y_n = j) = 1/n$. The r.v.'s $Y_1, ..., Y_H$ are independent since, for $i_1 = 1, i_2 \leqslant 2, ..., i_n \leqslant n$,

$$P(Y_1 = 1, Y_2 = i_2, ..., Y_n = i_n) = \frac{1}{n!} = \prod_{k=1}^{n} P(Y_{i_k} = i_k)$$

(see Vol. 1, E1.3.11).

The filtration is $\mathbb{F} = (F_k)_{1 \leqslant k \leqslant H}$, where $F_k = \sigma(Y_1, ..., Y_k)$. Let ν be a stopping time for $\mathbb{F}$:

$$P[\sigma(\nu) = 1] = \sum_{n \geqslant 1} E[\mathbf{1}_{(\nu=n, \sigma(n)=1)}]$$
$$= \sum_{n \geqslant 1} E[\mathbf{1}_{(\nu=n)} P(\sigma(n) = 1 | F_n)].$$

However,

$$Z_n = P[\sigma(n) = 1 | F_n] = \begin{cases} \dfrac{n}{H}, & \text{if } Y_n = 1 \\ 0, & \text{if } Y_n > 1 \end{cases}.$$

We thus try to maximize $E(Z_\nu) = P(\sigma(\nu) = 1)$, for ν a stopping time associated with $\mathbb{F}$. It is easily seen, by backward recurrence, that:

(a) $V(H) = \mathbf{1}_{[Y_H=1]}$;

(b) $V(n + 1)$ is independent of $(Y_1, ..., Y_n)$, hence

$$E[V(n + 1) | F_n] = E[V(n + 1)].$$

(c) $V(n) = \sup(Z_n, E[V(n + 1)])$.

Hence the optimal stopping time is $\nu = \inf\{n; Z_n \geqslant E[V(n+1)]\}$. Since Z_n is zero for $Y_n \neq 1$, we see that,

$$\nu = \left[\inf\left\{n; Y_n = 1 \text{ and } \frac{n}{H} \geqslant E[V(n + 1)]\right\}\right] \wedge H.$$

However, $E[V(n + 1)] > (n/H)$ implies

$$E[V(n)] \geqslant E[V(n+1)] > \frac{n}{H} > \frac{n-1}{H}.$$

Hence, let

$$r = \sup\left\{n;\ E[V(n)] > \frac{n-1}{H}\right\},$$

and let $\tau_r = \inf\{n;\ n > r,\ Y_n = 1\}$. Then $\nu = \inf(\tau_r, H)$ is optimal. In order to determine r, consider $u \in \{1, ..., H\}$ and $\tau_u = \inf\{n;\ n > u,\ Y_n = 1\}$; we have $P[\tau_u = u] = (1/u)$, and, for $H \geqslant k > u$,

$$P[\tau_u = k] = P[Y_n \neq 1, ..., Y_{k-1} \neq 1, Y_k = 1]$$

$$= \frac{1}{k}\,\frac{k-2}{k-1} \cdots \frac{u-1}{u} = \frac{u-1}{k(k-1)}.$$

From which it follows

$$E[Z_{\tau_u}] = \sum_{k=u}^{H} \frac{k}{H}\,\frac{1}{k}\,\frac{u-1}{k-1}$$

$$= \frac{u-1}{H}\left(\sum_{k=u}^{H} \frac{1}{k-1}\right);$$

$$E[Z_{\tau_{u+1}} - Z_{\tau_u}] = \frac{1}{H}\left[\sum_{k=u}^{H-1} \frac{1}{k} - 1\right].$$

We thus choose

$$r = \inf\left\{u;\ \frac{1}{u} + \frac{1}{u+1} + ... + \frac{1}{H-1} < 1\right\}.$$

We note that

$$\sum_{k=r}^{H-1} \frac{1}{k} \leqslant 1 < \sum_{k=r-1}^{H-1} \frac{1}{k}$$

and

$$\int_r^{H-1} \frac{dy}{y} = \operatorname{Log}(H-1) - \operatorname{Log} r \leqslant 1 \leqslant \int_{r-2}^{H-1} \frac{dy}{y}$$
$$= \operatorname{Log}(H-1) - \operatorname{Log}(r-2),$$

from which we deduce that r/H tends to $1/e$ for $H \to \infty$. Hence we choose to observe a proportion close to $1/e$ of the objects, then to take the first that we find better than the previous ones.

5.1.6. Optimal Stopping for a Markov Chain

Let $(\Omega,\mathcal{A},(P_x)_{x\in E},(X_n)_{n\geq 0})$ be a canonical Markov chain taking values in the measurable space $(E,\mathcal{E})$ with transition π, and let $\mathcal{F}_n = \sigma(X_0, ..., X_n)$. Let f be a bounded r.v. on $(E,\mathcal{E})$. We set ourselves the problem of optimal stopping with horizon H for the sequence $(f(X_n))$.

We define by recurrence the functions,

$$\begin{cases} f_0 = f \\ f_{n+1} = \sup[f, \pi(f_n)]. \end{cases}$$

It is easy to see, by backward recurrence, that $V^H(n) = f_{H-n}(X_n)$, hence that

$$V^H(0) = \sup_{0\leq\nu\leq H} E_x(f(X_\nu)) = f_H(x),$$

and

$$\nu^H = \inf\{n;\ f(X_n) = f_{H-n}(X_n)\}.$$

For the infinite horizon, we set $f^* = \lim \uparrow f_H$ and

$$f^*(x) = \sup_{0\leq\nu\leq\infty} E_x(f(X_\nu));$$

there exists an optimal stopping time if, and only if, $\nu^\infty = \inf\{n;\ f(X_n) = f^*(X_n)\}$ is finite, and ν^∞ is then optimal.

5.2. Control of Markov Chains

Should we replace a machine before it is too late? Here is how we could, by simplifying, model such a question.

A machine can be of use for at most k days. If it works one day without replacement (action C, for "continue") it may break down with probability $\theta \in]0,1[$, or continue intact to the following day with probability $1 - \theta$. Let x be the number of days during which it has been in use. The ageing of the machine is described as follows. For $x = k$, the machine is scrapped; it must be replaced, and the following day the age of the observed machine will be 1. For $x < k$, we have a choice between two actions: to replace, and what follows is similar, or to continue to use the machine, and the following day we shall observe either a broken machine (to which we attribute the age k) or a machine aged $x + 1$.

5.2.1. Controlled Markov Chains

Here is a model intended to describe Markov chains of which the evolution at each step can be controlled as a function of the past observations, called controlled Markov chains. The **state space** being $(E,\mathcal{E})$, we have an **action space** $(A,\mathcal{U})$, and we give for each $x \in E$ a set of possible actions $D(x) \in \mathcal{U}$ when the system is in state x. Moreover we are given a transition π from $(E \times A, \mathcal{E} \otimes \mathcal{U})$ into $(E,\mathcal{E})$, such that if the state of the system is x and the action undertaken is a, the state of the system at the following instant will be in Γ with probability $\pi(x,a;\Gamma)$.

Example. For the replacement problem $E = \{1,2, ..., k\}$, $A = \{C,R\}$, $D(k) = R$, and for $1 \leqslant x < k$ we have $D(x) = A$. Set

$$\pi(x,R;1) = 1 \quad \text{for all} \quad x,$$

$$\pi(x,C;k) = \theta = 1 - \pi(x,C;x + 1) \text{ for } 1 \leqslant x < k - 1,$$

$$\pi(k - 1,C;k) = 1$$

and the definition of $\pi(k,C;\cdot)$, as we shall see, will be unimportant, we can take $\pi(k,C;k) = 1$.

If we observe the state of the system each day and we decide on an action, the observation will be an element of $(\Omega,\mathcal{A}) = (E \times A, \mathcal{E} \otimes \mathcal{U})^{\mathbb{N}}$. Let $\omega \in \Omega$; set $\omega = (X_n(\omega),A_n(\omega))_{n\geqslant 0}$, (X_n,A_n) is the $(n+1)$th coordinate, X_n is the observation at time n, A_n the action taken on the same day. If $(u_n)_{n\geqslant 0}$ is a sequence, in the rest of this paragraph we denote $u^{(n)} = (u_0,u_1, ..., u_n)$. At each time n, a **strategy** is given by a measurable function δ_n from $(E,\mathcal{E})^{n+1}$ into $(A,\mathcal{U})$. Denoting $\mathbb{F} = (\mathcal{F}_n)$, with $\mathcal{F}_n = \sigma(X^{(n)})$, this implies that $(d_n(X^{(n)}))$ is an $\mathbb{F}$-adapted process taking values in $(A,\mathcal{U})$. We shall require that these strategies be **coherent**, i.e. only giving allowed actions: $\delta_n(x_0,x_1, ..., x_n)$ takes its values in $D(x_n)$. Let $\mathcal{D}$ be the set of coherent strategies. Given $x \in E$ and a strategy $\delta = (\delta_n)$, we define a probability $P_{\delta,x}$ on $(\Omega,\mathcal{A})$ by

$$P_{\delta,x}(X_0 = x) = 1$$

$$P_{\delta,x}(X_{n+1} \in \cdot, A_n = \delta_n(X^{(n)}) \mid F_n) = \pi(X_n, \delta_n(X^{(n)}); \cdot).$$

The existence of $P_{\delta,x}$ follows from Ionescu Tulcea's theorem ([4.1.1]) as does the existence of canonical Markov chains. The probability $P_{\delta,x}$ describes the evolution of a Markov chain with initial state x, controlled by δ. On $\sigma(X^{(n+1)}, A^{(n)})$, the probability $P_{\delta,x}$ is determined by $(\delta_0, ..., \delta_n)$.

We have a kind of Markov property. We denote by δ^x the translated strategy δ, $\delta^x = (\delta_p^x)_{p \geq 0}$ with $\delta_p^x(X_0, ..., X_p) = \delta_{p+1}(x, X_0, ..., X_p)$. Then, by definition, we see that for g a bounded r.v. on $E \times A$ and $n \geq 1$,

$$E_{\delta,x}[g(X_n, A_n)] = \int \pi(x, \delta_0(x); dy) E_{\delta^x, y}[g(X_{n-1}, A_{n-1})].$$

What criterion should be used to choose δ? We are certainly dealing with optimizing a gain. We shall deal with two criteria, the maximum gain with horizon H or the average maximum gain.

5.2.2. Optimal Control with Finite Horizon

Let us assume that at any time $n < H$, the action a taken in state x results in a gain $g(x,a)$. At time H, we sell off the system for the price $\delta(x)$ if it is in state x. The total gain is thus,

$$G = \sum_{n=0}^{H-1} g(X_n, A_n) + \gamma(X_H).$$

The **optimal average gain** (called in control theory the **value** of the problem with horizon H starting from x) is

$$W^H(x) = \sup\{E_{\delta,x}[G]; \delta \in \mathcal{D}\}$$

and a strategy δ^H is **optimal with horizon** H if, for every x,

$$W^H(x) = E_{\delta^H, x}[G].$$

The search for W^H and δ^H rests on **Bellman's principle** which follows, analogous to the idea used for optimal stopping. For $1 \leq n < H$, the average gain with horizon H is the average gain with horizon n replacing the selling price γ by W^{H-n}, the optimal average gain hoped for in the following $H - n$ time points. This principle is very general. Let us show how it can be applied here. Set $\ell_0(x) = \gamma(x)$,

$$\ell_{n+1}(x) = \sup_{a \in D(x)} \{g(x,a) + \int \pi(x,a;dy)\ell_n(y)\}.$$

Let us assume that for all n there exists a measurable function d_n from $(E,\mathcal{E})$ into $(A,\mathcal{U})$ such that $d_n(x)$ is in $D(x)$ for any x, and that,

$$\ell_{n+1}(x) = g(x,d_n(x)) + \int \pi(x,d_n(x);\, dy)\ell_n(y).$$

In particular, this is the case if $D(x)$ is finite for arbitrary x.

Let us now show that $d^H = (d_H(X_0),d_{H-1}(x_1), \ldots, d_1(X_{H-1}))$ is an optimal strategy with horizon H, and that $W^H = \ell_H$. This is true for $H = 0$, since $W^0 = \ell_0 = \gamma$ and every strategy is optimal. Assume the result is true for H and let us show it for $H + 1$. For every strategy δ, we have,

$$\begin{aligned} E_{x,\delta}(G) &= g(x,\delta_0(x)) + E_{x,\delta}\left[\sum_{n=1}^{H-1} g(X_n,A_n) + \gamma(X_H)\right] \\ &= g(x,\delta_0(x)) \\ &\quad + \int \pi(x,\delta_0(x);dy)E_{y,\delta^x}\left[\sum_{n=0}^{H-2} g(X_n,A_n)+\gamma(X_{H-1})\right] \\ &\leqslant g(x,\delta_0(x)) + \int \pi(x,\delta_0(x);\, dy)\,\ell_{H-1}(y) \\ &\leqslant \sup\{g(x,a) + \int \pi(x,a;dy)\ell_{H-1}(y);\ a \in D(x)\} \\ &= \ell_H(x). \end{aligned}$$

However, taking $\delta = d^H$ we see that equality holds; hence d^H is optimal.

Theorem 5.2.4. *Consider a controlled Markov chain with state space* $(E,\mathcal{E})$, *action space* $(A,\mathcal{U})$, *and with transition* π, *the permissible actions in state* x *being the elements of* $D(x) \subset A$. *Let* g *be a bounded* r.v. *on* $(E \times A,\ \mathcal{E} \times \mathcal{U})$ *and let* γ *be a bounded* r.v. *on* $(E,\mathcal{E})$. *For a gain with horizon* H

$$G = \sum_{n=0}^{H-1} g(X_n,A_n) + \gamma(X_H).$$

The optimal average gain $W^H(x)$ if the initial state is x may be calculated by recurrence from the relations

$$W^0(x) = \gamma(x)$$

$$W^{n+1}(x) = \sup\{g(x,a) + \int \pi(x,a;dy)W^n(x);\ a \in D(x)\}.$$

For each $n \leq H$, if there exists an r.v. d_n from $(E,\mathcal{E})$ *into* $(A,\mathcal{U})$ *such that $d_n(x)$ is in $D(x)$ for all x and such that,*

$$W^{n+1}(x) = g(x,d_n(x)) + \int \pi(x,d_n(x);dy)W^n(y),$$

then the strategy $d^H = (d_H(X_0),d_{H-1}(X_1), ..., d_1(X_{H-1}))$ is optimal with horizon H. In other words for any x,

$$E_{d^H,x}(G) = W^H(x).$$

Note. The strategy d^H is said to be **Markovian** since each action is determined by the state of the system at this time: it is useless to memorize the entire past.

5.2.3. Optimal Control on Average

To avoid the technical details, let us assume that E and A are finite. We look for an **average optimal gain** μ such that, for every x and every δ

$$\overline{\lim_{N\to\infty}} \frac{1}{N} \sum_{k=1}^{N-1} g(X_k,A_k) \leq \mu \qquad P_{\delta,x}\text{-a.s.}$$

and that, for an **optimal strategy on average** δ^0, we have, for any x,

$$\lim_{N\to\infty} \frac{1}{N} \sum_{k=1}^{N-1} g(X_k,A_k) = \mu \quad P_{\delta,x}\text{-a.s.}$$

We limit ourselves first of all to the case of the simplest strategies, said to be **stationary**, of the form $\delta = (d(X_n))_{n\geq 0}$, where d is a measurable function from $(E,\mathcal{E})$ into $(A,\mathcal{U})$, with $d(x) \in D(x)$ for every x. We shall also denote by d the strategy δ. Let S be the set of stationary strategies. For $d \in S$, the controlled chain is a stationary Markov chain with transition

$$(x,y) \longmapsto \pi(x,d(x);y) = \pi^d(x,y).$$

The following hypothesis is made: for every $d \in S$, the chain is recurrent aperiodic.

Example. The above hypothesis holds if there exists a state y and a $k \in \mathbb{N}$ such that $\inf\{\pi_x(x,a;y);\ x \in E,\ a \in A\} > 0$. This is the case for the replacement problem set in [5.2.1] with $y = k$.

Now let μ^d be a probability invariant under π^d, and let π_n^d be the nth iterate of π^d. Let us denote by g_d the function $x \longmapsto g(x,d(x))$. In the class of stationary strategies S, the average gain to be optimised is $\mu^d(g_d)$, the $P_{d,x}$-a.s. limit of $(1/N)\Sigma_{n=0}^{N-1}g(X_n,A_n)$ for $N \to \infty$, for any x. Set

$$\ell_d = \sum_{n \geqslant 0} (\pi_n^d g_d - \mu^d(g_d))$$

(this series converges from [4.3.5]). We have

$$\pi^d \ell_d = \sum_{n \geqslant 0} (\pi_{n+1}^d g_d - \mu^d(g_d)) = \ell_d - g_d + \mu^d(g_d).$$

The function ℓ_d is the solution of Poisson's equation,

$$\mu^d(g_d) = \pi^d \ell_d - \ell_d + g_d.$$

Let us show that we can find a strategy d_0 optimal on average in the class S. At the same time the proof will give an improved algorithm for stationary strategies which allows d_0 to be obtained. Let $d \in S$, construct d_1 such that,

$$g(x,d_1(x)) + \sum_y \pi(x,d_1(x);y)\ell_d(y)$$
$$= \sup_a [g(x,a) + \sum_y \pi(x,a;y)\ell_d(y)];$$

$$g_{d_1} + \pi^{d_1}\ell_d \geqslant g_d + \pi^d \ell_d = \mu^d(g_d) + \ell_d;$$

$$\mu^{d_1}(g_{d_1}) + \mu^{d_1}(\ell_d) \geqslant \mu^d(g_d) + \mu^{d_1}(\ell_d).$$

The strategy d_1 is thus better than d, strictly so unless,

$$g_d(x) + \pi^d \ell_d(x) = \sup_{a \in D(x)} \left[g(x,a) + \sum_y \pi(x,a;y)\ell_d(y) \right].$$

S is, however, finite and the algorithm will stop after a finite number of steps. Thus there exists a strategy d_0 such that,

$$\begin{aligned} g_{d_0} + \pi^{d_0}\ell_{d_0} &= \mu^{d_0}(g_{d_0}) + \ell_{d_0} \\ &= \sup_a \left[g(\cdot,a) + \sum_y \pi(\cdot,a;y)\ell_{d_0}(y) \right] \end{aligned}$$

and this strategy is optimal on average in the class S.

Let us show that d_0 is optimal on average amongst all the strategies of $\mathcal{D}$. Let $\ell_{d_0} = \ell$ and $k = \mu^{d^0}(g_{d_0})$,

$$\Phi(x,a) = g(x,a) - \ell(x) - k + \sum_y \pi(x,a;y)\ell(y).$$

We note that $\Phi(x,d_0(x)) = 0$ and $\Phi(x,a) \leqslant 0$. Consider the sequence,

$$Y_n = g(X_n,A_n) + \ell(X_{n+1}) - k - \ell(X_n) - \Phi(X_n,A_n).$$

Let $\delta = (\delta_n(X^{(n)}))$ be a strategy of $\mathcal{D}$,

$$\begin{aligned} E_{\delta,x}[Y_n | \mathcal{F}\,] &= g(X_n,\delta_n(X^{(n)})) - k - \ell(X_n) \\ &\quad - \Phi(X_n,A_n) + \sum_y \pi\,(X_n,\delta_n(X^{(n)});y)\ell(y) = 0. \end{aligned}$$

The r.v.'s Y_n are $\mathcal{F}_{n+1}$-measurable, there exists a constant λ majorizing $|Y_n|$ for all n. The sequence $M_n = (\Sigma_{p=1}^n Y_p)$ is thus a square integrable centered martingale, adapted to $(\mathcal{F}_{n+1})$, and its increasing process is majorized by $n\lambda^2$. From the law of large numbers of Corollary 2.6.30,

$$\frac{1}{N} \sum_{n=0}^{N-1} Y_n$$

tends $P_{\delta,x}$-a.s. to 0. In other words,

$$\frac{1}{N} \sum_{n=0}^{N-1} g(X_n,A_n) - k - \frac{1}{N} \sum_{n=0}^{N-1} \Phi(X_n,A_n)$$

tends $P_{\delta,x}$-a.s to 0. Since Φ is integrable, we obtain $P_{\delta,x}$-a.s,

$$\overline{\lim} \frac{1}{N} \sum_{n=0}^{N-1} g(X_n,A_n) \leqslant k$$

and d_0 is optimal on average.

A Study of the Case of a Controlled Markov Chain Dependent on a Parameter θ. Recall the replacement example mentioned above, the probability of breakdown being perhaps unknown. Assume given $\Theta \times E \times A$ and a transition π from $\Theta \times E \times A$ into E, such that $\inf\{\pi(\theta,x;y);\ \theta \in \Theta,\ x \in E\} > 0$ for a given state y. To every strategy $\delta \in \mathcal{D}$, every initial state x and every $\theta \in \Theta$ corresponds a probability $P_{\theta,\delta,x}$ on Ω. For each θ, we can find a stationary strategy d_θ optimal on average, with average gain $k(\theta)$. However θ is unknown, and we look for a strategy optimal on average, independent of θ.

Let $(\hat{\theta}_n)$ be a sequence of estimators of θ; $\hat{\theta}_n$ is a function of $X^{(n)}$ and $A^{(n-1)}$. We assume that it is **consistent**, i.e. for every θ, δ, x, $(\hat{\theta}_n)$ converges to θ, $P_{\theta,\delta,x}$-a.s. Then the strategy $\delta_0 = (d_{\hat{\theta}_n}(X_n))$ is optimal on average. In fact, for any $\delta \in \mathcal{D}$,

$\hat{\theta}_n$ coincides with θ from a certain point onwards ($P_{\theta,\delta,x}$-a.s.). Thus

$$\frac{1}{N} \sum_{n=0}^{N-1} \Phi(X_n,A_n)$$

tends to θ, $P_{\theta,\delta,x}$-a.s., and δ_0 is optimal.

How Is a Consistent Sequence of Estimators to Be Found? Let

$$L_n(\theta) = \prod_{p=1}^{n} \pi(\theta,X_{p-1},A_{p-1};X_p).$$

For every $x \in E$, $\delta \in \mathcal{D}$, $\theta \in \Theta$,

$$P_{\theta,\delta,x}[X_0 = x, \dots, X_n = x_n, A_0 = a_0, \dots, A_{N-1} = a_{N-1}]$$
$$= \prod_{p=1}^{n} \pi(\theta,x_{p-1},a_{p-1};x_p)$$

denoting $x = x_0$. Thus, on $(\Omega,\sigma(X^{(n)},A^{(n-1)}))$, $L_n(\theta)$ is a likelihood. We can study the maximum likelihood estimators. Let us assume that, for $\theta \neq \phi$ and every pair (x,a), there exists a y such that $\pi(\theta,x,a;y) \neq \pi(\phi,x,a;y)$. Since the logarithm function is strictly concave,

$$\sum_y \left[\text{Log}\,\frac{\pi(\phi,x,a;y)}{\pi(\theta,x,a;y)}\right]\pi(\theta,x,a;y)$$

$$< \text{Log}\,\sum \frac{\pi(\phi,x,a;y)}{\pi(\theta,x,a;y)}\,\pi(\theta,x,a;y) < 0$$

(summing over the y's such that $\pi(\theta,x,a;y) > 0$). Set $f(\theta,x,a,y) = -\text{Log}\,\pi(\theta,x,a;y)$. For $\phi \neq \theta$ and $(x,a) \in E \times A$,

$$\sum_y [f(\phi,x,a,y) - f(\theta,x,a,y)]\pi(\theta,x,a;y) > 0.$$

A function f on $\Theta \times E \times A \times E$ which satisfies this relation for all $\phi \neq \theta$ and every $(x,a) \in E \times A$ is a **contrast function** ([3.2.2]). A **minimum contrast estimator** $\hat{\theta}_n$ is defined by,

$$\sum_{p=0}^{n-1} f(\hat{\theta}_n(x_0, \dots, x_n, a_0, \dots, a_{n-1}), x_p, a_p, x_{p+1})$$

$$= \inf\left\{\sum_{p=0}^{n-1} f(\theta, x_p, a_p, x_{p+1});\ \theta \in \Theta\right\}.$$

Let us show that the sequence $(\hat{\theta}_n)$ is consistent. Let θ be the true value of the parameter and ϕ another value. Set

$$h(\phi,\theta,x,a) = \sum_y [f(\phi,x,a,y) - f(\theta,x,a,y)]\pi(\theta,x,a,y),$$

$$Y_n = f(\phi,X_n,A_n,X_{n+1}) - f(\theta,X_n,A_n,X_{n+1}) - h(\phi,\theta,X_n,A_n);$$

$$E_{\theta,\delta,x}[Y_n | X^{(n)}] = \sum_a 1_{\{a=\delta_n(X^{(n)})\}}$$

$$\times \left[\sum_y \pi(\theta,X^{(n)},a;y)[f(\phi,X^{(n)},a,y)\right.$$

$$\left. - f(\theta,X^{(n)},a,y)] - h(\phi,\theta,X^{(n)},a)\right].$$

Corollary 2.6.30 entails that, $P_{\theta,\delta,x}$-a.s.,

$$\frac{1}{N}(Y_0 + \dots + Y_{N-1})$$

$$= \frac{1}{N}\sum_{n=0}^{N-1} [f(\phi,X_n,A_n,X_{n+1}) - f(\theta,X_n,A_n,X_{n+1})]$$

$$- \frac{1}{N}\sum_{n=0}^{N-1} h(\phi,\theta,X_n,A_n)$$

tends to 0 if N tends to infinity.

However h is negative, thus majorized by an $\alpha < 0$. Thus $P_{\theta,\delta,x}$-a.s. for N large enough,

$$\frac{1}{N} \sum_{n=0}^{N-1} [f(\phi,X_n,A_n,X_{n+1}) - f(\theta,X_n,A_n,X_{n+1})]$$

is less than $\alpha/2$ and ϕ cannot be the minimum contrast estimator. We have taken Θ to be finite and, for n large enough, $\hat{\theta}_n = \theta$.

5.3. Sequential Statistics

Quality control (which is not really a control problem in the sense of [5.2]) is the inspection of a stock of objects produced by a machine. If a sample of n objects is observed from this stock, a natural decision is to declare the stock defective if the random number of defective objects sampled is greater than a certain number k, and to accept it otherwise. However if we are given such a rule, and if the n sample objects are sampled one by one, it is clear that we shall stop the inspection as soon as we have sampled k defective objects if this happens before the inspection of the nth object. The size of the observed sample will then be random.

This is an example of sequential statistics. Classical statistics is the tool of the statistician called on to analyze the results of an enquiry or of an experiment dealing with a given sample. Sequential statistics is that of a statistician who can choose the size of the sample, taking into account the cost of the experiments and the risk of errors. He certainly should choose to stop or to carry on the experiments in the light of previous results. Thus there are two types of decisions to be taken: when to stop the experiment? What decision to take for the statistical problem set (test, estimation)?

5.3.1. The Ideas of Sequential Statistics

Consider a statistical model $(\Omega,\mathcal{A},(P_\theta)_{\theta\in\Theta})$ dominated by P along the filtration $\mathbb{F} = (\mathcal{F}_n)_{n\in\mathbb{N}}$, in the sense of Definition 3.1.1. Θ is a measurable set and in general we shall omit reference to its σ-algebra $\mathcal{C}$. Assume that $(\theta,A) \longmapsto P_\theta(A)$ is a transition from $(\Theta,\mathcal{C})$ into (Ω,A). Let $(L_n(\theta))$ be a likelihood. If ν is a finite stopping time, we have (Proposition 3.2.5),

$$P_\theta = L_\nu(\theta)P \quad \text{on} \quad (\Omega,\mathcal{F}_\nu).$$

Let us now consider a statistical decision problem in the sense defined in [Vol. I, 8.1]. We are given a measurable set $(A,\mathcal{U})$ of actions and a cost C, an r.v. defined on $\Theta \times A$: $C(\theta,a)$ is the cost of action a if the parameter equals θ. At time ν, an action can only be chosen in the light of the past. A strategy is a measurable function d_ν from $(\Omega,\mathcal{F}_\nu)$ into $(A,\mathcal{U})$ and its average risk is, if $C(\theta,d_\nu)$ is P_θ-integrable

$$R(\theta,d_\nu) = E_\theta[C(\theta,d_\nu)].$$

We have here the double choice of ν and of d_ν.

5.3.2. Bayesian Sequential Statistics

Recall the Bayesian framework ([Vol. I, 8.2]) where Θ is equipped with a **prior** distribution α. Then the average cost of the action a is $\int\alpha(d\theta)C(\theta,a)$, assuming that these integrals exist. We call the **Bayesian risk** relative to α its minimum value

$$\rho(\alpha) = \inf\{\int\alpha(d\theta)C(\theta,a);\ a \in A\},$$

and if this lower bound is attained by $a = \delta(\alpha)$ we call $\delta(\alpha)$ the **Bayesian decision** relative to α: $\rho(\alpha) = C(\theta,\delta(\alpha))$. We then place ourselves on $\{\Theta \times \Omega, \mathcal{C} \times \mathcal{A}, Q\}$ with $Q(d\theta) = \alpha(d\theta) \times P_\theta$,

$$Q(C \times B) = \int\alpha(d\theta)1_C(\theta)P_\theta(B),$$

for $C \in \mathcal{C}$ and $B \in \mathcal{A}$. We again denote by $\mathcal{F}_\nu$ the σ-algebra $\{\Theta,\Phi\} \otimes \mathcal{F}_\nu$. Then the distribution of θ conditional on the observations prior to ν, i.e., conditional on $\mathcal{F}_\nu$ is called its **posterior distribution** at time ν and is denoted by α_ν. We have ([Vol. I, 8.2.6]),

$$\alpha_\nu = \frac{L_\nu(\cdot)}{\int L_\nu(u)\alpha(du)}\alpha.$$

Note that, on $\{\nu = n\}$, $\alpha_\nu = \alpha_n$. For every distribution α on Θ if there is a Bayesian decision $\delta(\alpha)$, and if we stop at time ν, the decision $\delta(\alpha_\nu)$ is Bayesian, i.e. it gives the minimum average risk amongst F_ν measurable decisions,

$$\int \alpha(d\theta)R(\theta,\delta(\alpha_\nu))$$
$$= \inf\{\int \alpha(d\theta)R(\theta,d_\nu);\ d_\nu\ \ F_\nu\text{-measurable}\}.$$

Thus in the Bayesian framework, we know the best decision to take when ν is given. Before seeing how ν is chosen, let us recall the two principle examples ([Vol. I, 8.2.5]).

Examples. (a) **A Test of Two Hypotheses "$\theta \in \Theta_0$" Against "$\theta \in \Theta_1$" for a Partion $\Theta = \Theta_0 \cup \Theta_1$ of Θ.** The set of actions is $\{0,1\}$ and we set, for $c_{10} > 0$ and $c_{01} > 0$,

$$C(\theta,0) = \begin{cases} 0 & \text{if } \theta \in \Theta_0 \\ c_{10} & \text{if } \theta \in \Theta_1 \end{cases},$$

$$C(\theta,1) = \begin{cases} 0 & \text{if } \theta \in \Theta_1 \\ c_{01} & \text{if } \theta \in \Theta_0 \end{cases}.$$

Let $\beta = \alpha(\Theta_1) = 1 - \alpha(\Theta_0)$. Then

$$\int d\alpha(\theta)C(\theta,0) = \beta c_{10},\quad \int d\alpha(\theta)C(\theta,1) = (1-\beta)c_{01};$$

$$\rho(\alpha) = \inf[(1-\beta)c_{01},\ \beta c_{10}];$$

$$d(\alpha) = \begin{cases} 0 & \text{if } \beta < c_{01}/(c_{01}+c_{10}) \\ 1 & \text{if } \beta > c_{01}/(c_{01}+c_{10}) \end{cases}$$

(either if equality)

Let

$$\beta_\nu = \alpha_\nu(\Theta_1) = \frac{\int_{\Theta_1} L_\nu(\theta)d\alpha(\theta)}{\int_{\Theta} L_\nu(\theta)d\alpha(\theta)}$$

the Bayesian decision d_ν is a **likelihood ratio test**

$$d_\nu = 1_{\{\lambda_\nu \geq (c_{01}/c_{10})\}} \quad \text{with} \quad \lambda_\nu = \frac{\int_{\Theta_1} L_\nu(\theta)d\alpha(\theta)}{\int_{\Theta_0} L_\nu(\theta)d\alpha(\theta)}.$$

(b) **Estimation.** Θ being an interval of $\mathbb{R}$ and $A = \Theta$, we take: $c(\theta,a) = [\theta - a]^2$. Then $\delta(\alpha)$ is the mean $\int \theta \, d\alpha(\theta)$ of α and $\rho(\alpha)$ its variance. Hence, after an observation of length ν, the Bayesian estimator of θ is $\hat{\theta}_\nu$, the mean of α_ν, and its Bayesian risk is the variance of α_ν. For an n-sample from $N(\theta,\sigma^2)$ and for a prior distribution $N(m,\tau^2)$, we obtain (Vol. 1, 8.2.2), assuming σ^2, τ^2, m known, the posterior distribution

$$N\left[\frac{\tau^2 \sum_{i=1}^{n} x_i + \sigma^2 m}{n\tau^2 + \sigma^2}, \frac{\sigma^2\tau^2}{n\tau^2 + \sigma^2}\right].$$

In order to estimate θ after an observation of length ν, we shall take,

$$\hat{\theta}_\nu = \frac{\tau^2 \sum_{i=1}^{\nu} x_i + \sigma^2 m}{\nu\tau^2 + \sigma^2}$$

and the Bayesian risk is

$$\int d\alpha(\theta) E_\theta\left[\frac{\sigma^2\tau^2}{\nu\tau^2 + \sigma^2}\right].$$

Optimal Choice of the Length of an Experiment. We are given an increasing sequence (γ_n) adapted to $\mathbb{F}$ and we assume that γ_n is the cost of the experiment up to time n ($\gamma_0 = 0$). If we stop at time ν by deciding d_ν, the total cost will be $\gamma_\nu + C(\theta,d_\nu)$. We know its minimum value, obtained by taking $d_\nu = \delta(\alpha_\nu)$. We then choose for ν the stopping time which minimizes $E_\alpha[\gamma_\nu + \rho(\alpha_\nu)]$. We are therefore dealing with an optimal stopping problem for the sequence $Z = (Z_n)$, where $Z_n = -\gamma_n - \rho(\alpha_n)$. If we restrict ourselves to the problem with horizon H of experiments of length $\leq H$, we know how to solve this problem by backward recurrence.

For a problem of the test of two hypotheses or of estimation, let us look at optimal stopping $\leq H$, when the experiments form a Markov chain $\{\Omega, A, (P_{\theta,x})_{x\in E}, X_n\}$ taking values in $(E,\mathcal{E})$ with transition $\pi(\theta,x;dy) = f(\theta,x,y)\pi(x;dy)$ with a given initial distribution. We denote $\pi(\alpha,\cdot;\cdot) = \int\alpha(d\theta)\pi(\theta,\cdot;\cdot)$. For initial state x and parameter θ, the probability on Ω is $P_{\theta,x}$,

$$L_n(\theta) = \prod_{1\leqslant k\leqslant n} f(\theta,X_{k-1},X_k).$$

Set

$$\hat{\alpha}_{x_0,\ldots,x_n}(d\theta) = \frac{\prod_{1\leqslant k\leqslant n} f(\theta,x_{k-1},x_k)\alpha(d\theta)}{\int \prod_{1\leqslant k\leqslant n} f(u,x_{k-1},x_k)\alpha(du)};$$

$$\hat{\alpha}_{x_0,\ldots,x_n} = T_{x_{n-1},x_n}(\hat{\alpha}_{x_0,\ldots,x_{n-1}}),$$

with

$$T_{x,y}(\alpha)(d\theta) = \frac{f(\theta,x,y)\alpha(d\theta)}{\int f(u,x,y)\alpha(du)}.$$

Then $\alpha_\nu = \hat{\alpha}_{X_0,\ldots,X_\nu}$ for each finite stopping time ν.

Let us assume **Markovian costs**, i.e. an observation equal to x costs $\Gamma(x)$. Γ is a positive r.v. on $(E,\mathcal{E})$ and $\gamma_n = \Sigma_{k=0}^n \Gamma(X_k)$. Let

$$V^H(n) = \operatorname*{ess\,inf}_{n\leqslant\nu\leqslant H} E_{\alpha,x}(\gamma_\nu + \rho(\alpha_\nu)|\, \mathcal{F}_n),$$

for initial state x;

$$V^H(H) - \gamma_H = \rho(\hat{\alpha}_{X_0,\ldots,X_H})$$

$$V^H(H-1) - \gamma_{H-1} = \inf[\rho(\hat{\alpha}_{X_0,\ldots,X_{H-1}}),$$

$$E_{\alpha,x}(\Gamma(X_H) + \rho(\hat{\alpha}_{X_0,\ldots,X_H})|\, \mathcal{F}_{H-1})].$$

However if ϕ is a positive r.v. on E^{n+1}, we see

$$E_{\alpha,x}[\phi(X_0, \ldots, X_n)|\, \mathcal{F}_{n-1}]$$

$$= \int\hat{\alpha}_{X_0,\ldots,X_{n-1}}(d\theta)\int\pi(\theta,X_{n-1};dy)\phi(X_0, \ldots, X_{n-1},y),$$

an expression denoted by

$$\int \pi(\hat{\alpha}_{X_0,\dots,X_{n-1}}, X_{n-1}; dy)\phi(X_0, \dots, X_{n-1}, y).$$

From which it follows that,

$$V^H(H-1) - \gamma_{H-1} = \inf\Big\{\rho(\hat{\alpha}_{X_0,\dots,X_{H-2}}),$$

$$\int \pi(\hat{\alpha}_{X_0,\dots,X_{H-1}}, X_{H-1}; dy)[\Gamma(y)$$

$$+ \rho(T_{X_{H-1},y}(\hat{\alpha}_{X_0,\dots,X_{H-1}}))]\Big\}.$$

We therefore define by recurrence, $c_0(\alpha,x) = \rho(\alpha)$ and, for $n \geqslant 1$,

$$c_n(\alpha,x) = \inf\left[\rho(\alpha), \int \pi(\alpha,x;dy)(\Gamma(y) + c_{n-1}(T_{x,y}(\alpha),y))\right].$$

Then $V^H(n) - \gamma_n = c_{H-n}(\hat{\alpha}_{X_1,\dots,X_n}, X_n)$, and the optimal stopping is

$$\nu^H = \inf\{n;\ \rho(\hat{\alpha}_{X_1,\dots,X_n}) = c_{H-n}(\hat{\alpha}_{X_1,\dots,X_n}, X_n)\}.$$

We have $c_H(\alpha,x) = \inf\{E_{\alpha,x}(\gamma_\nu + \rho(\alpha_\nu));\ 0 \leqslant \nu \leqslant H\}$. Thus,

$$\begin{aligned} c(\alpha,x) &= \lim_{H\to\infty} \downarrow c_H(\alpha,x) \\ &= \inf\{E_{\alpha,x}(\gamma_\nu + \rho(\alpha_\nu));\ \nu \text{ bounded}\}. \end{aligned}$$

Assume $\rho(\alpha_n)$ is bounded; if ν is a finite stopping time and if H increases to ∞, then $E_{\alpha,x}(\gamma_{\nu\wedge H})$ increases to $E_{\alpha,x}(\gamma_\nu)$, and $E_{\alpha,x}(\rho_{\alpha_{\nu\wedge H}})$ tends to $E_{\alpha,x}(\rho(\alpha_\nu))$. Then

$$c(\alpha,x) = \inf\{E_{\alpha,x}(\gamma_\nu + \rho(\alpha_\nu));\ \nu \text{ finite}\}.$$

When $\nu = \lim \uparrow \nu^H$ is finite; ν is optimal.

Examples. (a) *Sequential estimation of the mean of a Gaussian distribution.* An n-sample from $N(\theta,\sigma^2)$ is observed with prior distribution $N(m,\tau^2)$ (σ^2, τ^2, m are known). Then the Bayesian risk at time n is $\sigma^2\tau^2/(n\tau^2+\sigma^2) = \rho_n$. For the problem with horizon H, by taking the cost Γ to be > 0 we calculate

$$V(H) = H\Gamma + \rho_H$$

$$\beta_n = V(n) - n\Gamma = \inf[\rho_n, \Gamma + \rho_{n+1}].$$

The sequence $(\rho_n - \rho_{n+1})$ decreases to 0. Let $n_0 = \inf\{n; \rho_n - \rho_{n+1} \leqslant \Gamma\}$. For $H > n_0$, we have

$$V(n) = n\Gamma + \rho_n \quad \text{if} \quad H \geqslant n \geqslant n_0;$$

$$V(n) = n_0\Gamma + \rho_{n_0} \quad \text{if} \quad n \leqslant n_0 ;$$

and $\nu^H = n_0$. The optimal length of experiments for the problem with horizon H is the fixed time n_0. Hence for the infinite horizon also, we obtain n_0. For this sequential estimation problem the optimal length of experiment is n_0.

(b) *Test of two simple hypotheses* $\{\theta_0\}$ *and* $\{\theta_1\}$. Let us denote here $\alpha = \alpha\{\theta_1\} = 1 - \alpha\{\theta_0\}$ (the number α characterizes the measure α and we confuse the notations). We have,

$$\begin{aligned} c(\alpha,x) &= \inf\ [\ (\alpha), \inf_{\nu\geqslant 1} E_{\alpha,x}[\gamma_\nu + \rho(\alpha_\nu)] \\ &= \inf(\rho(\alpha), u_1(\alpha,x))\ ] \end{aligned}$$

with

$$\begin{aligned} u_1(\alpha,x) = \inf_{\nu\geqslant 1} [(1-\alpha)E_{\theta_0,x}(\gamma_\nu + c(\theta_0,d_\nu)) \\ + \alpha E_{\theta_1,x}(\gamma_\nu + c(\theta_1,d_\nu))]. \end{aligned}$$

The function $\alpha \longmapsto u_1(\alpha,x)$ is concave and equals $\Gamma(x)$ for $\alpha = 0$ and $\alpha = 1$. Thus, either $\rho(\alpha) \leqslant u_1(\alpha,x)$ for all α, or else there exist two numbers

$$A_1(x) < \frac{c_{01}}{c_{01} + c_{10}} < B_1(x)$$

such that $\rho(\alpha)$ is greater than $u_1(\alpha,x)$ for $A_1(x) < \alpha < B_1(x)$.

The maximum of $\rho(\alpha)$ is

$$\frac{c_{01}c_{10}}{c_{01} + c_{10}};$$

for $\Gamma(x)$ greater than this value, it is better not to experiment, and take $\nu = 0$.

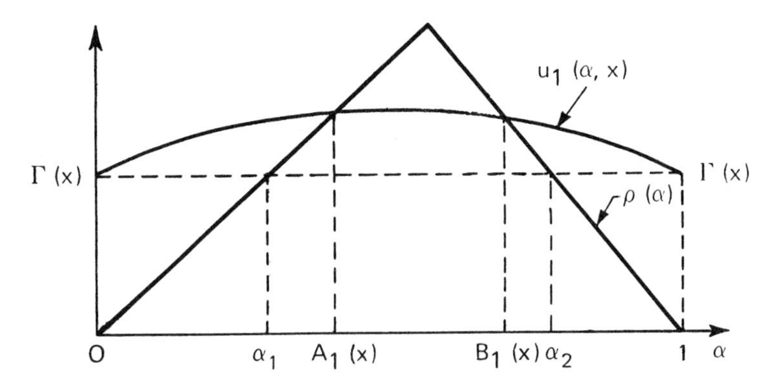

Let us assume $\Gamma(x)$ to be less than this value; $\rho(\alpha)$ equals $\Gamma(x)$ for $\alpha_1 = \Gamma(x)/c_{10}$ and for $\alpha_2 = 1 - (\Gamma(x)/c_{01})$. Thus

$$\frac{\Gamma(x)}{c_{10}} < A_1(x) < B_1(x) < 1 - \frac{\Gamma(x)}{c_{01}}.$$

For every x, we can thus find

$$A(x) = \frac{A_1(x)}{1 - A_1(x)} \quad \text{and} \quad B(x) = \frac{B_1(x)}{1 - B_1(x)}$$

such that

$$\frac{\Gamma(x)}{c_{10} - \Gamma(x)} \leqslant A(x) \leqslant \frac{c_{01}}{c_{10}} \leqslant B(x) \leqslant \frac{c_{01} - \Gamma(x)}{\Gamma(x)}.$$

and

$$\rho(\alpha) > u_1(\alpha,x) \Leftrightarrow A(x) < \frac{\alpha}{1-\alpha} < B(x).$$

The Bayesian test is then defined by

$$\nu = \inf\left\{n; \frac{L_n(\theta_1)}{L_n(\theta_0)} \notin \left]A(X_n),B(X_n)\right[\right\},$$

$$d_\nu = 1 \quad \text{for} \quad \frac{L_\nu(\theta_1)}{L_n(\theta_0)} \geqslant B(X_\nu),$$

$$d_\nu = 0 \quad \text{for} \quad \frac{L_\nu(\theta_1)}{L_\nu(\theta_0)} \leqslant A(X_\nu),$$

as long as the stopping time is finite. This is a sequential likelihood ratio test with "boundaries," A and B are functions of the state of the chain at time ν. Let us assume for example that, for θ_0 and for θ_1, the Markov chain has a positive recurrent point to which all other points lead. Let μ_{θ_0} and μ_{θ_1} be the invariant probabilities. Then, for all x,

$$\frac{1}{n} \operatorname{Log} \frac{L_n(\theta_1)}{L_n(\theta_0)} = \frac{1}{n} \sum_{k=1}^{n} \operatorname{Log} \frac{f(\theta_1, X_{k-1}, X_k)}{f(\theta_0, X_{k-1}, X_k)}$$

tends $P_{\theta_0,x}$-a.s., if $n \to \infty$, to

$$\int d\mu_{\theta_0}(x) \int \pi(\theta_0, x; dy) \operatorname{Log} \frac{f(\theta_1, x, y)}{f(\theta_0, x, y)} .$$

This limit is

$$\int d\mu_{\theta_0}(x) K[\pi(\theta_0, x; \cdot), \pi(\theta_1, x; \cdot)]$$

if K signifies Kullback information. Therefore if μ_{θ_0} and μ_{θ_1} charge

$$\{x; \pi(\theta_0, x; \cdot) \neq \pi(\theta_1, x; \cdot)\},$$

then $(\operatorname{Log}(L_n(\theta_1)/L_n(\theta_0))$ tends $P_{\theta_0,x}$-a.s. to $-\infty$, and $P_{\theta_1,x}$-a.s. to $+\infty$. Let us assume that Γ is minorized by a constant γ. For any initial state, the stopping time

$$\nu_1 = \inf\left\{n; \operatorname{Log} \frac{L_n(\theta_1)}{L_n(\theta_0)} \notin \left] \operatorname{Log} \frac{\gamma}{c_{10} - \gamma}, \operatorname{Log} \frac{c_{01} - \gamma}{\gamma} \right[\right\}$$

is finite $P_{\theta_0,x}$ and $P_{\theta_1,x}$-a.s. However the stopping time ν of the Bayesian test defined above is majorized by ν_1. ν is finite a.s. Of course, in the independent case, the probabilities $P_{\theta,x}$ do not depend on x. If the cost Γ is constant, similar results are obtained, but c_n, c, A and B do not depend on x.

5.3.3. Sequential Likelihood Ratio Tests

The sequential likelihood ratio test obtained in the Bayesian situation is a natural test and can be studied directly by abandoning the Bayesian point of view. A statistical model $(\Omega, \mathcal{A}, (P_{\theta_i})_{i=0,1})$ with a filtration $\mathbb{F} = (\mathcal{F}_t)_{t \in T}$ and $T = \mathbb{N}$ or $T = \mathbb{R}_+$ is always dominated by

$$Q = \frac{P_{\theta_1} + P_{\theta_2}}{2}.$$

Assume given a likelihood $(L_t(\theta_i))_{i=0,1}$.

Definition 5.3.5. A **sequential likelihood ratio test** (abbreviated SLRT) $S(A,B)$ of θ_0 against θ_1 is defined for A and B constants $0 < A < 1 < B$, by

$$\nu_{A,B} = \inf\left\{t;\ \frac{L_t(\theta_1)}{L_t(\theta_0)} \leqslant A \ \text{ or } \ \frac{L_t(\theta_1)}{L_t(\theta_0)} \geqslant B\right\},$$

$$d_{\nu_{A,B}} = 1 \quad \text{if} \quad L_{\nu_{A,B}}(\theta_1) \geqslant BL_{\nu_{A,B}}(\theta_0),$$

$$d_{\nu} \quad = 0 \quad \text{if} \quad L_{\nu_{A,B}}(\theta_1) \leqslant AL_{\nu_{A,B}}(\theta_0),$$

on condition that $\nu_{A,B}$ is a finite $\mathbf{F}$-stopping time.

Strength of a Sequential Likelihood Ratio Test. Let us denote $\lambda_t = L_t(\theta_1)/L_t(\theta_0)$ and $\nu = \nu_{A,B}$. The probabilities of the two types of error of the SLRT are

$$P_{\theta_0}(d_\nu = 1) = \alpha, \quad P_{\theta_1}(d_\nu = 1) = \beta;$$

the pair (α,β) is the **strength** of this test. We have

$$A \geqslant E_{\theta_0}[\lambda_\nu|\lambda_\nu \leqslant A] = \frac{P_{\theta_1}(\lambda_\nu \leqslant A)}{P_{\theta_0}(\lambda_\nu \leqslant A)} = \frac{\beta}{1-\alpha}$$

$$B \leqslant E_{\theta_0}[\lambda_\nu|\lambda_\nu \geqslant B] = \frac{P_{\theta_1}(\lambda_\nu \geqslant B)}{P_{\theta_0}(\lambda_\nu \geqslant B)} = \frac{1-\beta}{\alpha}.$$

The test

$$S\left[\frac{\beta}{1-\alpha}, \frac{1-\beta}{\alpha}\right]$$

has at least the strength (α,β). Therein lies the beauty of sequential statistics. A test with strength at least equal to (α,β) is obtained without making any calculations on the distributions used, without consulting statistical tables. Besides, if $m \leqslant \lambda_\nu - \lambda_{\nu^-} \leqslant M$, with $m \leqslant 0 \leqslant M$,

$$A + m \leqslant \frac{\beta}{1-\alpha} \leqslant A \quad \text{and} \quad B \leqslant \frac{1-\beta}{\alpha} \leqslant B + M.$$

If the process λ is continuous, or if the jumps are small enough, we obtain Wald's approximations:

$$A \simeq \frac{\beta}{1-\alpha}, \quad B \simeq \frac{1-\beta}{\alpha}.$$

In order to test two sample hypotheses with the help of a sample, in sequential statistics we can fix the error probabilities α and β which we do not wish to exceed. The possible choice of the length of the experiment allows us to deal with the test problem by giving a symmetric role to the two hypotheses, contrary to statistics on a fixed sample.

Study of the Duration of the SLRT in the Case of a Sample. Assume that the observations $(X_n)_{n\geq 1}$ take values in $(E,\mathcal{E})$, are independent and have distribution $f(\theta,\cdot)\pi(\cdot) = \pi(\theta;\cdot)$. Let us take $\mathcal{F}_0 = \{\Phi,\Omega\}$, $\mathcal{F}_n = \sigma(X_1, ..., X_n)$,

$$\operatorname{Log} \frac{L_n(\theta_1)}{L_n(\theta_0)} = \sum_{i=1}^{n} \operatorname{Log} \frac{f(\theta_1,X_i)}{f(\theta_0,X_i)}$$

$$= \sum_{i=1}^{n} Z_i = S_n.$$

The r.v.'s $(Z_i)_{i\geq 1}$ are independent and identically distributed, for any value of the parameter θ,

$$E_\theta(Z_1) = E_\theta\left[\operatorname{Log} \frac{f(\theta_1,X_1)}{f(\theta_0,X_1)}\right] = \int \operatorname{Log} \frac{f(\theta_1,x)}{f(\theta_0,x)} f(\theta,x)\pi(dx).$$

For $\theta = \theta_1$, $E_{\theta_1}(Z_1) = K[\theta_1,\theta_0]$ is the Kullback information of $\pi(\theta_1;\cdot)$ on $\pi(\theta_0;\cdot)$. For $\theta = \theta_0$, $E_{\theta_0}(Z_1) = -K(\theta_0,\theta_1)$. The SLRT $S(A,B)$ becomes, by setting $a = \operatorname{Log} A$, $b = \operatorname{Log} B$ $(a < 0 < b)$,

$$\nu = \inf\{n; S_n \notin]a,b[\}, \quad d_\nu = 1_{(S_\nu \geq b)}.$$

This is the problem of gambler's ruin again ([2.3.2]). The stopping time ν has moments of all orders when Z_1 is nonzero P_θ-a.s., i.e.,

$$\pi(\theta; \{f(\theta_1,\cdot) \neq f(\theta_0,\cdot)\})$$

is nonzero. This is the only case where, under P_θ, the hypotheses θ_0 and θ_1 are distinguishable. Assume,

$$M(u) = \int \pi(\theta;dx)\left[\frac{f(\theta_1,x)}{f(\theta_0,x)}\right]^u$$

is finite for any $u \in \mathbb{R}$, in other words, $E_\theta(e^{uZ_1})$ is finite. The function $u \longmapsto M(u)$ is strictly convex since its second derivative

$$u \longmapsto M''(u) = E_\theta\left[Z_1^2 e^{uZ_1}\right]$$

is strictly positive. Noting that $M'(0)$ equals $E_\theta(Z_1)$, we see that, if $E_\theta(Z_1)$ is nonzero, there exists a number $u(\theta) \neq 0$ such that $M(u(\theta))$ equals 1. The process

$$(e^{u(\theta)S_n})_{n \geq 0}$$

is then a martingale (by setting $S_0 = 0$) adapted to $\mathbb{F}$ on $(\Omega, \mathcal{A}, P_\theta)$; in fact

$$\begin{aligned} E_\theta[e^{u(\theta)S_{n+1}} | \mathcal{F}_n] &= e^{u(\theta)S_n} E_\theta[e^{u(\theta)Z_{n+1}} | \mathcal{F}_n] \\ &= [e^{u(\theta)S_n}] M(u(\theta)) = e^{u(\theta)S_n}. \end{aligned}$$

The stopping theorem (Theorem 2.4.19) then gives

$$E_\theta[e^{u(\theta)S_\nu}] = 1.$$

With Wald's approximation already made, where we neglect the overstepping of the boundary, we obtain

$$e^{u(\theta)a} P_\theta[S_\nu \leq a] + e^{u(\theta)b} P_\theta[S_\nu \geq b] \simeq 1;$$

$$P_\theta(S_\nu \leq a) \simeq \frac{1 - B^{u(\theta)}}{A^{u(\theta)} - B^{u(\theta)}},$$

$$P_\theta(S_\nu \geq b) \simeq \frac{1 - A^{u(\theta)}}{B^{u(\theta)} - A^{u(\theta)}}.$$

The average duration of the game follows from Wald's theorem 2.3.9. $(S_n - nE_\theta(Z_1))$ is a martingale and,

$$E_\theta(\nu) = \frac{E_\theta(S_\nu)}{E_\theta(Z_1)} \simeq \frac{a[1 - B^{u(\theta)}] - b[1 - A^{u(\theta)}]}{E_\theta(Z_1)[A^{u(\theta)} - B^{u(\theta)}]}.$$

For $\theta = \theta_0$, $u(\theta)$ equals 1 and,

$$P_{\theta_0}(S_\nu \geq b) = \beta \simeq \frac{1 - A}{B - A};$$

$$E_{\theta_0}(\nu) \simeq \frac{1}{K(\theta_0,\theta_1)} \left[\frac{b(A - 1) + a(1 - B)}{B - A}\right].$$

For $\theta = \theta_1$, $u(\theta)$ equals -1 and,

$$P_{\theta_1}(S_\nu \leq a) = \alpha \simeq A \frac{B - 1}{B - A},$$

$$E_{\theta_1}(\nu) \simeq \frac{1}{K(\theta_1,\theta_0)} \left[bB \frac{1 - A}{B - A} + aA \frac{B - 1}{B - A}\right].$$

The above holds without assuming $M(u)$ finite for all u, if there exists a nonzero u for which $M(u)$ equals 1. Thus it always holds for $\theta = \theta_0$ or $\theta = \theta_1$.

If $E_\theta(Z_1)$ is zero, we again use Wald's theorem. (S_n) and $((S_n^2) - nE_\theta(Z_1^2))$ are centered martingales and,

$$E_\theta[S_\nu] = 0 \simeq aP_\theta(S_\nu \leq a) + bP_\theta(S_\nu \geq b),$$

$$P_\theta(S_\nu \leq a) \simeq \frac{b}{b - a}, \quad P_\theta(S_\nu \geq b) \simeq \frac{a}{a - b},$$

$$E_\theta(S_\nu^2) = E_\theta(\nu)E_\theta(Z_1^2), \quad E_\theta(\nu) \simeq -\frac{ab}{E_\theta(Z_1^2)}.$$

Note. For a sample from a distribution, we can show that if another sequential test $\hat{d}_{\nu'}$ is such that $P_{\theta_0}(\hat{d}_{\nu'} = 1) \leq \alpha$, $P_{\theta_1}(\hat{d}_{\nu'} = 0) \leq \beta$, then it is on average longer than the SLRT:

$$E_{\theta_0}(\nu') \geq E_{\theta_0}(\nu), \quad E_{\theta_1}(\nu') \geq E_{\theta_1}(\nu).$$

The SLRT is therefore the most economic possible amongst tests giving at most the same errors.

Example. For an exponential model,

$$f(\theta,x) = C(\theta)f(x)\exp[D(\theta)g(x)];$$

$$\text{Log}\,\frac{f(\theta_1,x)}{f(\theta_0,x)} = \text{Log}\,\frac{C(\theta_1)}{C(\theta_0)} + [D(\theta_1) - D(\theta_0)]g(x)$$

$$S_n = n \text{ Log } \frac{C(\theta_1)}{C(\theta_0)} + [D(\theta_1) - D(\theta_0)] \sum_{i=1}^{n} g(X_i).$$

If, for example, $D(\theta_1)$ is greater than $D(\theta_0)$, we obtain

$$\nu = \inf \left\{ n; \sum_{i=1}^{n} g(X_i) \notin]\alpha_1 + n\beta, \alpha_2 + n\beta[\right\}$$

with

$$\alpha_1 = \frac{a}{D(\theta_1) - D(\theta_0)}, \quad \alpha_2 = \frac{b}{D(\theta_1) - D(\theta_0)}; \quad a < 0 < b$$

$$\beta = \frac{1}{D(\theta_0) - D(\theta_1)} \text{ Log } \frac{C(\theta_1)}{C(\theta_0)}.$$

The **inspection diagram** has the following appearance:

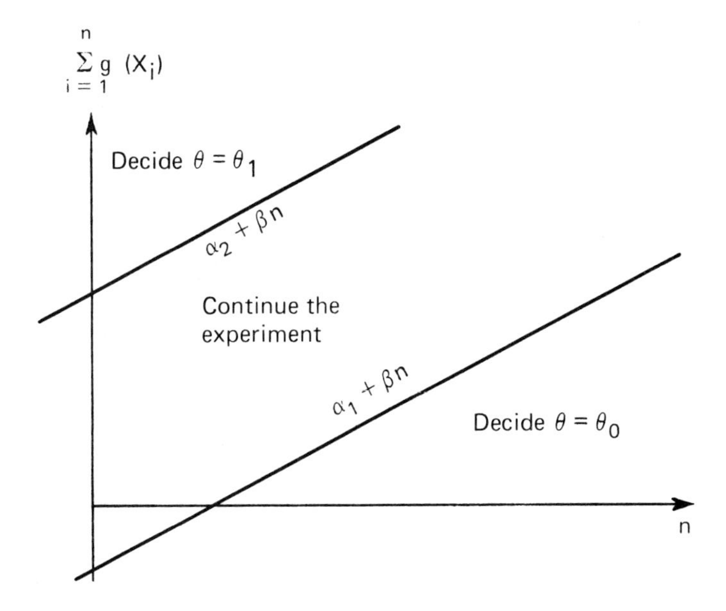

For example, in the Gaussian case,

$$f(\theta,x) = \frac{1}{\sqrt{2\pi}} \exp\left[-\frac{(x-\theta)^2}{2}\right]$$

and

$$Z_n = (\theta_1 - \theta_0)\left[X_n - \frac{\theta_1 + \theta_2}{2}\right].$$

For $\theta \neq (\theta_1+\theta_0)/2$, we have

$$u(\theta) = \frac{\theta_1 + \theta_0 - 2\theta}{\theta_1 - \theta_0}$$

which allows Wald's approximations to be made.

5.4. Large Deviations and Likelihood Tests

The asymptotic theories which we have met up till now rest on a law of large numbers of the form

$$Z_n = \frac{X_1 + \dots + X_n}{n} \xrightarrow{\text{a.s.}} m$$

and a central limit theorem which studies the convergence of $\sqrt{n}(Z_n - m)$.

We can also study the rate of convergence to 0 of $P[|Z_n - m| > a]$ for an $a > 0$. A theorem of this form is a **large deviation theorem**. The usual formulation of these results is,

$$\overline{\lim_{n\to\infty}} \frac{1}{n} \operatorname{Log} P[|Z_n - m| > a] = -h(a) < 0.$$

This means that, for every $\varepsilon > 0$ and for n large enough,

$$P[|Z_n - m| > a] \leqslant e^{-n[h(a)-\varepsilon]}.$$

We then say that (Z_n) **tends to m at an exponential rate.** We can sometimes obtain a minorization,

$$\underline{\lim}_{n\to\infty} \frac{1}{n} \operatorname{Log} P[|Z_n - m| > a] = -h(a)$$

which means that for $\varepsilon > 0$ and n large enough

$$P[|Z_n - m| > a] \geqslant e^{-n[h(a)+\varepsilon]}$$

and the sequence (Z_n) converges to m at a rate $(e^{-nh(a)})$.

5.4.1. Large Deviations for a Sample

Let F be a distribution on $\mathbb{R}$, integrable with mean m, F being different from Dirac's measure at m. Set $\phi(t) = \int e^{tx} dF(x)$. The set $\{t;\ \phi(t) < \infty\}$ is an interval (α_F, β_F) with or without its endpoints. Let $\psi = \operatorname{Log} \phi$. The **Cramer transform** h_F of F is defined [Vol. I, 3.3.5] by

$$h_F(a) = \sup\{at - \psi(t);\ t \in (\alpha_F, \beta_F)\}.$$

Let us state the results of [Vol. I, 3.3.5] and Theorem 4.4.22(c) of Volume I. The function ψ' increases strictly on $]\alpha_F, \beta_F[$ and ψ is convex. Extend ψ' and ψ at α_F (resp. β_F) by taking their left limits (resp. right limits), whether finite or not. The derivative of $t \longmapsto at - \psi(t)$ is $a - \psi'$. From which it follows that

$$h_F(a) = \begin{cases} a\beta_F - \psi(\beta_F) & \text{for } a \geq \psi'(\beta_F) \\ a\alpha_F - \psi(\alpha_F) & \text{for } a \leq \psi'(\alpha_F) \\ a\psi'^{-1}(a) - \psi[\psi'^{-1}(a)] & \text{for } \psi'(\alpha_F) < a < \psi'(\beta_F). \end{cases}$$

On $]\psi'(\alpha_F), \psi'(\beta_F)[$, h_F' is the inverse function of ψ'. The point 0 is always in $[\alpha_F, \beta_F]$, and $\psi'(0) = m$. From which it follows that $h_F(m) = 0$ and $\psi'^{-1}(m) = 0$. For $a > m$,

$$h_F(a) = \sup_{u>0} (ua - \psi(u)).$$

Geometric Construction Let $a \in]\psi'(\alpha_F), \psi'(\beta_F)[$, $\psi'(h_F'(a)) = a$, implies that the line of the equation

$$y = a(x - h_F'(a)) + \psi(h_F'(a)) = ax - h_F(a)$$

is tangent to the graph of ψ. From which we obtain the following construction: draw the graph of ψ, draw the tangent to ψ with slope a. The intersection of this tangent and the y axis is $-h_F(a)$. (See diagram on following page.)

Chernov's Theorem 5.4.6. *Let (X_n) be a sample from a distribution F on $\mathbb{R}$. Assume F is integrable with mean m and with Cramer transform h_F. Set $S_n = X_1 + \ldots + X_n$ and*

$$I = \{t;\ \int e^{tx} dF(x) < \infty\}.$$

Assume that $\mathring{I}$ is a neighborhood of 0.

(a) *For $a > m$, $P[S_n \geq na] \leq e^{-nh_F(a)}$; for $a < m$, $P[S_n \leq na] \leq e^{-nh_F(a)}$. In both cases,* $h_F(a)$ is > 0.

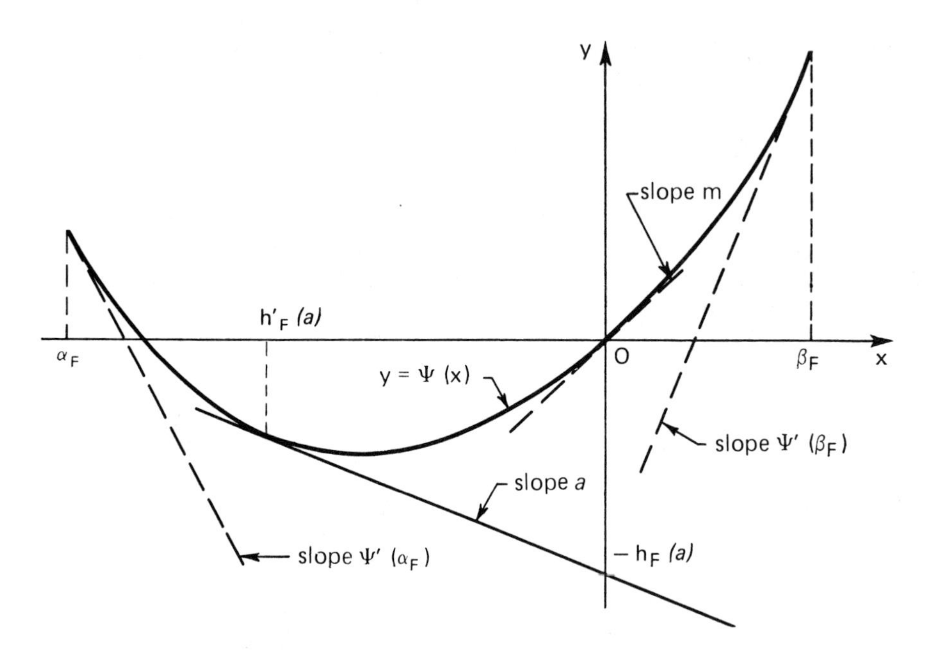

(b) *Setting* $\psi(t) = \text{Log} \int e^{tx} dF(x)$, $\beta_F = \sup\{\psi'(t); t \in I\}$ *and* $\alpha_F = \inf\{\psi'(t); t \in I\}$, *for* $m < a < \beta_F$,

$$\lim_{n\to\infty} \frac{1}{n} \text{Log } P[S_n \geqslant na] = -h_F(a);$$

for $\alpha_F < a < m$,

$$\lim_{n\to\infty} \frac{1}{n} \text{Log } P[S_n \leqslant na] = -h_F(a).$$

Note. Since h_F is continuous at a, the limits obtained above are the same replacing the inequalities by strict inequalities.

Proof. Assume $a > m$. The case $a < m$ follows by replacing (X_n) by $(-X_n)$.

(a) $$P[S_n \geqslant na] \leqslant \inf\{E(e^{-una+uS_n}), u \in]0,\beta_F]\}$$

$$= \exp[-n \sup\{ua - \psi(u)\}; \; u \in]0,\beta_F]]$$

$$= \exp[-nh_F(a)].$$

(b) It remains to show,

$$\underline{\lim}_{n\to\infty} \frac{1}{n} \operatorname{Log} P(S_n > na) \geqslant -h_F(a).$$

For this, a change of probabilities is made. Let $t \in]0,\beta_F[$; set $\phi(t) = \int e^{tx} dF(t)$ and $f_t(x) = (\phi(t))^{-1}e^{tx}$. We consider $F_t = f_t F$. This is a probability on $\mathbb{R}$. Let us denote on $(\Omega, A) = (\mathbb{R}, \mathcal{B}_{\mathbb{R}})^{\mathbb{N}}$, $P = F^{\otimes \mathbb{N}}$ and $P_t = F_t^{\otimes \mathbb{N}}$; X_n is the nth coordinate. On $\sigma(X_1, ..., X_n)$, we have

$$P_t = (\phi(t))^{-n} e^{tS_n} P \quad \text{and} \quad P = \phi^n(t) e^{-tS_n} P_t.$$

From which it follows that,

$$P[S_n \geqslant na] = E_t[1_{(S_n \geqslant na)} \phi^n(t) e^{-tS_n}]$$

$$= (\phi(t))^n e^{-nat} E_t\left[\exp\left\{-nt\left(\frac{S_n}{n} - a\right)\right\} 1_{(S_n \geqslant na)}\right].$$

Let $\varepsilon > 0$; we have,

$$\frac{1}{n} \operatorname{Log} P[S_n \geqslant na]$$
$$\geqslant -at + \operatorname{Log} \phi(t) + \frac{1}{n} \operatorname{Log} E_t\left[e^{-nt\varepsilon} 1_{(na \leqslant S_n \leqslant na+\varepsilon)}\right].$$

The mean of F_t is $\psi'(t)$. If a is in $]n,\psi'(\beta_F)[$, we can take $a < \psi'(t) < a + \varepsilon$. For n large enough,

$$P_t\left[a \leqslant \frac{S_n}{n} \leqslant a + \varepsilon\right] \geqslant \frac{1}{2}.$$

Then

$$\frac{1}{n} \operatorname{Log} P[S_n \geqslant na] \geqslant -at + \psi(t) - t\varepsilon - \frac{1}{n} \operatorname{Log} 2$$
$$\underline{\lim}_{n\to\infty} \frac{1}{n} \operatorname{Log} P[S_n \geqslant na] \geqslant -at + \psi(t) - t\varepsilon.$$

This is true for all t such that $\psi'^{-1}(a) < t < \psi'^{-1}(a + \varepsilon)$. By letting t tend to $\psi'^{-1}(a)$ we obtain in the right side of the inequality $h_F(a) - t\varepsilon$. Then by letting ε tend to zero, we obtain the lower bound $h_F(a)$.

5.4.2. Neyman-Pearson Test of Two Distributions on $\mathbb{R}$ Using Samples

Let F_0 and F_1 be two distributions on $\mathbb{R}$, f_0 and f_1 their densities with respect to $F = (1/2)(F_0 + F_1)$. A sample is observed from these distributions $(\Omega, \mathcal{A}, P_i) = (\mathbb{R}, \mathcal{B}_{\mathbb{R}}, F_i)$ for $i = 0,1$, with X_n the nth coordinate. Set

$$U = \text{Log}\,\frac{f_1}{f_0}, \quad S_n = U(X_1) + \ldots + U(X_n).$$

A Newman-Pearson test with rejection region

$$D_n = \left\{ \sum_{i=1}^{n} \text{Log}\,\frac{f_1}{f_0}(X_i) \geq d_n \right\} = \{S_n \geq d_n\}$$

is always the most powerful possible at its level ([Vol. I, 8.3.2]). Let us denote this level by $\alpha_n = P_0(S_n \geq d_n)$, and by $\beta_n = P_1(S_n < d_n)$ the type II error.

Let us study what happens for $d_n = nc$, c being fixed. Assume U is integrable for F_0 and F_1;

$$\int U\, dF_0 = \int \text{Log}\,\frac{f_1}{f_0}\, dF_0 = -K(F_0, F_1),$$

$$\int U\, dF_1 = K(F_1, F_0).$$

For

$$\phi_0(t) = E_0[e^{tU(X_n)}] = \int f_1^t f_0^{1-t} dF,$$

we have

$$\phi_0(0) = \phi_0(1) = 1, \quad \phi_0'(0) = -K(F_0, F_1),$$

$$\phi_0'(1) = E_0[U(X_n)e^{U(X_n)}] = K(F_1, F_0).$$

For

$$\phi_1(t) = E_1[e^{tU(X_n)}] = \int f_1^{1+t} f_0^t dF$$

we have

$$\phi_1(t) = \phi_0(1 + t).$$

Therefore we can apply Chernov's theorem and majorize $\alpha_n = P_0(S_n \geqslant nc)$ and $\beta_n = P_1(S_n < nc)$ if $-K(F_0,F_1) < c < K(F_1,F_0)$. On $]-K(F_0,F_1), K(F_1,F_0)[$, let us consider the Cramer transforms h_0 and h_1 of the distributions of $U(X_n)$ for P_0 and P_1 respectively; h_0' is the inverse function of ϕ_0'/ϕ_0 and h_1' that of ϕ_1'/ϕ_1. The relation $\phi_1(t) = \phi_0(1 + t)$ for $-1 < t < 0$ implies $h_0' - h_1' = 1$. At the point $y = K(F_1,F_0)$ we calculate $h_1'(y) = h_1(y) = 0$ and $h_0'(y) = 1$, $h_0(y) = y$. From which it follows that for all $c \in]-K(F_0,F_1), K(F_1,F_0)[$: $h_0(c) - h_1(c) = c$.

Theorem 5.4.7. *Let $F_0 = f_0F$ and $F_1 = f_1F$ be two distinct distributions on $\mathbb{R}$. Assume that the Kullback informations $K(F_0,F_1)$ and $K(F_1,F_0)$ are finite, and $c \in]-K(F_0,F_1), K(F_1,F_0)[$. Consider a sample (X_n) from these distributions, and for each n a test of F_0 against F_1 with rejection region*

$$D_n = \left\{\sum_{i=1}^{n} \operatorname{Log} \frac{f_1}{f_0}(X_i) \geqslant cn\right\}.$$

Let h be the Cramer transform of the distribution $(\operatorname{Log}(f_1/f_0))F_0$. We have $h(c) > 0$ and

(a) $$\alpha_n = P_0(D_n) \leqslant e^{-nh(c)} \quad and \quad \frac{\operatorname{Log} \alpha_n}{n} \to -h(c),$$

(b) $$\beta_n = P_1(D_n^c) \leqslant e^{-n(h(c)-c)} \quad and \quad \frac{\operatorname{Log} \beta_n}{n} \to c - h(c).$$

Consequence. Let $0 < \alpha < K(F_1,F_0)$. A Neyman-Pearson test of level $e^{-n\alpha}$ ([Vol. 1, 8.3.2]) has a critical function lying between $1_{(S_n \geqslant nc_n)}$ and $1_{(S_n > nc_n)}$ for a constant c_n. By part (a) of Theorem 5.4.7 and the note which follows Theorem 5.4.6, $(c_n) \to c$ with $h(c) = \alpha$. From part (b) of Theorem 5.4.7, we have $c < \alpha$, and the Type II error β_n is then such that

$$\left[\frac{\operatorname{Log} \beta_n}{n}\right] \to c - \alpha.$$

In order to see this more clearly, set $\psi = \operatorname{Log} \phi_0$, and note that, if t varies from 0 to 1, $h'(t)$ varies from $-K(F_0,F_1) = \psi'(0)$ to $K(F_1,F_0) = \psi'(1)$; moreover: $\alpha = h(c)$. By using the same construction as [5.4.1], we carry out the following: draw the graph of ψ and the tangent to this graph from the point $(0,-\alpha)$. The slope of this tangent is c and its contact point is the x-axis point $h'(c)$; $c - \alpha$ is the point with x coordinate 1

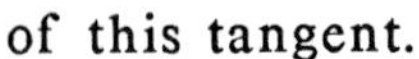
of this tangent.

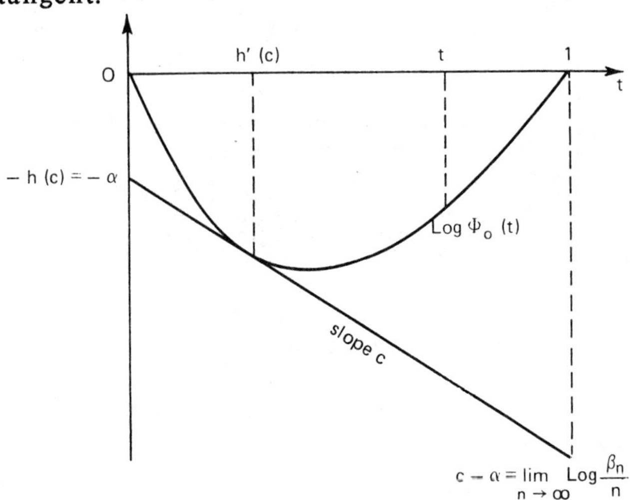

Therefore, for a Neyman-Pearson test of level $e^{-n\alpha}$, the Type II error $\beta_n(\alpha)$ is close to $e^{-n(c-\alpha)}$, and the number $c - \alpha$ can be constructed by the above method. By considering the tangent at the point (0,0) with slope $-K(F_0,F_1)$, we see that $0 < \alpha - c < K(F_0,F_1)$. For every $\alpha \in]0,K(F_1,F_0)[$, the error $\beta_n(\alpha)$ decreases more quickly than $e^{-nK(F_0,F_1)}$.

Example. Taking $F_i = N(\theta_i,1)$, i = 0,1, we obtain

$$\psi_0(t) = \text{Log } \phi_0(t) = -\frac{1}{2}(\theta_1 - \theta_0)^2 t(1-t).$$

We can then, with an n-sample, choose a test the Type I and Type II errors of which are majorized by

$$\exp\left\{-\frac{n}{4}(\theta_1 - \theta_0)^2\right\}.$$

5.4.3. Likelihood Ratio Tests

We are given a model $\{\Omega,A,(P_\theta)_{\theta\in\Theta}\}$, dominated along the filtration $\mathbb{F} = (F_t)_{t\in T}$ with $T = \mathbb{N}$ on $\mathbb{R}_+$, and let $(L_t(\theta))_{t\in T}$ be a likelihood. Assume Θ is the union of two disjoint parts Θ_0 and Θ_1, and we test "$\theta \in \Theta_0$" against "$\theta \in \Theta_1$." Let us assume that at each time t maximum likelihood estimators $\hat\theta_{t,0}$, $\hat\theta_{t,1}$, and $\hat\theta_t$ exist for $\theta \in \Theta_0$, $\theta \in \Theta_1$, $\theta \in \Theta$:

$$L_t(\hat{\theta}_{t,0}) = \sup\{L_t(\theta);\ \theta \in \Theta_0\};$$

$$L_t(\hat{\theta}_{t,1}) = \sup\{L_t(\theta);\ \theta \in \Theta_1\};$$

$$L_t(\hat{\theta}_t) = \sup\{L_t(\theta);\ \theta \in \Theta\}.$$

Two forms of **likelihood ratio tests** are then natural at each time t (cf. [Vol. I, 7.3.2]); their rejection regions are respectively of the form,

$$D_t = \left\{\frac{L_t(\hat{\theta}_{t,1})}{L_t(\hat{\theta}_{t,0})} > d_t\right\}$$

and

$$D_t' = \left\{\frac{L_t(\hat{\theta}_{t,1})}{L_t(\hat{\theta}_t)} > d_t'\right\}.$$

We can also envisage the following **sequential Chernov test.** Let $A(\theta)$ be the set Θ_0 or Θ_1, which contains θ, and let $\hat{\hat{\theta}}_t$ be the maximum likelihood estimator in $\Theta \backslash A(\hat{\theta}_t)$:

$$\hat{\hat{\theta}}_t = \hat{\theta}_{t,0} \quad (\text{resp.} = \hat{\theta}_{t,1}),$$

when $\hat{\theta}_t \in \Theta_1$ (resp. Θ_0). Let $c > 0$; we define a sequential test by stopping the experiment at time

$$\nu_c = \inf\{t;\ \ell_t(\hat{\theta}_t) - \ell_t(\hat{\hat{\theta}}_t) \geq c\}$$

(assumed finite) and by then deciding $\theta \in A(\hat{\theta}_{\nu_c})$; ℓ_t still means $\text{Log}\, L_t$. Thus let us consider on $(\mathbb{R}, \mathcal{B}_{\mathbb{R}})$ a family $(F_\theta)_{\theta \in \Theta}$ with Θ finite. It is dominated by a distribution F, $F_\theta = f_\theta F$. For $\theta \neq \theta'$ denote by $h(\theta,\theta';\cdot)$ the Cramer transform of

$$\text{Log}\,\frac{f_{\theta'}}{f_\theta}(F_\theta),$$

and by $K(\theta,\theta')$ the Kullback information of F_θ with respect to $F_{\theta'}$.

$$K(\Theta_0,\Theta_1) = \inf\{K(\theta,\theta');\ \theta \in \Theta_0,\ \theta' \in \Theta_1\},$$

$$h(\Theta_0,\Theta_1,\cdot) = \inf\{h(\theta,\theta',\cdot);\ \theta \in \Theta_0,\ \theta \in \Theta_1\}.$$

Proposition 5.4.8. *A sample is observed from a family $(F_\theta)_{\theta \in \Theta}$ of distributions on $\mathbb{R}$, Θ having q elements. We assume, for every pair (θ,θ'), $0 < K(F_\theta,F_{\theta'}) < \infty$.*

(a) *If $(\hat{\theta}_n)$ is a maximum likelihood estimator, we can find an $a > 0$ and $b > 0$ such that, for all n,*

$$P_\theta[\text{there exists an } m;\ m > n,\ \hat{\theta}_m \neq \theta] \leq ae^{-bn};$$

the sequence $(\hat{\theta}_n)$ converges to θ at an exponential rate.

(b) *With the above notations a likelihood ratio test with rejection region $\{\ell_n(\hat{\theta}_{n,1}) - \ell_n(\hat{\theta}_{n,0}) > nc\}$ has, for $c \in]-K(\Theta_0,\Theta_1), K(\Theta_1,\Theta_0)[$, Type I and II errors α_n and β_n such that*

$$\overline{\lim}\ \frac{\operatorname{Log} \alpha_n}{n} < 0 \quad \text{and} \quad \overline{\lim}\ \frac{\operatorname{Log} \beta_n}{n} < 0.$$

These errors tending to 0 at an exponential rate.

(c) *For $0 < c$ the Chernov sequential test has an exponentially bounded duration ν_c,* i.e.

$$\overline{\lim}\ \frac{1}{n} \operatorname{Log} P_\theta(\nu_c > n) < 0,$$

and its error probability is majorized by qe^{-c}.

Proof. For $\theta \neq \theta'$, $K(\theta,\theta')$ is finite, and $K(\Theta_0,\Theta_1)$ and $K(\Theta_1,\Theta_0)$ are strictly positive numbers. Let $c \in]-K(\Theta_0,\Theta_1),K(\Theta_1,\Theta_0)[$; $h(\Theta_0,\Theta_1,c)$ and

$$\begin{aligned} h(\Theta_1,\Theta_0,c) &= h(\Theta_0,\Theta_1,c) - c \\ &= \inf\{h(\theta',\theta,c);\ \theta \in \Theta_0,\ \theta' \in \Theta_1\} \end{aligned}$$

are both strictly positive. Applying Theorem 5.4.7, we have for every $\theta \in \Theta$,

$$\begin{aligned} P_\theta[\hat{\theta}_n \neq \theta] &\leq \sum_{\theta' \neq \theta} P_\theta[\ell_n(\theta') > \ell_n(\theta)] \\ &= qe^{-nh(\theta,\Theta\setminus\{\theta\},0)} \end{aligned}$$

which implies (a). Therefore we can also find an $a_0 > 0$ and $b_0 > 0$, such that, for $\theta \in \Theta_0$,

$$P_\theta(\theta \neq \hat{\theta}_{n,0}) \leq a_0 e^{-b_0 n}.$$

Then

$$P_\theta(\ell_n(\hat\theta_{n,1}) - \ell_n(\hat\theta_{n,0}) \geq cn) \leq P_\theta(\hat\theta_{n,0} \neq \theta)$$

$$+ P_\theta(\ell_n(\hat\theta_{n,1}) - \ell_n(\theta) \geq cn)$$

$$\leq a_0 e^{-b_0 n} + \sum_{\theta' \in \Theta_1} P_\theta(\ell_n(\theta') - \ell_n(\theta) \geq cn)$$

$$\leq a_0 e^{-b_0 n} + q e^{-nh(\Theta_0,\Theta_1,c)} .$$

For $\theta' \in \Theta_1$, we can find an $a_1 > 0$ and $b_1 > 0$ such that $P_{\theta'}(\hat\theta_{n,1} \neq \theta) \leq a_1 e^{-b_1 n}$. Then,

$$P_{\theta'}(\ell_n(\hat\theta_{n,1}) - \ell_n(\hat\theta_{n,0}) \leq cn) \leq P_{\theta'}(\hat\theta_{n,1} \neq \theta')$$

$$+ P_{\theta'}(\ell_n(\theta') - \ell_n(\hat\theta_{n,0}) \leq cn)$$

$$\leq a_1 e^{-b_1 n} + q e^{-n[h(\Theta_0,\Theta_1,c)-c]} .$$

From which part (b) of the proposition follows.

Let us show part (c). We have

$$P_\theta(\nu_c \geq n) \leq P_\theta(\hat\theta_n \neq \theta) + \sum_{\theta' \neq \theta} P_\theta(\ell_n(\theta') - \ell_n(\theta) \geq 0)$$

$$\leq 2q e^{-nh(\theta,\Theta\backslash\{\theta\},\ 0)} .$$

Moreover the error probability is

$$\sum_{\theta' \notin A(\theta)} P_\theta(\hat\theta_{\nu_c} = \theta').$$

However

$$P_\theta(\hat\theta_{\nu_c} = \theta') = E_{\theta'}\left[1_{(\hat\theta_{\nu_c}=\theta')} \frac{L_{\nu_c}(\theta)}{L_{\nu_c}(\theta')}\right].$$

And, on $\{\hat\theta_{\nu_c} = \theta'\}$

$$\frac{L_{\nu_c}(\hat\theta_{\nu_c})}{L_{\nu_c}(\theta)} \geq \frac{L_{\nu_c}(\hat\theta_{\nu_c})}{L_{\nu_c}(\hat{\hat\theta}_{\nu_c})} \geq e^c,$$

from which

$$P_\theta[\hat{\theta}_{\nu_c} = \theta'] \leqslant e^{-c}.$$

Extensions. The preceding proposition may be easily extended to a dominated model of Markov chains, $\pi(\theta,x;\cdot) = f(\theta,x,\cdot)\pi(x;\cdot)$, by assuming Θ finite and, for every $\theta \in \Theta$, $0 < m_\theta \leqslant \text{Log } f(\theta,\cdot,\cdot) \leqslant M_\theta < \infty$ for constants m_θ and M_θ. We then apply Theorem 2.6.32 of large deviations relative to martingales with bounded jumps, to martingales

$$M_{\theta,\theta',n} = \sum_{i=1}^{n} \left[\text{Log } \frac{f(\theta',X_{i-1},X_i)}{f(\theta,X_{i-1},X_i)} - K(\theta,\theta',X_{i-1})\right],$$

$K(\theta,\theta',\cdot)$ being Kullback information defined in [4.4.2].

Bibliographic Notes

The general ideas of control are those of non-stochastic control, linked to differential equations, in particular those of Bellman. In view of its strategic and economic importance, control is a subject with a vast literature.

Here are a few basic works, in the spirit of this chapter. For optimal stopping, Chow-Robbins-Siegmund, Shiryayev, Neveu [4]. For control of Markov chains on a countable space Derman, Howard; and in a more general framework Ross, Kusher, Gikhman-Skorokhod [4].

Section [5.2.3] follows the works of Mandl and Georgin.

For sequential statistics, we refer to Wald, Wetherill, Ferguson, DeGroot and above all, Ghosh and Govindarajulu which contain all the useful references.

The use of large deviations in statistics has not been the object of a systematic work. We refer to Asterisque [2] where a fairly complete bibliography is given and to Chernoff. This chapter takes up some ideas from Duflo-Florens Zmirou.

Chapter 6
COUNTING PROCESSES

Objectives

In this chapter we shall study the simplest continuous time processes, the trajectories of which are increasing and continuous on the right.

The first goal is the study of processes $(N_t)_{t \geq 0}$ where N_t is the number of events of a certain type observed before t. These "counting" processes are essential in reliability or for the study of queues. We can associate with these processes a "compensator" analogous to the compensator defined in [2.2.3] for sequences of r.v.'s.

We are given an increasing process taking values in $[0,\infty[$, with trajectories continuous on the right; on each trajectory we can define Stieltjes integrals. The study of these stochastic integrals is useful for counting processes and serves as an introduction to Chapter 8 where other stochastic integrals will be defined (similar to [1.2] where the integral with respect to the Poisson process had introduced the integral with respect to second order spatial processes with orthogonal increments).

6.1. Renewal Processes and Random Walks

6.1.1. Renewal Processes

Consider a type of machine which can function without breakdown during a random time τ with distribution F on $[0,\infty]$. If at each breakdown the machine is replaced by a new machine identical to the preceding one, the time of the nth breakdown is,

$$T_n = T_0 + \tau_0 + \tau_1 + \ldots + \tau_n$$

where T_0 is the initial time with distribution G and (τ_n) a sequence of independent r.v.'s with distribution F. T_0 is independent of the sequence (τ_n).

Definition 6.1.1. We call a **renewal process** an increasing sequence $(T_n)_{n\geq 0}$ of positive r.v.'s on $(\Omega,\mathcal{A},P)$, such that

$$T_n = T_0 + \sum_{k=1}^{n} \Delta T_k,$$

where the r.v.'s (ΔT_n) are positive, independent and identically distributed (denote their distribution by F), and that these r.v.'s are independent of T_0 (the distribution of which is denoted G): T_n is the nth **renewal time**. We assume $F \neq \delta_0$.

The **counting process** associated with this renewal process is $(N_t)_{t\geq 0}$, with,

$$N_t = \sum_{n\geq 0} 1_{(T_n \leq t)} \qquad \text{(number of breakdowns before } T\text{)}.$$

For each ω, $(N_t(\omega))$ is the distribution function of the measure

$$N(\omega,\cdot) = \sum_{n\geq 0} \delta_{T_n(\omega)} \, ;$$

the counting process is associated with the random measure $N = \Sigma_{n\geq 0}\delta_{T_n}$.

Examples. (a) For $G = \delta_0$ and F an exponential distribution (λ) with parameter $\lambda > 0$, the process $(N_t)_{t\geq 0}$ is a **Poisson process** [1.2.1]. This is proved in [Vol. I, E4.4.12] and will also be proved in [6.3].

(b) We are given a Markov chain of which a is a recurrent point. If a is the initial state, the sequence of passage times to a is a renewal process ([4.2.1]).

(c) We have

$$E(N_t) = \sum_{n=0}^{\infty} G*F^{*n}([0,t]).$$

The function $t \longmapsto E(N_t)$ is linear if and only if

$$\sum_{n=0}^{\infty} G*F^{*n} = \alpha L,$$

with L Lebesgue measure on $\mathbb{R}_+$ and $\alpha \in \mathbb{R}_+$. The Laplace transform of αL is $u \longmapsto (\alpha/u)$. If $\hat{G}$ and $\hat{F}$ are Laplace transforms of G and F, then the Laplace transform of $\Sigma_{n=0}^{\infty} G*F^{*n}$ is

$$\sum_{n=0}^{\infty} \hat{G}\,\hat{F}^n = \frac{\hat{G}}{1-\hat{F}}.$$

Thus $t \longmapsto E(N_t)$ is linear if and only if

$$\hat{G}(u) = \alpha \frac{1-\hat{F}(u)}{u}.$$

It is easy to check that the only probability having such Laplace transform has density

$$t \longmapsto \frac{1}{m} F([t,\infty[), \quad \alpha = \frac{1}{m}.$$

For F an exponential distribution of parameter λ,

$$\frac{1}{m} F([t,\infty[) = \lambda e^{-\lambda t} \quad \text{and} \quad G = F.$$

Breakdowns Subsequent to Time *t*. Let the observations begin at time t. We observe the breakdowns at times $(T_n^{(t)})_{n \geq 0}$ with $T_n^{(t)} + t = T_{N_t+n}$. For $n \geq 1$, we have on $\{N_t = k\}$:

$$T_n^{(t)} - T_{n-1}^{(t)} = \Delta_{T_{n+k}} = \tau_{n+k}.$$

Let $\Gamma_1, ..., \Gamma_n$ be n Borel sets of $\mathbb{R}_+$ and $k \in \mathbb{N}$:

$$P[N_t = k, T_0^{(t)} \in \Gamma_0, T_1^{(t)} - T_2^{(t)} \in \Gamma_1, ..., T_n^{(t)} - T_{n-1}^{(t)} \in \Gamma_n]$$

$$= P[N_t = k, T_k = t \in \Gamma_1, \tau_{k+1} \in \Gamma_2, ..., \tau_{k+n} \in \Gamma_n]$$

$$= P[N_t = k, T_0^{(t)} \in \Gamma_1] \prod_{i=2}^{k} F(\Gamma_i)$$

$$= P[T_0^{(t)} \in \Gamma_1] \prod_{i=2}^{k} F(\Gamma_i).$$

Thus $(T_n^{(t)})_{n \geqslant 0}$ is a new renewal process and the r.v.'s $(T_n^{(t)} - T_{n-1}^{(t)})_{n \geqslant 1}$ have the distribution F. The counting process of this renewal process is $(N_{t+u} - N_t)_{u \geqslant 0}$. If $u \longmapsto E(N_u)$ is linear, then $u \longmapsto E(N_{t+u} - N_t)$ is linear. If G has the density

$$t \longmapsto \frac{1}{m} F([t,\infty[),$$

$T_0^{(t)}$ has the same distribution for any t.

6.1.2. Random Walks and Renewal

Consider on a probability space $(\Omega, \mathcal{A}, P)$ a filtration $\mathbb{F} = (\mathcal{F}_n)_{n \geqslant 0}$ and sequence of r.v.'s $(X_n)_{n \geqslant 0}$ adapted to $\mathbb{F}$ and such that, for every $n \geqslant 1$, X_n is independent of $\mathcal{F}_{n-1}$. We assume that for $n \geqslant 1$, these r.v.'s have the same distribution concentrated either on $\mathbb{R}$ or on $[0,\infty]$, denoted by F. The distribution of X_0 is denoted by G.

The associated random walk is then (S_n), with $S_n = X_0 + ... + X_n$. This is a Markov chain with initial distribution G and transition probability $(x,\Gamma) \longmapsto F(\Gamma - x)$ ([4.2.3]). The strong Markov property takes a particular form here. Let us denote by ϕ the Fourier transform of F. Let T be an $\mathbb{F}$-stopping time, p and integer and $(u_j)_{1 \leqslant j \leqslant p} \in \mathbb{R}^p$,

$$1_{(T<\infty)} E\left[\exp i \sum_{j=1}^{p} u_j X_{T+j} \mid \mathcal{F}_T\right]$$

$$= \sum_{k=0}^{\infty} 1_{(T=k)} E\left[\exp i \sum_{j=1}^{p} u_j X_{k+j} \mid \mathcal{F}_k\right]$$

$$= \sum_{k \geqslant 0} 1_{(T=k)} \prod_{j=1}^{p} \phi(u_j) = 1_{(T<\infty)} \prod_{j=1}^{p} \phi(u_j).$$

Therefore, if T is finite, $(S_{n+T} - S_T)$ is a random walk independent of $\mathcal{F}_T$ and with the same distribution as $(S_n - S_0)$.

The above applies to renewal processes for which F is concentrated on $[0,\infty]$.

The **ladder indices** of the random walk (S_n) are the times where it takes values larger than previously. The lower bound of an empty set of $\mathbb{N}$ being $+\infty$, these ladder indices are defined by,

$$T_0 = 0, \quad T_1 = \inf(n;\ S_n > S_0),\ \ldots,$$

$$T_k = \inf(n;\ S_n > S_{T_{k-1}}).$$

These are stopping times. Let μ be the distribution of T_1. We have

$$P(T_k - T_{k-1} = q |\ \mathcal{F}_{T_{k-1}})\mathbf{1}_{(T_{k-1}<\infty)} = P\left[\bigcap_{n=1}^{q-1} (S_{T_{k-1}+n} \leqslant S_{T_{k-1}})\right.$$

$$\left.\cap\ (S_{q+T_{k-1}} > S_{T_{k-1}})|\ \mathcal{F}_{T_{k-1}}\right]\mathbf{1}_{(T_{k-1}<\infty)}$$

$$= \mu(q)\mathbf{1}_{(T_{k-1}<\infty)}.$$

Therefore the sequence of ladder indices is a renewal process with initial state 0 associated with μ. We can set in Definition 6.1.1,

$$\Delta T_k = (T_k - T_{k-1})\mathbf{1}_{(T_k<\infty)} + \tau_k\mathbf{1}_{(T_{k-1}=\infty)}$$

for a sequence (τ_k) of independent r.v.'s with distribution μ, independent of (X_n).

6.1.3. Total Number of Renewals

In everything that follows we use the framework of Definition 6.1.1, assuming T_0 is zero.

The distribution of T_n is then F^{*n}, and we have

$$E(N_t) = \sum_{n=1}^{\infty} P(T_n \leqslant t) = \sum_{n=1}^{\infty} F^{*n}(t).$$

However

$$F^{*n}(t) = P(T_n \leqslant t) \leqslant P\left[\sum_{k=1}^{n} (\Delta T_k \leqslant t)\right] \leqslant (F(t))^n.$$

Let $\varepsilon > 0$ and $\delta > 0$ such that $F(]\delta,\infty]) = 1 - F(\delta) > \varepsilon$. For integer k, $k\delta > t$, we have

$$P(T_k > t) = 1 - F^{*k}(t) > \varepsilon^k.$$

Thus

$$P(T_{nk} < t) \leqslant P(T_k \leqslant t, T_{2k} - T_k \leqslant t, \ldots, T_{nk} - T_{(n-1)k} \leqslant t)$$

$$\leqslant (F^{*n}(t))^n ;$$

$$E(N_t) \leqslant k \sum_{n=1}^{\infty} P(T_{nk} \leqslant t) \leqslant \frac{kF^{*k}(t)}{1 - F^{*n}(t)} < \infty .$$

Let $\nu = \inf\{n; T_n = \infty\} = \inf\{n; \Delta T_n = \infty\}$. Denoting $F(\{\infty\}) = F(\infty)$, we have

$$P(\nu = k) = P[\Delta T_1 < \infty, \ldots, \Delta T_{k-1} < \infty, \Delta T_k = \infty]$$

$$= (1 - F(\infty))(F(\infty))^{k-1},$$

$$E(\nu) = \sum_{k\geqslant 1} kF(\infty)^{k-1}(1 - F(\infty)) = \frac{1}{1 - F(\infty)},$$

$$\lim_{t\to\infty} E(N_t) = E(\nu).$$

Proposition 6.1.2. *For an arbitrary renewal process the following renewal function U is finite for any t:*

$$U(t) = \sum_{n\geqslant 1} F^{*n}(t).$$

Taking $T_0 = 0$, *two cases are distinguished*:

$$F(\infty) > 0 \ \text{and} \ \lim_{t\to\infty} E(N_t) = \frac{1}{1 - F(\infty)};$$

$$F(\infty) = 0 \ \text{and} \ (N_t) \xrightarrow[t\to\infty]{\text{a.s.}} \infty.$$

In what follows, we study the case $F(\infty) = 0$. Certain theorems are clear by assuming F to be integrable with mean m. From the law of large numbers:

$$\frac{T_n}{n} \xrightarrow{\text{a.s.}} m \quad \text{and} \quad \frac{T_{N_t}}{N_t} \xrightarrow{\text{a.s.}} m.$$

By using the inequality $T_{N_t} \leqslant t < T_{N_t+1}$, we have:

$$\frac{N_t}{t} \xrightarrow{\text{a.s.}} \frac{1}{m}.$$

If F has variance σ^2, we can show that

$$\sqrt{t}\left(\frac{N_t}{t} - \frac{1}{m}\right) \xrightarrow{\mathcal{D}} N\left(0, \frac{\sigma^2}{m^2}\right).$$

Application to Random Walks. With the notations of [6.1.2], set $S_0 = 0$, and

$$M = \sup_{n \geqslant 1} (S_n).$$

Let

$$N = \sum_{k=1}^{\infty} 1_{(T_k < \infty)};$$

$(T_1 < \infty) = (M > 0)$. Thus, if $P(M \leqslant 0) > 0$, we have $E(N) = 1/P(M \leqslant 0)$; and $P(M < \infty) = 1$. If $P(M \leqslant 0) = 0$, all the ladder indices are finite a.s. Then the sequence $(S_{n+T_k} - S_{T_k})$ is independent of F_{T_k}, with the same distribution as (S_n): its first ladder index is $T_{k+1} - T_k$. Thus the r.v.

$$S_{T_{k+1}} - S_{T_k}$$

is independent of F_{T_k} with the same distribution as S_{T_1}. Moreover, S_{T_k}/k tends a.s. to $E(S_{T_1})$ if $k \to \infty$ (whether or not this value is finite). In particular, we then have $P(M = \infty) = 1$.

When F has mean m, we have $P(M < \infty) = 1$ if $m < 0$ and $P(M < \infty) = 0$ if $m \geqslant 0$. This is easily deduced from the law of large numbers if $m \neq 0$. The case $m = 0$ is found in [Vol. 1, E4.4.13], and will be a consequence of Theorem 6.1.4.

6.1.4. Sets Attained by a Random Walk

In this section we generalize the results of [4.2.3] to a random walk associated with an arbitrary distribution F defined on $\mathbb{R}$. Assuming that the initial distribution of the walk is F, a Borel set of $\mathbb{R}$ is reached by the walk if it is charged by the measure $U = \Sigma_{n \geqslant 1} F^{*n}$. Recall that if μ is a measure on $\mathbb{R}$, its support S_μ is the set of $x \in \mathbb{R}$ of which U charges every neighborhood. We have

$$S_U = \cup S_{F^{*n}}.$$

Proposition 6.1.3. (a) *If F is concentrated on $\mathbb{R}_+$, two cases are possible*:

- *F is arithmetic with step $d > 0$*: i.e. *S_U is contained in $d\mathbb{N}$ and contains all multiples nd of d for n greater than a certain number*;
- *For every $\varepsilon > 0$, there exists an $x_\varepsilon > 0$ such that $y \geqslant x$ implies $U(y - \varepsilon, y + \varepsilon) > 0$.*

(b) *If F charges $]0,\infty[$ and $]-\infty,0[$, two cases are possible*:

- *F is arithmetic of step $d > 0$*, i.e. $S_U = d\mathbb{Z}$;
- *F is non-arithmetic*, i.e., $S_U = \mathbb{R}$

Proof. (a) Let F and G be two distributions on $\mathbb{R}$ We have

$$a \in S_F,\ \ b \in S_G \Rightarrow a + b \in S_{F*G}.$$

Let us consider an r.v. X with distribution F and an r.v. Y with distribution G, X and Y independent. For every $\varepsilon > 0$,

$$P[|(X + Y) - (a + b)| \leqslant \varepsilon] \geqslant P[|X - a| \leqslant \varepsilon/2]$$

$$+ P[|Y - b| \leqslant \varepsilon/2] > 0.$$

(b) Let us study the support S_U of U. Let a and b be in S_U, $0 < a < b$. For every integer m and n the points $ma + nb$ are in S_U. Let $m \in \mathbb{N}$ such that $m(b - a) > a$. The points ma, $(m - 1)a + b$, ..., $(m - k)a + kb$, ..., $(m + 1)a$, mb form an increasing sequence of points spaced at most $(b - a)$.

Let $d = \inf\{|b - a|;\ 0 < a < b,\ (a,b) \in S_U^2\}$ and let $\varepsilon > d$. If S_U intersects $\mathbb{R}_+$, take a and b in S_U such that $0 < a < b$ with $b - a \leqslant \varepsilon$ and $x_\varepsilon = ma$ for $m(b - a) > a$. S_U contains $x_\varepsilon + (b - a)\mathbb{N}$. For $d > 0$, we can take $b - a = d$. In the alternative case, there would indeed be another pair (a',b') of the same type such that $d < b' - a' < b - a < 2d$ and S_U would contain pairs of positive points at distances $(b - a) - (b' - a') < d$. For every $c \in S_U \cap \mathbb{R}_+$, $x + c \in S_U$, thus $c \in d\mathbb{N}$ and $S_U \cap \mathbb{R}_+ \subset d\mathbb{N}$. The proof concludes thus in the case where F is defined on $\mathbb{R}_+$.

(c) Assume that F charges $]0,\infty[$ and $]-\infty,0[$. If $S_U \cap \mathbb{R}_+ \subset d\mathbb{N}$, S_U contains all the multiples nd of d for $n \geqslant n_0$. Then let $b < 0$, $b \in S_U$. For n large enough $b + nd > n_0 d$ and this number is a multiple of d: $b \in d\mathbb{Z}$. Hence F is a distribution concentrated on $d\mathbb{Z}$ and Proposition 4.1.10 applies; $S_U = d\mathbb{Z}$ (Nothing is changed in [4.2.3] by taking $d > 0$ to be real instead of d an integer > 0.)

In the non-arithmetic case, we apply (b) to $S_U \cap]0,\infty[$ and to $S_U \cap]-\infty,0[$. For every $\varepsilon > 0$, there exists an x_ε such that for $|y| > x_\varepsilon$ the ball $B(y,\varepsilon)$ with center y and radius ε is charged by U. Let $t \in \mathbb{R}$ and let y be such that $|y + (t/2)| > x_\varepsilon$ and $|y - (t/2)| > x_\varepsilon$. The measure U charges the balls $B((t/2) + y,\varepsilon)$ and $B((t/2) - y,\varepsilon)$ hence also the ball $B(t,2\varepsilon)$. This being true for all $\varepsilon > 0$, $t \in S_U$ and $S_U = \mathbb{R}$.

If F is a non-arithmetic distribution on $\mathbb{R}$ which is integrable and centered we obtain for the associated random walk a recurrence property analogous to Proposition 4.2.11. However, it deals only with recurrence in open sets of $\mathbb{R}$. If for example F and the initial distribution are concentrated on $\mathbb{Q}$, the set $\mathbb{R}\backslash\mathbb{Q}$ is not reached by the random walk.

Theorem 6.1.4. *Let F be an integrable, centered, non-arithmetic distribution on $\mathbb{R}$ which is non zero. For any initial distribution, a random walk associated with F returns almost surely infinitely often to each open set of $\mathbb{R}$.*

Proof. (a) First of all let us show the recurrence of the random walk with initial distribution F in every open set, i.e. in every ball $B(x,\varepsilon)$ with center x and radius $\varepsilon > 0$. Let

$$T_{x,\varepsilon} = \inf\{n;\ S_n \in B(x,\varepsilon)\},$$

$$N_{x,\varepsilon} = \sum_{n\geq 0} 1_{(S_n \in B(x,\varepsilon))} \cdot$$

From Proposition 6.1.3, for every x and every $\varepsilon > 0$, $P(T_{x,\varepsilon} < \infty) > 0$.

It is sufficient to show that, for every $\varepsilon > 0$, $P(N_{0,\varepsilon} = \infty) = 1$. To prove it, let us assume that $P(N_{x,2\varepsilon} < \infty) > 0$. Since

$$(S_{n+T_{-x,\varepsilon}} - S_{T_{-x,\varepsilon}})_{n\geq 1}$$

has, conditional on $(T_{-x,\varepsilon} < \infty)$, the same distribution as (S_n), we have,

$$P[T_{-x,\varepsilon} < \infty,\ \Sigma 1_{\{(S_{n+T_{-x,\varepsilon}} - S_{T_{-x,\varepsilon}}) \in B(x,2\varepsilon)\}} < \infty] > 0.$$

However

$$(S_{n+T_{-x,\varepsilon}} - S_{T_{-x,\varepsilon}}) \notin B(x,2\varepsilon)$$

implies $S_{n+T_{-x,\varepsilon}} \notin B(0,\varepsilon)$ and $P(N_{0,\varepsilon} < \infty) > 0$.

Let us show next that $U(B(0,\varepsilon)) = \infty$ for every $\varepsilon > 0$. By the weak law of large numbers,

$$P[|S_n/n| < \infty] \xrightarrow{n\to\infty} 1$$

and in the inequality

$$\frac{1}{N} U(B(0,N\varepsilon)) \geqslant \frac{1}{N} \sum_{n<N} F^{*n}(B(0,N\varepsilon))$$

$$\geqslant \frac{1}{N} \sum_{n<N} F^{*n}(B(0,n\varepsilon))$$

the right hand term tends to 1 for $N \to \infty$. However, for all $\varepsilon > 0$,

$$U(B(x,\varepsilon)) = E(\mathbf{1}_{(T_{x,\varepsilon}<\infty)} \Sigma \mathbf{1}_{(S_{n+T_{x,\varepsilon}} \in B(x,\varepsilon))})$$

$$\leqslant 1 + \Sigma P(T_{x,\varepsilon} < \infty, S_{n+T_{x,\varepsilon}} - S_{T_{n,\varepsilon}} \in B(0,2\varepsilon))$$

$$\leqslant 1 + U(B(0,2\varepsilon)).$$

The ball $B(0,N\varepsilon)$ can be covered by $(N\varepsilon/\eta + 1)$ balls of radius $\eta > 0$. Hence,

$$U(B(0,N\varepsilon)) \leqslant (N\varepsilon/\eta + 1)[U(B(0,2\eta)) + 1].$$

Dividing by N and letting $N \to \infty$, we obtain

$$\varepsilon/\eta[U(B(0,2\eta)) + 1] \geqslant 1.$$

This is true for all $\varepsilon > 0$ and $\eta > 0$. From which it follows that

$$U(B(0,2\eta)) = \infty, \quad \text{for every} \quad \eta > 0.$$

Finally, let $\sigma_\varepsilon = \sup(n; |S_n| < \varepsilon)$,

$$1 = P(N_{0,\varepsilon} = \infty) + P(T_{0,\varepsilon} = \infty) + \sum_{m \geq 0} P(\sigma_\varepsilon = m)$$

and:

$$1 \geq \sum_{m \geq 0} P(\sigma_\varepsilon = m)$$

$$\geq \sum_{m \geq 0} P(|S_m| < \varepsilon, |S_{n+m} - S_m| \geq 2\varepsilon \text{ for all } n > 0)$$

$$= U(B(0, \varepsilon))P(T_{0,2\varepsilon}) = \infty).$$

$U(B(0, \varepsilon)) = \infty$ implies $P(T_{0,2\varepsilon} = \infty) = 0$. For $\eta < \varepsilon$, we have,

$$P(|S_m| < \eta, |S_{n+m}| > 2\varepsilon \quad \text{for all } n > 0)$$

$$\leq P(|S_m| < \eta, |S_{n+m} - S_m| > \varepsilon - \eta \quad \text{for all } n > 0)$$

$$\leq P(|S_m| < \eta)P(T_{0,\varepsilon-\eta} = \infty) = 0.$$

Letting η tend to ε, we obtain $P(\sigma_\varepsilon = m) = 0$. From which it follows that: $1 = P(N_{0,\varepsilon} = \infty)$.

(b) If (S_n) is a random walk with an arbitrary initial distribution, associated with F, $(S_n - S_0)$ is a random walk with initial distribution F and is independent of S_0. From which, for all $y \in \mathbb{R}$ and every $\varepsilon > 0$,

$$P(\Sigma 1_{(S_n \in B(y, \varepsilon))} = \infty \mid S_0 = x)$$

$$= P(\Sigma 1_{(S_n - S_0 \in B(y-x, \varepsilon))} = \infty) = 1.$$

We also have $P(N_{y,\varepsilon} = \infty) = 1$.

6.1.5. The Renewal Theorem

The Renewal Theorem 6.1.5. *Let* (τ_n) *be a sequence of positive independent* r.v.'s *on a probability space* (Ω, A, P), *with the same distribution* F. *Assume* $0 < E(\tau_1) = m < \infty$. *Let*

$$T_n = \sum_{k=1}^{n} \tau_k, \quad N_t = \sum_{n=1}^{\infty} 1_{(T_n \leq t)},$$

$$U(t) = \sum_{n \geq 1} F^{*n}(t) = E(N_t).$$

(a) *If* F *is aperiodic, we have for all* $a > 0$,

$$\lim_{t\to\infty}[U(t+a) - U(t)] = \frac{a}{m}.$$

(b) *If F is periodic with period d, we have a similar result for $a \in d\mathbb{N}$.*

Proof. (1) Consider on a probability space $(\Omega', \mathcal{A}', P')$ a sequence (τ'_n) of independent r.v.'s with distribution F, and T'_0 independent of (τ'_n) having density $(1-F)/m$ with respect to Lebesgue measure. Then (Example (c) of [6.1.1]), setting

$$N'_t = \sum_{n\geq 0} \mathbf{1}_{(T'_n \leq t)},$$

we have

$$E[N'_{t+a} - N'_t] = \frac{a}{m}.$$

We must therefore prove that the renewal process, with $T_0 = 0$, is asymptotically close to the above.

(2) We use the coupling method already introduced in [4.3.4]. Assume F is aperiodic and let us work in the space $(\Omega, \mathcal{A}, P) \times (\Omega', \mathcal{A}', P')$ where we denote $T_n(\omega,\omega') = T_n(\omega)$, $T'_n(\omega,\omega') = T'_n(\omega')$, $\hat{P} = P \otimes P'$ and $\hat{E}$ the expectation under $\hat{P}$. The sequences (T_n) and (T'_n) are independent. The proof which follows resembles that of [4.3.5]. We study at which instant the two sequences are close enough, and starting from this time the evolutions are similar.

From Theorem 6.1.4, the random walk $(T'_n - T_n)$ returns infinitely often to $[0,\eta[$ for every $\eta > 0$ (in the case where F is arithmetic we can take $\eta = 0$).

Let $\eta > 0$ and let ν and ν' be the a.s. finite times given by

$$\nu = \inf\{n;\ 0 \leq T'_n - T_n \leq \eta\},$$

$$\nu' = \inf\{n;\ 0 \leq T_n - T'_n \leq \eta\}.$$

For every $\alpha > 0$, we have

$$\hat{E}\left[\sum_{n\leq\nu} \mathbf{1}_{[t<T_n\leq t+a]}\right]$$

$$= \hat{E}[(N_{t+a} - N_t)\mathbf{1}_{(t\leq T_\nu)}]$$

$$\leq \alpha\hat{P}[t \leq T_\nu] + \hat{E}[(N_{t+a} - N_t)\mathbf{1}_{(N_{t+a}-N_t\geq\alpha)}].$$

However, conditional on $\{N_u;\ u \leqslant t\}$, $(N_{t+a} - N_t)_{a \geqslant 0}$ is the counting process of a renewal process, the initial distribution of which is the conditional distribution of $T_{N_t+1} - t$, and the increments of which have the distribution F. It follows easily from the study of breakdowns subsequent to t made in [6.1.1].

However for a renewal process $(\breve{T}_n)$ with arbitrary initial distribution,

$$\breve{N}_a = \sum_{n=0}^{\infty} 1_{(\breve{T}_n \leqslant a)} \leqslant 1 + \sum_{n=1}^{\infty} 1_{(\breve{T}_n - \breve{T}_0 \leqslant a)}.$$

Thus $N_{t+a} - N_t$ is majorized in distribution by $N_a + 1$. From which,

$$\hat{E}[(N_{t+a} - N_t)1_{(N_{t+a}-N_t \geqslant \alpha)}]$$

$$\leqslant E[(N_a + 1)1_{(N_a+1 \geqslant \alpha)}].$$

For $\varepsilon > 0$, we fix α such that this majorant is $\leqslant \varepsilon/2$. For t large enough, we obtain

$$\hat{E}\left[\sum_{n \leqslant \nu} 1_{(t<T_n \leqslant t+a)}\right] \leqslant \varepsilon,$$

$$\lim_{t\to\infty} \hat{E}\left[\sum_{n \leqslant \nu} 1_{[t<T_n \leqslant t+a]}\right] = 0.$$

Similarly it can be shown

$$\lim_{t\to\infty} \hat{E}\left[\sum_{n \leqslant \nu} 1_{[t<T'_n \leqslant t+a]}\right] = 0;$$

(T_n) and (T'_n) are random walks adapted to $(\mathcal{F}_n) = (\sigma\{(T_p)_{p \leqslant n}, (T'_q)_{q \leqslant n}\})$, and ν is a stopping time for this filtration. From [6.1.2], the r.v.'s $(T'_{p+\nu} - T'_\nu)$ and $(T_{p+\nu} - T_\nu)$ are independent and identically distributed, from which it follows that

$$\hat{P}[t < T_{p+\nu} \leqslant t + a]$$

$$= \hat{P}[t + T'_\nu - T_\nu < T'_{p+\nu} \leqslant t + a + T'_\nu - T_\nu]$$

$$\geqslant \hat{P}[t + \eta < T'_{p+\nu} \leqslant t + a].$$

Likewise,

$$\hat{P}[t < T_{p+\nu'} \leqslant t + a] \leqslant \hat{P}[t < T'_{p+\nu'} \leqslant t + a + \eta].$$

We have

$$\frac{a}{m} = \lim_{t\to\infty} \hat{E}(N'_{t+a} - N'_t) = \lim_{t\to\infty} \hat{E}\left[\sum_{n\geqslant 1} \mathbf{1}_{(t<T'_n\leqslant t+a)}\right]$$

$$= \lim_{t\to\infty} \hat{E}\left[\sum_{p\geqslant 1} \mathbf{1}_{(t<T'_{p+\nu'}\leqslant t+a)}\right].$$

From which

$$\underline{\lim}_{t\to\infty} [U(t + a) - U(t)]$$

$$= \underline{\lim}_{t\to\infty} \hat{E}\left[\sum_{p\geqslant 1} \mathbf{1}_{(t<T_{p+\nu}\leqslant t+a)}\right] \geqslant \frac{a - \eta}{m};$$

$$\overline{\lim}_{t\to\infty} [U(t+a) - U(t)] = \overline{\lim}_{t\to\infty} \hat{E}\left[\sum_{p\geqslant 1} \mathbf{1}_{(t<T_{p+\nu'}\leqslant t+a)}\right]$$

$$\leqslant \frac{a + \eta}{m}.$$

This being true for all $\eta > 0$, part (a) of the theorem follows. Part (b) follows more simply by taking $\eta = 0$, $\nu = \nu'$.

Examples. (1) By reconsidering the example of Markov chains with a positive recurrent point a with period d, we have with the notations of [4.2] and [4.3]

$$\pi_{nd}(a,a) = E\left[\sum_{k\geqslant 1} \mathbf{1}_{\{T_a^k=nd\}}\right] \xrightarrow{n\to\infty} \frac{1}{E_a(T_a)}.$$

This result is also a consequence of Orey's theorem 4.3.17.

(2) Let us consider the random walk of [6.1.2] associated with a distribution F having a strictly positive mean m. Let us take here $X_0 = 0$. We first prove that $E(T_1) < \infty$. Let k be a constant such that $m' = E(X_1 \wedge k) > 0$ and $X'_n = X_n \wedge k$,

$$S'_n = X'_1 + \dots + X'_n, \quad T'_1 = \inf\{n;\ S'_n > 0\}.$$

We have $T_1 \leqslant T_1'$ and $S'_{T_1} \leqslant k$. Applying Wald's inequality to the stopping time $T_1' \wedge n$,

$$m' E[T_1' \wedge n] = E[S'_{T_1' \wedge n}] \leqslant k.$$

Letting $n \to \infty$ we obtain,

$$E(T_1) \leqslant E(T_1') \leqslant \frac{k}{m'}.$$

Then applying Wald's theorem again, we have,

$$E(S_{T_1}) = mE(T_1).$$

We have $P(X_1 > 0) = P(T_1 = 1) > 0$, thus,

$$\sum_{k=1}^{\infty} P(T_k = n) \xrightarrow[n \to \infty]{} \frac{1}{E(T_1)}.$$

Assuming the distribution of S_{T_1} to be aperiodic, we have, for $h > 0$, if $x \to \infty$,

$$\sum_{k=1}^{\infty} P(x < S_{T_k} \leqslant x + h) \to \frac{h}{E(S_{T_1})}.$$

Let us denote $M_n = \sup(S_p;\ 1 \leqslant p \leqslant n)$,

$$\sum_{n=1}^{\infty} P(x < M_n \leqslant x + h)$$

$$= \sum_{h=1}^{\infty} \sum_{k=1}^{\infty} P(x < S_{T_k} \leqslant x + h;\ T_k \leqslant n < T_{k+1})$$

$$= \sum_{k=1}^{\infty} E[(T_{k+1} - T_k) 1_{(x < S_{T_k} \leqslant x+h)}]$$

$$= E(T_1) \sum_{k=1}^{\infty} P(x < S_{T_k} \leqslant x + h).$$

Hence

$$\sum_{n=1}^{\infty} P(x < M_n \leqslant x + h) \to \frac{h}{m},$$

if $x \to \infty$.

6.2. Counting Processes

6.2.1. Counting Point Processes

On a probability space $(\Omega, \mathcal{A}, P)$, we observe successive random times (successive breakdowns...) $(T_n)_{n \geq 0}$. We assume $T_0 = 0$ and $T_n \leq T_{n+1}$, the inequality being strict for $T_n < \infty$. At each instant T_n, if it is finite, we make an observation Z_n, a measurable function from $(\Omega, \mathcal{A})$ into $(E, \mathcal{E})$. For every $\Gamma \in \mathcal{E}$, we consider the number of observations prior to t taking values in Γ,

$$N_t(\Gamma) = \sum_{n \geq 1} 1_{(T_n \leq t, Z_n \in \Gamma)} \cdot$$

Denote $N_t = N_t(E)$; N is a **counting point process.** We shall also say **point process** on its own. The observations (Z_n) are the **marks** of N. When E is reduced to a point, we have an unmarked point process or counting process. We then study (N_t) alone.

The most useful examples are the following. We can always consider a canonical version where $(\Omega, \mathcal{A}) = (E \times \mathbb{R}, \mathcal{E} \otimes \mathcal{B}_{\mathbb{R}^+})$ and (Z_n, T_n) is the nth coordinate.

(a) **Renewal processes studied in [6.1].** We also meet **renewal processes with repair**: if, after each breakdown, the strictly positive random duration till replacement or till the machine functions as new has distribution G, and if these durations and the lifetimes of the successive machines are independent. T_{2n-1} is the breakdown time of the nth machine, T_{2n} that of the start-up of the $(n+1)$th; $T_{2n} - T_{2n-1} = U_{2n-1}$ has distribution G, $T_{2n+1} - T_{2n} = U_{2n}$ has distribution F, and these r.v.'s $(U_n)_{n \geq 0}$ are independent.

(b) **Markovian jump processes.** We are given an r.v. q on E taking values in $]0,\infty[$, and a transition π from $(E, \mathcal{E})$ into $(E, \mathcal{E})$. We assume the sequences (T_n) and (Z_n) to be adapted to a filtration $\mathbb{G} = (\mathcal{G}_n)$, $T_0 = 0$. Finally, conditional on $\mathcal{G}_n$:

- $T_{n+1} - T_n$ has an exponential distribution with parameter $q(Z_n)$, denoted $e(q(Z_n))$.
- Z_{n+1} has distribution $\pi(Z_n, \cdot)$
- $T_{n+1} - T_n$ and Z_{n+1} are independent.

Set $T_\infty = \sup T_n$, $\tau_n = T_n - T_{n-1}$.

For $t \in [T_n, T_{n+1}[$, set $X_t = Z_n$. Then $(\Omega, \mathcal{A}, P, (X_t)_{0 \leqslant t < T_\infty})$ is a **jump process associated with** q **and** π, T_∞ is its **explosion time.** We often assume T_∞ to be a.s. infinite. In particular this is the case if q is bounded. Let A majorize q and let $a > 0$. We have,

$$\sum_n P[T_{n+1} - T_n \geqslant a | \mathcal{G}_n] = \sum_n \exp[-q(Z_n)a] \geqslant \sum_n e^{-aA} = \infty.$$

From the Borel-Cantelli lemma 2.7.33, the event $\{T_{n+1} - T_n \geqslant a\}$ occurs a.s. infinitely often and T_∞ is a.s. infinite.

Here is a construction of such a process. Let $\tilde{\pi}$ be the transition probability of $E \times \mathbb{R}_+$ into itself defined by $\tilde{\pi}(x,t;\cdot) = \pi(x,\cdot) \otimes e(q(x))$. We consider the canonical Markov chain associated with $\tilde{\pi}$: $\{\Omega, \mathcal{A}, (P_x)_{x \in E}, (Z_n, \tau_n)_{n \geqslant 0}\}$, denoting by P_x the distribution of the chain with initial distribution $\delta_x \otimes \delta_0$. We then say that $\{\Omega, \mathcal{A}, (P_x)_{x \in E}, (X_t)_{0 \leqslant t < T_\infty}\}$ is the canonical jump process associated with q and π... although we are dealing with a family of processes, which are not canonical in the sense of Definition 0.2.4. For a probability ν on E, we denote $P_\nu = \int \nu(dx) P_x$.

Taking $E = \mathbb{N}$, and $\pi(n,n+1)$ equal to 1 for all n, $Z_0 = 0$, we obtain the Poisson process. Still with $E = \mathbb{N}$, we have a **birth and death process** when

$$\pi(0,1) = 1; \quad \pi(n,n+1) = 1 - \pi(n,n-1)$$

for $n > 1$. In other words being in state n, there can be a birth at a random time α_n or a death at a random time β_n. Assuming α_n and β_n are independent of one another and independent of the past, α_n exponential with parameter λ_n and β_n with parameter μ_n, the jump takes place at time $\alpha_n \wedge \beta_n$, exponential with parameter $\lambda_n + \mu_n$,

$$P[\alpha_n < \beta_n] = \pi(n,n+1) = \frac{\lambda_n}{\lambda_n + \mu_n} = 1 - \pi(n,n-1),$$

$$q(n) = \lambda_n + \mu_n .$$

(These formulae hold for $n = 0$ by taking $\mu_0 = 0$.) These processes give good models for describing queues. Let A_n be the arrival time of the nth customer, B_n the time at which the

nth service finishes. Assume the sequences $(\Delta A_n = A_n - A_{n-1})_{n \geqslant 1}$ and $(\Delta B_n = B_n - B_{n-1})_{n \geqslant 1}$ are independent and formed from r.v.'s independent of each other. Here the r.v.'s $A_n - A_{n-1}$ are exponential with parameter λ. X_t is the number of customers in the shop (who are waiting or being served). Here are some schemes (M is the code for "exponential distribution," whilst, in [4.1.3] G is the code for "general distribution").

M/M/1 Queue. The r.v.'s $B_n - B_{n-1}$ are exponential with parameter μ; $\lambda_0 = \lambda$, $\mu_0 = 0$; $\lambda_n = \lambda$, $\mu_n = \mu$ for $n \geqslant 1$.

M/M/s Queue. There are s servers; if there are n customers in the line, $s \wedge n$ servers work at the same time independently of one another and their service lasts an exponential time with parameter μ. One of these services is completed at a time exponentially distributed with parameter $(s \wedge n)\mu$. Thus: $\lambda_n = \lambda$ and $\mu_n = (s \wedge n)\mu$.

M/M/s/k Queue. This is the same mechanism as M/M/s, but the capacity of the room is limited to k. Then $E = \{0,1, ..., k\}$ and

$$\lambda_n = \begin{cases} \lambda & \text{for } 0 \leqslant n < k \\ 0 & \text{for } n = k \end{cases} \qquad \mu_n = (s \wedge n)\mu.$$

M[x]/M/1 Queue (group arrivals). The customers arrive in groups, the time interval between the arrivals of two groups of k customers is exponential with parameter $\lambda(k)$ and the intervals are independent of each other;

$$\lambda = \sum_{k \geqslant 1} \lambda(k) < \infty.$$

We again have $E = \mathbb{N}$, however:

$$q(0) = \lambda, \quad q(n) = \lambda + \mu \quad \text{for} \quad n \geqslant 1$$

$$\begin{cases} \pi(n,n+k) = \dfrac{\lambda(k)}{\lambda + \mu} \\ \pi(n,n-1) = \dfrac{\mu}{\lambda + \mu} \end{cases} \text{for } n \geqslant 1, \ \pi(0,k) = \frac{\lambda(k)}{\lambda}.$$

(c) **Semi-Markov processes.** We can extend the definition of a Markovian jump process by removing the hypothesis of exponential distribution for the time. We are given π, a transition from $(E,\mathcal{E})$ into $(E \times \mathbb{R}_+, \mathcal{E} \otimes \mathcal{B}_{\mathbb{R}+})$, and we assume,

$$P[T_{n+1} - T_n \geq t, Z_{n+1} \in \Gamma | Z_0, ..., Z_n, T_1, ..., T_n]$$
$$= \pi(Z_n; \Gamma \times [t,\infty[).$$

The sequence $(Z_n,T_n)_{n\geq 1}$ is a **semi-Markov chain.**
By defining

$$X_t = \sum_{n\geq 0} Z_n \mathbf{1}_{(T_n \leq t < T_{n+1})},$$

we call (X_t) a **semi-Markovian process** on E.

M/G/1 Queue. Assume that the distribution of ΔA_n is exponential with parameter λ. Consider the exit time T_n of the nth customer and the number Z_n of customers who remain in the shop just after this customer has left. Let N_{n+1} be the number of arrivals during ΔB_{n+1}, the service time of the $(n+1)$th customer,

$$P(N_{n+1} = j | \Delta B_{n+1} = t) = e^{-\lambda t} \frac{(\lambda t)^j}{j!}$$

when Z_n is nonzero, we have

$$\Delta B_{n+1} = T_{n+1} - T_n,$$

$$P[Z_{n+1} = j, T_{n+1} - T_n > t | Z_0, ..., Z_n, T_1, ..., T_n]$$
$$= \int_t^\infty e^{-\lambda u} \frac{(\lambda u)^{j+1-Z_n}}{(j + 1 - Z_n)!} d\beta(u).$$

For Z_n zero, a customer must arrive first of all in order for there to be service

$$P[Z_{n+1} = j, T_{n+1} - T_n > t | Z_0, ..., Z_n, T_1, ..., T_n]$$
$$= \int_0^\infty \lambda e^{-\lambda s} ds \int_{t-s}^\infty e^{-\lambda u} \frac{(\lambda u)^j}{j!} d\beta(u).$$

We therefore construct a semi-Markov chain (Z_n,T_n).

G/M/1 Queue. Here it is the services ΔB_n which have an exponential distribution with parameter μ. We consider Z_n, the number of customers in the shop just before the arrival of the nth customer (hence just before A_n). Let M_{n+1} be the number of customers who have time to be served in the time interval $[A_n, A_{n+1}[$:

$$P(M_{n+1} = j | \Delta A_{n+1} = t) = e^{-\mu t} \frac{(\mu t)^j}{j!},$$

$$Z_{n+1} = (Z_n + 1 - M_{n+1})_+ .$$

As for the sequence (W_n) for the G/G/1 queue, we see that (Z_n) is a Markov chain. For $0 < j \leqslant Z_n + 1$,

$$P[Z_{n+1} = j, A_{n+1} - A_n > t | Z_0, \ldots, Z_n, T_1, \ldots, T_n]$$

$$= \int_t^\infty e^{-\mu u} \frac{(\mu u)^{Z_n+1-j}}{(Z_n + 1 - j)!} d\alpha(u).$$

$$P[Z_{n+1} = 0; A_{n+1} - A_n > t | Z_0, \ldots, Z_n, T_1, \ldots, T_n]$$
$$= \int_t^\infty e^{-\mu u} \left(\sum_{k \geqslant Z_n+1} \frac{(\mu u)^k}{k!} d\alpha(u) \right).$$

Thus (Z_n, A_n) is a semi-Markov chain.

6.2.2. Filtrations and Martingales

We deal with the obvious generalizations of Definitions 2.2.1 and 2.2.2 when the time t varies in $\mathbb{R}_+$.

As for a sequence of experiments, observations through time are often used to predict the future (a prediction problem) or are used for statistical problems. If the phenomenon under study is described by (Ω, A, P) at time t, a decision can only be made on the basis of events observed before t of which we assume that they form a σ-algebra F_t. If for example the observation is the process defined above, we shall often take $F_t = \sigma(X_s: s \leqslant t)$. Certainly, for $s \leqslant t$, an event observed before s is before t, and $F_s \subset F_t$.

Definition 6.2.5. (a) A **filtration** $\mathbb{F} = (F_t)_{t \in \mathbb{R}_+}$ on (Ω, A, P) is an increasing family of sub-σ-algebras of A: F_t is interpreted as the set of observable events at time t.

(b) A process $(\Omega, A, P, (X_t)_{t \in \mathbb{R}_+})$ is **adapted** to $\mathbb{F}$ if, for every t, X_t is observable at time t, i.e., if X_t is measurable on (Ω, F_t).

Definition 6.2.6. On (Ω, A, P) equipped with a filtration $\mathbb{F} = (F_t)_{t \in \mathbb{R}_+}$ a real process $X = (X_t)_{t \in \mathbb{R}_+}$ adapted to $\mathbf{F}$ is:

(a) A **submartingale** if, for every t and h in $\mathbb{R}_+$,

$$E(X_t^+) < \infty \quad \text{and} \quad E(X_{t+h} | F_t) \geq X_t.$$

(b) A **supermartingale** if, for every t and h in $\mathbb{R}_+$,

$$E(X_t^-) < \infty \quad \text{and} \quad E(X_{t+h} | F_t) \leq X_t.$$

(c) A **martingale** if, for every t and h in $\mathbb{R}_+$,

$$E(|X_t|) < \infty \quad \text{and} \quad E(X_{t+h} | F_t) = X_t.$$

If there is any confusion, the filtration $\mathbb{F}$ is specified and we speak of the $\mathbb{F}$-martingale or of the martingale adapted to $\mathbf{F}$. Likewise for sub or supermartingales.

6.2.3. Compensator of a Point Process

Let T be the random time of a breakdown, assumed strictly positive. Let F be its distribution. We denote $F(t) = P(T \leq t)$. When F has a density f we speak of the failure rate at time t,

$$h(t) = \lim_{u \to 0} \frac{1}{u} P[t < T \leq t + u \mid T > t] = \frac{f(t)}{1 - F(t)}.$$

If this rate is a constant equal to λ, F is the exponential distribution with parameter λ. If h increases there is wear and tear. If h decreases we are in a "breaking in" period (cf. Vol. I, E4.2.3). Without assuming that F has a density, we have, for every $u > 0$, denoting $\int_a^b$ for $\int_{]a,b]}$:

$$P[t < T \leqslant t + u|T > t] = \frac{1}{P(T > t)} \int_t^{t+u} dF(s)$$

$$\int_t^{t+u} dF(s) = \int_t^{t+u} \frac{1}{1 - F(s-)} \left[\int_{[s,\infty[} dF(u)\right] dF(s)$$

$$= E\left[\int_t^{t+u} \frac{dF(s-)}{1 - F(s-)} 1_{(T \geqslant s)}\right].$$

At each instant t, we have observed $\mathcal{F}_t = \sigma\{1_{(T \leqslant u)}; u \leqslant t\}$ which has an atom, the event $\{T > t\}$ and the trace of $\mathcal{F}_t$ on $\{T \leqslant t\}$ is that of $\sigma(T)$. Let us set $N_t = 1_{(T \leqslant t)}$ and

$$\tilde{N}_t = \int_0^{t \wedge T} \frac{dF(s)}{1 - F(s-)}.$$

When F has a density,

$$\tilde{N}_t = \int_0^{t \wedge T} h(s)ds = -\mathrm{Log}(1 - F(t \wedge T)).$$

The process $N - \tilde{N} = (N_t - \tilde{N}_t)_{t \geqslant 0}$ is a martingale adapted to $(\sigma(T \wedge t)_{t \geqslant 0})$. In fact we have, for $t \geqslant 0$ and $u \geqslant 0$,

$$E[N_{t+u} - N_t | \mathcal{F}_t] = E[\tilde{N}_{t+u} - \tilde{N}_t | \mathcal{F}_t],$$

since on $\{T > t\}$ this is the equality written above, and on $\{T \leqslant t\}$ the two terms are zero.

Now let us consider $\mathcal{F}_0$, a sub σ-algebra of $\mathcal{A}$, and, for $n \in \mathbb{N}$ and $t \in \mathbb{R}_+$,

$$\mathcal{G}_n = \sigma(T_1, T_2, \dots, T_n, Z_1, \dots, Z_n) \vee \mathcal{F}_0,$$

$$\mathcal{F}_t = \sigma(N_s(\Gamma); s \leqslant t, \Gamma \in \mathcal{E}) \vee \mathcal{F}_0,$$

$$\mathbb{F} = (\mathcal{F}_t)_{t \geqslant 0}.$$

On $\{T_n < \infty\}$, let us set $\tau_{n+1} = T_{n+1} - T_n$. We are given a transition probability F_n from $(\Omega, \mathcal{G}_n)$ into $(]0,\infty] \times E, \mathcal{B}_{]0,\infty]} \otimes \mathcal{E})$, the distribution of (τ_{n+1}, Z_{n+1}) conditional on $\mathcal{G}_n$. Then

$$P[\tau_{n+1} \geqslant t | \mathcal{G}_n] = F_n(\cdot; [t,\infty[\times E)$$

is denoted $H_n(t)$. The trace of $\mathcal{F}_n$ on $\{T_n \leqslant t < T_{n+1}\}$ is that of $\mathcal{G}_n$, since on $\{T_n \leqslant t < T_{n+1}\}$ we have observed only the events of G_n and $\{t < T_{n+1}\}$ (cf. Vol. I, E6.2.9). Let $A \in G_n$; the same calculation as above gives

$$P[A \cap \{T_n \leqslant t < T_{n+1} \leqslant t + u, Z_{n+1} \in \Gamma\}]$$

$$= E\left[1_A 1_{(T_n \leqslant t)} \int 1_{]t-T_n, t+u-T_n]}(s) 1_{(T_{n+1} \geqslant s)} \cdot \frac{1_\Gamma(x)}{H_n(s)} dF_n(s,x)\right],$$

$$N_t^{n+1}(\Gamma) = 1_{[T_{n+1} \leqslant t, Z_{n+1} \in \Gamma]}$$

and

$$\tilde{N}_t^{n+1}(\Gamma) = \int_0^{(t-T_n)^+ \wedge (T_{n+1}-T_n)} \frac{1_\Gamma(x)}{H_n(s)} dF_n(s,x).$$

We have, on $\{T_n \leqslant t < T_{n+1}\}$:

$$E[N_{t+u}^{n+1}(\Gamma) - N_t^{n+1}(\Gamma) | F_t]$$

$$= E[\tilde{N}_{t+u}^{n+1}(\Gamma) - \tilde{N}_t^{n+1}(\Gamma) | F_t].$$

Since this relation is clear on $\{T_n \leqslant t < T_{n+1}\}^c$, the process $(N_t^{n+1}(\Gamma) - \tilde{N}_t^{n+1}(\Gamma))_{t \geqslant 0}$ is an $\mathbb{F}$-martingale.

Theorem 6.2.7. *With the above notations, denoting* $\tilde{N}_0(\Gamma) = 0$ *and*

$$\tilde{N}_t(\Gamma) = \tilde{N}_{T_n}(\Gamma) + \int 1_{(s<t-T_n; x\in\Gamma)} \frac{dF_n(s,x)}{H_n(s)}$$

on $\{T_n < t \leqslant T_{n+1}\}$, *the process* $\{N_t(\Gamma) - \tilde{N}_t(\Gamma)\}_{t\geqslant 0}$ *stopped at* T_n *is, for each n, a centered* $\mathbb{F}$*-martingale. We shall say that N is a point process compensated by* $\tilde{N}$. *When, for every t,* $N_t(\Gamma)$ *is integrable, the process* $(N_t(\Gamma) - \tilde{N}_t(\Gamma))_{t\geqslant 0}$ *is itself a centered martingale.*

Proof. The process stopped at T_n is none other than

$$\sum_{p=1}^n N_t^p(\Gamma) - \sum_{p=1}^n \tilde{N}_t^p(\Gamma);$$

from which the first part of the theorem follows. If $N_t(\Gamma)$ is integrable for all t, we can apply Beppo-Levi's theorem and let n tend to ∞ in the relations

$$E\left[\sum_{p=1}^{n} (N^p_{t+u}(\Gamma) - N^p_t(\Gamma)| \mathcal{F}_t\right]$$

$$= E\left[\sum_{p=1}^{n} (\tilde{N}^p_{t+u}(\Gamma) - \tilde{N}^p_t(\Gamma))| \mathcal{F}_t\right].$$

Thus $(N_t(\Gamma) - \tilde{N}_t(\Gamma))$ is an $\mathbb{F}$-martingale.

Examples. (with the notation of [6.2.1]): (a) *Renewal process.*

$$A(t) = \int_0^t \frac{dF(u)}{1 - F(u-)} \quad \text{for } T_n < t \leqslant T_{n+1},$$

we have

$$\tilde{N}_t = A(\Delta T_1) + \dots + A(\Delta T_n) + A(t - T_n);$$

$\tilde{N}_t = \lambda t$, for a Poisson process.

(b) *Markovian jump process.* We can consider the **number of jumps landing in** Γ, i.e.,

$$N_t(\Gamma) = \sum_{n \geqslant 1} \mathbf{1}_{(T_n \leqslant t, Z_n \in \Gamma)} \cdot$$

We then take $\mathcal{F}_0 = \sigma(Z_0)$; for $T_n < t \leqslant T_{n+1}$ and for every initial distribution, we have,

$$\tilde{N}_t(\Gamma) = \tilde{N}_{T_n}(\Gamma) + \int_0^{t-T_n} q(Z_n)\pi(Z_n,\Gamma)du$$

$$= \int_0^t q(X_s)\pi(X_s,\Gamma)ds.$$

When E is countable, we also often consider, for $(i,j) \in E^2$, the number of jumps from i to j before t

$$N^{ij}_t = \sum_{n \geqslant 1} \mathbf{1}_{(T_n \leqslant t)(Z_{n-1}=i,\ Z_n=j)}$$

compensated by $\tilde{N}^{ij}_t = q(i)\pi(i,j)\tau_i(t)$ where

$$\tau_i(t) = \int_0^t \mathbf{1}_{\{i\}}(X_s)ds$$

is the time spent in i before t.

(c) *Birth and death process (and queues).* We introduce the number of births $N_t^+ = \Sigma_{i\geq 0} N_t^{i,i+1}$ and the number of deaths $N_t^- = \Sigma_{i\geq 1} N_t^{i,i-1}$. For an M/M/1 queue for example, we obtain,

$$\tilde{N}_t^+ = \lambda t, \qquad \tilde{N}_t^- = \mu \int_0^t 1_{(X_s \neq 0)} ds.$$

(d) *Semi-Markovian processes.* Having defined $N_t(\Gamma)$ as for jump processes, we easily obtain with $F_0 = \sigma(Z_0)$ for every initial distribution

$$\tilde{N}_t(\Gamma) = \int_0^t \frac{\pi(X_s;\Gamma,ds)}{\pi(X_s;E \times [s,\infty[)}.$$

6.2.4. Preliminaries on Stieltjes' Integral

Let A: $t \longmapsto A_t$ be an increasing function, continuous on the right, zero at 0, from $\mathbb{R}_+$ into $\mathbb{R}$. We associate with it a measure μ_A on $\mathbb{R}_+$ such that $\mu_A([0,t]) = A_t$.

Definition 6.2.8. A function V: $t \longmapsto V_t$ from $\mathbb{R}_+$ into $\mathbb{R}$ is of **bounded variation** if for every t,

$$T_t^V = \sup\left\{\sum_{i=0}^{n} |V_{t_i+1} - V_{t_i}|;\ 0 = t_0 < t_1 <\ldots< t_n = t\right\}$$

is finite. The function T^V is the **total variation** of V.

The study of these functions and the Stieltjes' integral appear in Kolmogorov-Fomine (Chapter 6) or Rudin (Chapter 8). It can be proved that V: $t \to V_t$, a function from $\mathbb{R}_+$ into $\mathbb{R}$ continuous on the right, is of bounded variation if, and only if, $V - V(0)$ is the difference of two functions A and B from $\mathbb{R}_+$ into $\mathbb{R}$, which are continuous on the right, increasing and zero at 0. Let the measure $\mu_V = \mu_A - \mu_B$ on $]0,\infty[$ and let $\mu_V\{0\} = V(0)$. Set, for $0 \leq u < t$,

$$\begin{aligned}\int_{]u,t]} f(s)d\mu_V(s) &= \int_u^t f(s)dV_s \\ &= \int_{]u,t]} f(s)dV(s).\end{aligned}$$

The total variation T^V of V is equal to $A + B$. We have

$$\left|\int_u^t f(s)dV_s\right| \leqslant \int_u^t |f(s)|dT_s^V.$$

Finally $\Delta V_t = V_t - V_{t-}$ denotes the jump of V at t, zero except for an at most countable set of t,

$$\sum_{s\leqslant t} |\Delta V_s| \leqslant T_t^V.$$

Integration by Parts Formula 6.2.9. If U and V are two functions of bounded variation,

$$U_tV_t = U_0V_0 + \int_0^t U_{s-}dV_s + \int_0^t V_s dU_s$$
$$= U_0V_0 + \int_0^t U_{s-}dV_s + \int_0^t V_{s-}dU_s + \sum_{s\leqslant t} \Delta U_s \Delta V_s$$

where $\Delta U_s = U_s - U_{s-}$; $\Delta V_s = V_s - V_{s-}$.

Proof. It is sufficient to calculate

$$\mu_U \otimes \mu_V[0,t]^2 = \mu_U \otimes \mu_V\{0\}$$
$$+ \mu_U \otimes \mu_V\{(x,y); 0 < x < y; 0 < y \leqslant t\}$$
$$+ \mu_U \otimes \mu_V\{(x,y); 0 < y \leqslant x; 0 < x \leqslant t\}.$$

Consequence. Let us assume that A is increasing, continuous on the right and zero at 0. Then,

$$A_t^2 = \int_0^t A_{s-}dA_s + \int_0^t A_s dA_s$$
$$\int_0^t A_{s-}dA_s \leqslant \frac{A_t^2}{2} \leqslant \int_0^t A_s dA_s$$

and by recurrence,

$$\int_0^t A_{s-}^{n-1}dA_s \leqslant \frac{A_t^n}{n!} \leqslant \int_0^t A_s^{n-1}dA_s\,.$$

Change of time formula 6.2.10. Let A be an increasing function, continuous on the right, from $\mathbb{R}_+$ into $\mathbb{R}_+$ such that $A_0 = 0$. For $0 < t < A_\infty = \sup A_t$, let us set $\tau(t) = \inf\{s; A_s > t\}$. Then, for every positive r.v. f on $\mathbb{R}_+$, we have,

$$\int_0^t f(s)dA_s = \int_0^{A_t} f(\tau(s))ds.$$

Proof. For $f = 1_{]0,u]}$ with $0 < u \leqslant t$, the equality is written

$$A_u = \int_0^{A_t} 1_{\{s \leqslant A_u\}} ds.$$

Hence it is true. The result is extended to step functions then to positive r.v.'s in the usual way.

Proposition 6.2.11. Solution of an Integral Equation. *Let $V = (V_t)_{t \geqslant 0}$ be a function continuous on the right, of bounded variation from $\mathbb{R}_+$ into $\mathbb{R}$, such that $V_0 = 0$. Then the integral equation*

$$Z_t = Z_0 + \int_0^t Z_{s-} dV_s$$

has a unique solution satisfying $\sup_{s \leqslant t} |Z_s| < \infty$ *for all t. If, for all t, we denote*

$$V_t = \sum_{s \leqslant t} \Delta V_s + V_t^c$$

(V^c is the continuous part of V), the solution is

$$Z_t = Z_0 \prod_{s \leqslant t} (1 + \Delta V_s) \exp \int_0^t dV_s^c .$$

Proof. (a) Note first of all that if V is of bounded variation, then so also is the function e^V: $t \longmapsto \exp V_t$. In fact for $0 = t_0 < \ldots < t_{n+1} = t$, we have,

$$\sum_{i=1}^{n} |\exp V_{t_{i+1}} - \exp V_{t_i}| \leqslant \sup_{s \leqslant t} (\exp V_s) \sum_{i=0}^{n} |V_{t_{i+1}} - V_{t_i}|$$

and

$$T_t^{e^V} \leqslant (\exp T_t^V) T_t^V .$$

(b) Set $X_t = Z_0 \prod_{s \leqslant t} (1 + \Delta V_s)$ and

$$Y_t = \exp \int_0^t dV_s^c .$$

The functions X and Y are of bounded variation, since their logarithms are of bounded variation. The integration by parts formulae gives,

$$Z_t \leqslant X_t Y_t = Z_0 + \int_0^t X_{s-} Y_s dV_s^c + \sum_{s \leqslant t} Y_s X_{s-} \Delta V_s$$
$$= Z_0 + \int_0^t Z_{s-} dV_s.$$

(c) The difference $\widetilde{Z}$ between two bounded solutions on $[0,t]$ satisfies

$$\widetilde{Z}_t = \int_0^t \widetilde{Z}_{s-} dV_s .$$

Denoting $\ell = \sup_{s \leqslant t} |\widetilde{Z}_{s-}|$ we have, for $s < t$, $|\widetilde{Z}_s| \leqslant \ell T_s^V$ and

$$|\widetilde{Z}_t| \leqslant \ell \int_0^t T_{s-}^V dT_s^V \leqslant \frac{\ell (T_t^V)^2}{2};$$

$$|\widetilde{Z}_t| \leqslant \frac{\ell}{2} \int_0^t (T_{s-}^V)^2 dT_s^V \leqslant \frac{1}{6} (T_t^V)^3.$$

By continuing, we obtain

$$|\widetilde{Z}_t| \leqslant \frac{1}{n!} (T_t^V)^n ,$$

which tends to 0 if n tends to $+\infty$: $\widetilde{Z}$ is zero.

6.2.5. Stochastic Integral with Respect to an Increasing Process

$(\Omega, \mathcal{A}, P)$ and a filtration $\mathbb{F} = (\mathcal{F}_t)_{t \in \mathbb{R}_+}$ are fixed.

Definition 6.2.12. (a) An $\mathbb{F}$**-increasing process** A is a real valued process adapted to $\mathbb{F}$ satisfying the following property: for every $\omega \in \Omega$, the trajectories $t \longmapsto A_t(\omega)$ are increasing, continuous on the right and zero at 0.

(b) The $\mathbb{F}$-increasing process A is said to be **integrable, square integrable**, ... if that is the case for the r.v.'s A_t, for any t.

(c) Another integrable $\mathbb{F}$-increasing process $\widetilde{A}$ is a **compensator** of A, an integrable $\mathbb{F}$-increasing process, if the process $A - \widetilde{A}$ is an $\mathbb{F}$-martingale.

Note. With this definition, A may have several compensators, in particular itself. However the examples given in [6.2.3] are sufficient to justify the interest in this notion.

In what follows we consider integrable $\mathbb{F}$-increasing processes A and $\tilde{A}$, such that $V = A - \tilde{A}$ is an $\mathbb{F}$-martingale ($\tilde{A}$ is a compensator of A). For which type of processes $C = (C_s)_{s\geq 0}$ does Stieltjes' integral ($\int_0^t C_s dV_s$) make sense, and do we have the property "$C \cdot V = (\int_0^t C_s dV_s)_{t\geq 0}$ is a centered $\mathbb{F}$-martingale?"

Let $s < t$ and $\Gamma \in \mathcal{F}_s$. Set $C_u(\omega) = \mathbf{1}_{]s,t]\times\Gamma}(u,\omega)$. It is easily seen that $C.V$ is a martingale. The set Q of subsets of $]0,\infty] \times \Omega$ such that, for $\mathbf{1}_Q(t,\omega) = C_t(\omega)$, the process $C.V$ has the property, is stable under differences and under increasing limits. It contains the σ-algebra $\mathcal{P}$ generated by the sets $]s,t] \times \Gamma$ for $s < t$ and $\Gamma \in \mathcal{F}_s$ $(0 < s < t < \infty)$.

Let C be a step r.v. on $(]0,\infty[\times \Omega, \mathcal{P})$. $C.V$ is again a centered martingale, thus

$$E\left[\int_0^t C_s dA_s\right] = E\left[\int_0^t C_s d\tilde{A}_s\right].$$

This last equality also holds for every positive r.v. C on $(]0,\infty[\times \Omega, \mathcal{P})$. If C is an r.v. on $(]0,\infty[\times \Omega, \mathcal{P})$, when

$$E\left[\int_0^t |C_s| dA_s\right] < \infty$$

we also have

$$E\left[\int_0^t |C_s| d\tilde{A}_s\right] < \infty,$$

and $C.V$ is also a martingale.

Definition 6.2.13. The σ-algebra $\mathcal{P}$ of $]0,\infty[\times \Omega$ generated by the sets $]s,t] \times \Gamma$, where $\Gamma \in \mathcal{F}_s$ and $0 < s < t$, is the **σ-algebra of predictable sets.** An r.v. C defined on $(]0,\infty[\times \Omega, \mathcal{P})$ is a **predictable process** (we denote $C(t,\omega) = C_t(\omega)$). We also say $\mathbb{F}$-predictable if we want to specify $\mathbb{F}$.

Examples of Predictable Processes. Real processes C, adapted to $\mathbb{F}$ and *continuous on the left*, are predictable. Similarly for *deterministic processes* (measurable functions C from $\mathbb{R}_+$ into $\mathbb{R}$). It is easy to see for deterministic processes since $\mathcal{B}_{\mathbb{R}_+} = \sigma(]s,t], 0 < s < t)$, and also for left continuous processes C by writing, for $t > 0$,

$$C_t = \lim_n \left[\sum_{0\leq q<2^n} C_{q/2^n} \mathbf{1}_{\{(q/2^n)<t\leq(q+1)/2^n\}}\right].$$

Proposition 6.2.14. *For every positive predictable process and every t, $\int_0^t C_s dA_s$ is an* r.v. *and*

$$E\left[\int_0^t C_s dA_s\right] = E\left[\int_0^t C_s d\tilde{A}_s\right].$$

We say that C is a A-integrable, when

$$E\left[\int_0^t |C_s| dA_s\right] = E\left[\int_0^t |C_s| d\tilde{A}_s\right]$$

is finite for every t. Then the process

$$C\left(A - \tilde{A}\right) = \left(\int_0^t C_s dA_s - \int_0^t C_s d\tilde{A}_s\right)_{t \geqslant 0}$$

is a centered $\mathbb{F}$-martingale.

Now let us consider two integrable increasing processes A and B, $\mathbb{F}$-adapted and right continuous, compensated by $\tilde{A}$ and $\tilde{B}$. We have, by the integration by parts formula,

$$A_t^2 = 2\int_0^t A_{s-}\, dA_s + \sum_{s \leqslant t} \Delta A_s^2 .$$

If A is square integrable, (A_{t-}) is A-integrable and

$$E\left(\sum_{s \leqslant t} \Delta A_s^2\right) < \infty.$$

If C and D are two bounded $\mathbb{F}$ -adapted processes, continuous on the left, we have

$$\begin{aligned}
&(C \cdot (A - \tilde{A}))_t (D \cdot (B - \tilde{B}))_t \\
&\quad = \int_0^t C_s D_s (A - \tilde{A})_{s-} d(B - \tilde{B})_s \\
&\quad + \int_0^t C_s D_s (B - \tilde{B})_{s-} d(A - \tilde{A})_s \\
&\quad + \sum_{s \leqslant t} C_s D_s \Delta(A - \tilde{A})_s \Delta(B - \tilde{B})_s .
\end{aligned}$$

(This can be verified by the integration by parts formula by taking first of all C and D to be positive, then by separating the positive and negative parts). In particular,

$$(A - \tilde{A})_t^2 = 2\int_0^t (A - \tilde{A})_{s-} d(A - \tilde{A})_s + \sum_{s \leqslant t} \Delta(A - \tilde{A})_s^2 .$$

Proposition 6.2.15. *Let A, $\widetilde{A}$, B and $\widetilde{B}$ be four square integrable $\mathbb{F}$-increasing processes such that $A - \widetilde{A}$ and $B - \widetilde{B}$ are $\mathbb{F}$-martingales. Then*

$$\sum_{s \leq t} E(\Delta A_s - \Delta \widetilde{A}_s)^2 < \infty.$$

For every bounded predictable processes C and D, $M = (M_t)_{t \geq 0}$ is a centered $\mathbb{F}$-martingale, by setting

$$M_t = \int_0^t C_s d(A - \widetilde{A})_s \int_0^t D_s d(B - \widetilde{B})_s - \sum_{s \leq t} C_s D_s (\Delta A_s - \Delta \widetilde{A}_s)(\Delta B_s - \Delta \widetilde{B}_s).$$

Uniqueness of a Predictable Compensator. For A and $\widetilde{A}$ square integrable, we have,

$$(A - \widetilde{A})_t^2 = \int_0^t (A_s - \widetilde{A}_s) d(A - \widetilde{A})_s + \int_0^t (A_{s-} - \widetilde{A}_{s-}) d(A - \widetilde{A})_s .$$

Assuming A and $\widetilde{A}$ predictable, this means $E(A_t - \widetilde{A}_t)^2 = 0$ and $P(A_t = \widetilde{A}_t) = 1$. By the right continuity of the processes

$$P[\omega;\ A_t(\omega) = \widetilde{A}_t(\omega) \ \text{ for all } \ t] = 1.$$

Proposition 6.2.16. *A square integrable increasing process A has at most one square integrable predictable compensator (two predictable compensators have identical trajectories except on a negligible set).*

The Case of an Unmarked Counting Process N Compensated by a Continuous $\widetilde{N}$. This is the case for a Poisson process, for the renewal process with continuous F and for the process N^{ij} associated with jump processes. Then for every predictable process C,

$$\sum_{s \leq t} C_s^2 (\Delta N_s - \Delta \widetilde{N}_s)^2 = \sum_{T_n \leq t} C_{T_n}^2 = \int_0^t C_s^2 \, dN_s ,$$

$$\left[\int_0^t C_s d(N - \widetilde{N})_s\right]^2 = \int_0^t C_s^2 dN_s + 2\int_0^t C_s^2 (N - \widetilde{N})_{s-} d(N - \widetilde{N})_s .$$

From this we deduce the following theorem.

Theorem 6.2.17. (a) *If an integrable unmarked counting process N adapted to $\mathbb{F}$ is compensated by a continuous $\tilde{N}$, then for $\mathbb{F}$-predictable C,*

$$E\left[\int_0^t |C_s| dN_s\right] = E\left[\int_0^t |C_s| d\tilde{N}_s\right]$$

and

$$E\left[\int_0^t C_s^2 dN_s\right] = E\left[\int_0^t C_s^2 d\tilde{N}_s\right].$$

If, for all t, these two quantities are finite, $C\cdot(N - \tilde{N})$ and $[C\cdot(N - \tilde{N})]^2 - C^2\cdot\tilde{N}$ are centered $\mathbb{F}$-martingales.

(b) *If*

$$N^1 = \left(\sum_{n\geq 1} 1_{(T_n^1 \leq t)}\right)_{t\geq 0}$$

and

$$N^2 = \left(\sum_{n\geq 1} 1_{(T_n^2 \leq t)}\right)_{t\geq 0}$$

are two point process without common jump adapted to $\mathbb{F}$ with continuous compensators $\tilde{N}^1$ and $\tilde{N}^2$, if C and D are predictable and if

$$\int_0^t (|C_s| + C_s^2) dN_s^1 + \int_0^t (|D_s| + D_s^2) dN_s^2$$

is integrable for all t, then $(C\cdot(N^1 - \tilde{N}^1))\,(D\cdot(N^2 - \tilde{N}^2))$ is a centered $\mathbb{F}$-martingale.

6.3. Poisson Processes

A Poisson process on $\mathbb{R}_+$ the intensity of which is a measure μ has various equivalent characteristics. We shall show the last of these after the statement of the following definition.

Definition 6.3.18. Let F be a function from $\mathbb{R}_+$ into $\mathbb{R}_+$, increasing, continuous, zero at 0 and only at 0. A process $N = (\Omega, \mathcal{A}, (N_t)_{t\geq 0}, P)$ is a **Poisson process with intensity F** if,

- it is a process taking values in $\mathbb{N}$, with increasing trajectories, continuous on the right, zero at 0 and with jumps equal to 1;
- it satisfies one of the three following equivalent conditions.

(a) By associating, with $\Gamma \in \mathcal{B}_{\mathbb{R}_+}$, $N(\Gamma) = \int 1_\Gamma(t)dN_t$, we define a process

$$(N(\Gamma))_{\Gamma \in \mathcal{B}_{\mathbb{R}_+}}$$

which is a spatial Poisson process [1.2.1] with intensity

$$F : \Gamma \longmapsto \int 1_\Gamma(s)dF(s).$$

(b) N is a process with independent increments ([0.2.3]) and, for $s < t$, $N_t - N_s$ has a Poisson distribution with intensity $F(t) - F(s)$.

(c) $(N_t - F(t))$ is a centered martingale.

Proof of Characteristic. (c) Consider a counting process N adapted to $\mathbb{F}$ with deterministic compensator $\tilde{N}$. Let $\lambda > 0$ and $0 \leqslant s < t$

$$\begin{aligned} e^{-\lambda N_t} - e^{-\lambda N_s} &= \sum_{s<u\leqslant t} (e^{-\lambda N_u} - e^{-\lambda N_{u-}}) \\ &= \sum_{s<u\leqslant t} e^{-\lambda N_{u-}}[e^{-\lambda} - 1][N_u - N_{u-}] \\ &= (e^{-\lambda} - 1)\int_s^t e^{-\lambda N_{u-}} dN_u \end{aligned}$$

$$= (e^{-\lambda} - 1)\left[\int_s^t e^{-\lambda N_{u-}} d(N_u - \tilde{N}_u) + \int_s^t e^{-\lambda N_{u-}} d\tilde{N}_u\right];$$

$$E[e^{-\lambda N_t} - e^{-\lambda N_s} \mid \mathcal{F}_s] = (e^{-\lambda} - 1)E\left[\int_s^t e^{-\lambda N_{u-}} d\tilde{N}_u \mid \mathcal{F}_s\right].$$

Let $(Z_t)_{t\geqslant s}$ be a right continuous version of

$$(E[e^{-\lambda(N_c - N_s)} \mid \mathcal{F}_s])_{t\geqslant s}.$$

We have

$$Z_t = 1 + (e^{-\lambda} - 1)\int_s^t Z_{u-} d\tilde{N}_u,$$

and Proposition 6.2.11 implies

$$Z_t = E[e^{-\lambda(N_t - N_s)} \mid F_s]$$
$$= \prod_{s<u\leq t} (1 + (e^{-\lambda} - 1)\Delta\tilde{N}_u)\exp[(e^{-\lambda}-1)(\tilde{N}_t^c - \tilde{N}_s^c)].$$

This is a constant, hence $(N_t)_{t\geq 0}$ is a process with independent increments and, for continuous $\tilde{N}$, it is a Poisson process. Even better, $(N_t - N_s)$ is independent of F_s, thus N_t is independent of F_0.

Corollary 6.3.19. *Let N be a counting $\mathbb{F}$-point process with deterministic compensator $\tilde{N}$. Then, for $0 \leq s < t$, $N_t - N_s$ is independent of F_s. N is a process with independent increments.*

If $\tilde{N}$ is continuous, N is a Poisson process.

Note. From the preceding proof, the equivalence of the following properties can be seen:

(a) N is an integrable $\mathbb{F}$-counting processes with a continuous and deterministic compensator $\tilde{N}$.

(b) $Z = (Z_t)_{t\geq 0}$ is for every $\lambda > 0$, an $\mathbb{F}$-martingale, by denoting

$$Z_t = \exp[-\lambda N_t + (1 - e^{-\lambda})\tilde{N}_t].$$

We shall obtain **exponential martingales** of this form in Chapter 8.

Proposition 6.3.20. *Let N^1 and N^2 be two integrable counting processes adapted to the same filtration $\mathbb{F}$. Assume that N^1 and N^2 have no jumps in common and have deterministic and continuous compensators $\tilde{N}^1$ and $\tilde{N}^2$ with respect to $\mathbb{F}$. Then N^1 and N^2 are independent Poisson processes.*

Proof. Let us set

$$X_t = e^{-\lambda N_t^1 - \mu N_t^2} = 1 + \int_0^t (e^{-\lambda} - 1)X_{s-}\, dN_s^1$$
$$+ \int_0^t (e^{-\mu} - 1)X_{s-} dN_s^2 .$$

For $r < t$, we have

For $r < t$, we have

$$\frac{X_t}{X_r} = 1 + \int_r^t (e^{-\lambda} - 1) \frac{X_{s-}}{X_r} dN_s^1 + \int_r^t (e^{-\mu} - 1)\frac{X_{s-}}{X_r} dN_s^2,$$

$$Z_t = E\left[\frac{X_t}{X_r}\right] = 1 + \int_r^t (e^{-\lambda} - 1)Z_{s-}d\tilde{N}_s^1 + \int_r^t (e^{-\mu} - 1)Z_{s-}d\tilde{N}_s^2 .$$

Proposition 6.2.11 then gives

$$E[e^{-\lambda(N_t^1-N_r^1) -\mu(N_t^2-N_r^2)}] = \exp[(e^{-\lambda} - 1)(\tilde{N}_t^1 - \tilde{N}_r^1)]\exp[(e^{-\mu} - 1)(\tilde{N}_t^2 - \tilde{N}_r^2)].$$

6.4. Statistics of Counting Processes

We shall be content with dealing with two very simple cases. We shall return to the asymptotic statistics of counting processes in [8.3].

6.4.1. The Compensator's Role as Predictor

With the notations of [6.2.3], let us assume $\tilde{N}$ continuous: $\tilde{N}$ increases regularly whereas N increases by jumps at "unpredictable" times. For any predictable N-integrable process C, $C \cdot \tilde{N}$ is a continuous process, and $E[(C \cdot \tilde{N})_t] = E[(C \cdot N)_t]$. If C and $\tilde{N}$ are deterministic this implies that $(C \cdot \tilde{N})_t$ is an "unbiased predictor of $(C \cdot N)_t$". In any case $C \cdot \tilde{N}$ is more regular than $C \cdot N$ and is a help in its study (as the compensator of discrete sequences did in Chapter 2). Conversely, $\tilde{N}$ depends on the distribution of the process, and $(C \cdot N)_t$ is an unbiased estimator of $(C \cdot \tilde{N})_t$. Let us look at a statistical example.

Test of Identity of Distributions of Two Samples. Let F be a distribution concentrated on $]0,\infty]$, and $(U_1, ..., U_n)$ an n-sample from F. Then

$$N_t = \sum_{i=1}^{n} 1_{(U_i \leqslant t)}$$

is, for the filtration $\mathbb{F} = (\sigma\{U_i \wedge t;\ i = 1, \ldots, n\})_{t \geqslant 0}$, compensated by

$$\widetilde{N}_t = \int_0^t \frac{dF(s)}{1 - F(s-)} (n - N_{s-}).$$

(This follows easily from the fact that it is true when $n = 1$.) Let $T < \infty$, be such that

$$\int_0^T \frac{dF(s)}{1 - F(s-)} < \infty.$$

Let F^1 and F^2 be two distributions on $]0,\infty]$,

$$(U_i^1)_{1 \leqslant i \leqslant n_1} \quad \text{and} \quad (U_j^2)_{1 \leqslant j \leqslant n_2}$$

an n_1-sample from F^1 and an n_2-sample from F^2. For Y^1 and Y^2 bounded and predictable,

$$\left\{\int_0^t Y_s^i dN_s^i - \int_0^t Y_s^i \frac{dF^i(s)}{1 - F^i(s-)}(n_i - N_{s-}^i)\right\}_{t \leqslant T}$$

is a centered $\mathbb{F}$-martingale. Taking Y predictable and $Y_s^1 = (n_2 - N_{s-}^2)Y_s$, $Y_s^2 = (n_1 - N_{s-}^1)Y_s$, we obtain that, if $F^1 = F^2 = F$,

$$\left\{\int_0^t Y_s(n_2 - N_{s-}^2)dN_s^1 - \int_0^t Y_s(n_1 - N_{s-}^1)dN_s^2 = W_{n_1 n_2}(t)\right\}_{t \leqslant T}$$

is a centered $\mathbb{F}$-martingale.

If F is continuous and $F^1 = F^2 = F$, then N^1 and N^2 do not have any jumps in common and

$$\left\{W_{n_1 n_2}^2(t) - \int_0^t (Y_s)^2((n_2 - N_{s-}^2) + (n_1 - N_{s-}^1))\frac{dF(s)}{1 - F(s)}(n_1 - N_{s-}^1)(n_1 - N_{s-}^2)\right\}_{t \leqslant T}$$

is a centered $\mathbb{F}$-martingale.

This leads to testing "$F_1 = F_2$" against "$F^1 \neq F^2$" by using rejection regions of the form $\{|W_{n_1,n_2}(t)| \geqslant a\}$. To various values of Y correspond various rank tests. For $Y = 1$ we obtain,

$$W_{n_1,n_2}(t) = \sum_{n=1}^{n_2} \sum_{m=1}^{n_1} 1_{(t \geq U_n^2 > U_m^1)} - \sum_{n=1}^{n_2} \sum_{m=1}^{n_1} 1_{(t \geq U_m^1 > U_n^2)}:$$

this is **Wilcoxon's statistic** [Vol. 1, 4.5.2 and E4.5.1].

6.4.2. The Statistics of Jump Processes

Let E be a countable set equipped with the σ-algebra of the set of its subsets, let π be a transition from E into E, and q an r.v. on E taking values in $]0,\infty[$. With the notations of 6.2.1(b), assume that we are in the non-explosive case, and consider the canonical jump process associated with q and π $(\Omega, \mathcal{A}, (P_x)_{x \in E}, (X_t)_{t \geq 0})$. Let us denote by $\{\Omega, \mathcal{A}, (Q_x)_{x \in E}, (X_t)_{t \geq 0}\}$ the canonical process associated with q identically equal to 1 and a transition $\hat{\pi}$ such that $\pi(i,j) > 0$ implies $\hat{\pi}(i,j) > 0$.

For each n, Q_x dominates P_x on $(\Omega, \mathcal{G}_n)$, with $\mathcal{G}_n = \sigma(Z_0, ..., Z_n, T_1, ..., T_n) = \sigma(Z_0, ..., Z_n, \tau_1, ..., \tau_n)$ (by denoting $\tau_i = T_i - T_{i-1}$ for $i \geq 1$). In fact for $t_1, ..., t_n$ in $\mathbb{R}_+$ and ν an arbitrary initial distribution,

$$\begin{aligned} P_\nu[Z_0 = x_0, ..., Z_n = x_n, \ \tau_1 \geq t_1, ..., \tau_n \geq t_n] \\ = \nu(x_0)\pi(x_0,x_1) \ ... \ \pi(x_{n-1},x_n) \\ \times \exp[-q(x_0)t_1 - q(x_1)t_2 - ... - q(x_{n-1})t_n] \\ = Q_\nu[1_{(Z_0=x_0,...,Z_n = x_n, \ \tau_1 \geq t_1 \ , \ ..., \ \tau_n \geq t_n)} L_{T_n}] \end{aligned}$$

by setting

$$L_{T_n} = \frac{\pi(Z_0,Z_1)...\pi(Z_{n-1},Z_n)}{\hat{\pi}(Z_0,Z_1)...\hat{\pi}(Z_{n-1},Z_n)} (q(Z_1) \ ... \ q(Z_n))$$

$$\times \exp[(1 - q(Z_0))\tau_1 \ ... \ (1 - q(Z_n))\tau_n]$$

$$= \exp\left[\sum_{(i,j)\in E^2}\left(\operatorname{Log}\frac{\pi(i,j)}{\overset{\wedge}{\pi}(i,j)}\right)N^{ij}_{T_n} + \sum_{i\in E} N^{i\cdot}_t \operatorname{Log} q(i)\right.$$

$$\left. + \sum_{i\in E}(1 - q(i))\tau_i(T_n)\right],$$

where N^{ij}_t is the number of jumps from i to j before t, $\tau_i(t)$ the time spent in i before t and $N^{i\cdot}_t = \Sigma_{j\in E} N^{ij}_t$. For $t \in \mathbb{R}_+$, let us set:

$$L_t = \exp\left[\sum_{(i,j)\in E^2}\left(\operatorname{Log}\frac{\pi(i,j)}{\overset{\wedge}{\pi}(i,j)}\right)N^{ij}_t + \sum_{i\in E} N^{i\cdot}_t \operatorname{Log} q(i)\right.$$

$$\left. + \sum_{i\in E}(1 - q(i))\tau_i(t)\right].$$

Then, if (T_n) tends to ∞, we prove $P_\nu = L_t Q_\nu$ on $\mathcal{F}_t$. Consider $A \in \mathcal{F}_t$ and let us denote by $\hat{E}_\nu$ the expectation relative to Q_ν:

$$P_\nu\,[A \cap \{T_n \leqslant t < T_{n+1}\}] = E_\nu[1_{A\cap(T_n\leqslant t)}P_\nu(T_{n+1}>t)|\mathcal{G}_n)]$$

$$= E_\nu(1_{A\cap(T_n\leqslant t)}\exp[-q(Z_n)(t - T_n)])$$

$$= \hat{E}_\nu(1_{A\cap(T_n\leqslant t)}\exp[-q(Z_n)(t - T_n)]L_{T_n})$$

$$= \hat{E}_\nu(1_{A\cap(T_n\leqslant t)}L_{T_n}\exp[(1-q(Z_n))(t-T_n)]\;\cdot$$

$$\cdot\; Q_\nu(T_{n+1} > t \mid \mathcal{G}_n)$$

$$= \hat{E}_\nu(1_{A\cap(T_n\leqslant t<T_{n+1})}L_t).$$

Summing over n, we obtain $P_\nu = L_t \cdot Q_\nu$ on $(\Omega,\mathcal{F}_t)$.

Proposition 6.4.21. *Let E be a countable space and let Θ be a set of parameters. We are given q, a function from $\Theta \times E$ into $]0,\infty[$, and π, a function from $\Theta \times E^2$ into $[0,1]$ such that, for every $(\theta,i) \in \Theta \times E$, $\Sigma_j \pi(\theta,i,j) = 1$. Denote by $\{\Omega,\mathcal{A},(P_{\theta,x})_{x\in E}, (X_t)_{t\geqslant 0}\}$ the canonical jump process associated with $q(\theta,\cdot)$ and $\pi(\theta,\cdot,\cdot)$. Set:*

$$L_t(\theta) = \exp\left[\sum_{(i,j)\in E^2}(\operatorname{Log}\pi(\theta,i,j))N^{ij}_t\right.$$

$$\left. + \sum_{i\in E} N^{i\cdot}_t \operatorname{Log} q(\theta,i) + \sum_{i\in E}(1 - q(\theta,i))\tau_i(t)\right].$$

We then have, for every initial distribution γ, *a model* $(P_{\theta,\nu})_{\theta\in\Theta}$ *dominated at each instant* T_n *with likelihood* $L_{T_n}(\theta)$. *If* T_n *tends to* ∞, $P_{\theta,\nu}$-a.s. *for every* θ, *then the model is dominated at each instant* t, *with likelihood* $L_t(\theta)$.

Proof. It is sufficient to take in the above $\hat{\pi}(i,j) > 0$ for every $(i,j) \in E^2$ and the dominating measure μ_ν defined by $\mu_\nu = [\hat{\pi}(Z_0,Z_1)...\hat{\pi}(Z_{n-1},Z_n)]^{-1}Q_\nu$ on $(\Omega, \mathcal{G}_n)$. When this likelihood has been calculated, we can without difficulty adapt to jump processes the Markov chain statistics studied in [4.4.1,2 and 3].

Bibliographic Notes

Renewal processes and point processes are essential for the study of failures (reliability) and of queues. Various simple probability models of these very interesting phenomena are found in Gnedenko-Beliaev-Soloviev, Cinlar, and Ross.

The renewal theorem is proved in Feller (Volume 2) and in Gikhman and Skorokhod ([2] and [3]) with applications. The proof given here is Lindvall's.

In [6.2] we deal only with the time aspect of point processes. Bremaud gives a clear account of the subject with many applications to queues. Jacod's very complete book, is more difficult; Bremaud-Jacod presents a unified view. Kolmogorov-Folmine and Rudin study the Stieltjes' integral of deterministic functions of bounded variation.

A more complete study of jump processes and of semi-Markovian processes can be worthwhile, as in the study of Markov chains. This will be covered briefly in Chapter 8. However at the moment Cinlar, Cox-Smith, Gikhman- Skorokhod, Ross, Takacs... can be read. The study of queues has been developed in order to optimize the use of computers; see Kleinrock.

Snyder, Cox-Lewis, Basawa-Rao cover some statistical problems of Poisson processes or jump processes. The interest in Aalen's point of view will not be very clear until after reading [8.3] which gives some of the asymptotic properties; see Aalen, Bremaud, Gill, and Rebolledo [1]. Using the compensator we can study the absolute continuity of point processes and obtain likelihoods which are the exponential martingales analogous to those of [6.4.2]; see

Bremaud, Duflo-Florens Zmirou, Jacod, Kutoiants, Liptzer-Shiryayev; Bremaud, Liptzer-Shiryayev also deal with filtering.

Finally there are many other aspects of point processes which we do not deal with; spatial point processes, stationarity, Cox processes (or doubly stochastic processes); see Neveu [4], Grandell, and Lewis.

Chapter 7
PROCESSES IN CONTINUOUS TIME

Objectives

Point processes have already led us to study a random evolution at times $t \in \mathbb{R}_+$. However the study has been simplified by the fact that it depends only on a sequence of r.v.'s. We are now going to deal with continuous time processes, such as Brownian motion, where the time t is an element of $\mathbb{R}_+$. Everything that is going to be said can easily be translated by taking, instead of $\mathbb{R}_+$, an interval of $\mathbb{R}_+$.

The principal tools, stopping times and martingales, are the same as those in discrete time processes. However the technique is more delicate. We are led to study the regularity of the trajectories. Finally we deal with convergence in distribution of processes, in particular convergence to a Brownian motion.

7.1. Stopping Times

7.1.1. Definitions and Operations

The history of the observations is described, as in discrete time, by a probability space $(\Omega, \mathcal{A}, P)$ and a filtration $\mathbb{F} = (\mathcal{F}_t)_{t \in \mathbb{R}_+}$ (Definition 6.2.5). The σ-algebra $\mathcal{F}_t$ is the set of events prior to t (including t). Only the events in $\mathcal{F}_\infty = \vee \mathcal{F}_t$ can be observed.

As for discrete time, it is important to be able to stop an experiment at a random time T. However the decision to stop can depend only on prior observations. An event is observable before T, if, for $T \leqslant t$ it is observable before t.

Definition 7.1.1. (a) An $\mathbb{F}$**-stopping time** T is a r.v. from (Ω,A) into $[0,\infty]$ such that, for all $t \geqslant 0$, $\{T \leqslant t\}$ is in F_t. $\mathbb{F}$ is to be understood when there is no confusion.

(b) If T is a stopping time, the **σ-algebra** F_T **of events prior to** T is

$$F_T = \{A;\ A \in F_\infty,\ A \cap \{T \leqslant t\} \in F_t \text{ for all } t \geqslant 0\}.$$

Properties 7.1.2.

(a) *A constant* t *is a stopping time.*

(b) F_T *is a σ-algebra.*

(c) *If* T_1 *and* T_2 *are stopping times, then* $T_1 \vee T_2 = \sup(T_1,T_2)$ *and* $T_1 \wedge T_2 = \inf(T_1,T_2)$ *are stopping times. The events* $\{T_1 < T_2\}$, $\{T_1 \leqslant T_2\}$, $\{T_1 = T_2\}$ *are in* $F_{T_1} \cap F_{T_2}$. *For* $T_1 \leqslant T_2$, *we have* $F_{T_1} \subset F_{T_2}$.

(d) *Let* (T_n) *be a sequence of stopping times;* $\sup T_n$ *is a stopping time.*

(e) *If* D *is a countable subset of* $[0,\infty]$, *a function* T *from* Ω *into* D *is a stopping time if and only if, for* $d \in D$, $\{T = d\} \in F_d$.

Proof. Properties (a) and (b) are easy. Let us prove (c):

$$\{T_1 \wedge T_2 \leqslant t\} = \{T_1 \leqslant t\} \cup \{T_2 \leqslant t\}$$

and

$$\{T_1 \vee T_2 \leqslant t\} = \{T_1 \leqslant t\} \cap \{T_2 \leqslant t\},$$

$$\{T_1 < T_2\} \cap \{T_2 \leqslant t\} = \bigcup_{q<t} (\{T_1 \leqslant q\} \wedge \{q < T_2 \leqslant t\})$$

($\cup_{q<t}$ is the union over the rationals q strictly less than t),

$$\{T_1 < T_2\} \cap \{T_1 \leqslant t\} = \bigcup_{q<t} (\{T_1 \leqslant q\} \cap \{q < T_2\}) \\ \cup (\{T_1 \leqslant t\} \cap \{T_2 > t\}).$$

Thus $\{T_1 < T_2\} \in \mathcal{F}_{T_1} \cap \mathcal{F}_{T_2}$. Similarly $\{T_1 < T_2\}^c = \{T_2 \leqslant T_1\}$, $\{T_2 < T_1\}$ and $\{T_1 = T_2\}$ are in $\mathcal{F}_{T_1} \cap \mathcal{F}_{T_2}$. Property (d) follows from

$$\{\sup T_n \leqslant t\} = \cap \{T_n \leqslant t\}.$$

And (e) follows since

$$\{T = d\} = \{T \leqslant d\} \setminus \bigcup_{d'<d} \{T \leqslant d'\}$$

and

$$\{T \leqslant t\}^c = \{T > t\} = \bigcup_{d>t} \{T = d\},$$

the union being taken for $d \in D$.

Questions. Let (T_n) be a sequence of stopping times. Can we state "inf T_n is a stopping time"? Unfortunately it is false that $\{\inf T_n \leqslant t\}$ is the union $\cup\{T_n \leqslant t\}$. All that we can state is $\{\inf T_n < t\} = \cup\{T_n < t\}$. Can we then characterize the stopping time property by: "for every t, $\{T < t\} \in \mathcal{F}_t$"? This is also false. Let us take, for example, an event A distinct from ϕ and Ω, and $\mathcal{F}_t = \{\phi,\Omega\}$ for $t \leqslant 1$, $\mathcal{F}_t = \sigma(A)$ for $t > 1$ and $T = 1 + 1_A$. We have $\{T < t\} \in \mathcal{F}_t$ for all t, but $\{T \leqslant 1\} \notin \mathcal{F}_1$.

Let T be a stopping time. We then always have

$$\{T < t\} = \cup\left\{T \leqslant t - \frac{1}{n}\right\} \in \mathcal{F}_t.$$

Every r.v. T satisfies, for all t,

$$\{T \leqslant t\} = \bigcap_n \left\{T < t + \frac{1}{n}\right\} \in \bigcap_{s>t} \mathcal{F}_s.$$

We denote $\mathcal{F}_t^+ = \cap_{s>t}\mathcal{F}_s$ and $\mathbb{F}^+ = (\mathcal{F}_t^+)_{t\geqslant 0}$.

Proposition 7.1.3. *Assume the filtration $\mathbb{F}$ is continuous on the right, i.e., $\mathbb{F} = \mathbb{F}^+ = (\mathcal{F}_t^+)$ with $\mathcal{F}_t^+ = \cap_{s>t}\mathcal{F}_s$. An r.v. T taking values in $[0,\infty]$ is a stopping time if and only if, for every t, $\{T < t\}$ is in $\mathcal{F}_t$. Then $\{T = t\} \in \mathcal{F}_t$ and*

$$\mathcal{F}_T = \{A;\ A \in \mathcal{F}_\infty,\ A \cap \{T < t\} \in \mathcal{F}_t,\ \textit{for all } t\}.$$

For any sequence (T_n) of stopping times, inf T_n *is a stopping* *time and* $\mathcal{F}_{\inf T_n} = \cap \mathcal{F}_{T_n}$.

Let us show the last property by denoting $T = \inf T_n$. Property 7.1.2(b) proves $\mathcal{F}_T \subset \cap_n \mathcal{F}_{T_n}$. Let $A \in \cap_n \mathcal{F}_{T_n}$; $A \in \mathcal{F}_\infty$ and, for all t:

$$A \cap \{T < t\} = A \cap \left\{\bigcup_n (T_n < t)\right\}$$
$$= \bigcup_n \{A \cap (T_n < t)\} \in \mathcal{F}_t.$$

7.1.2. Entrance Times

Consider a process $(\Omega,\mathcal{A},P,(X_t)_{t\in\mathbb{R}_+})$ taking values in $(E,\mathcal{E})$ adapted to $\mathbb{F}$.

Some pleasant properties in the case of sequences of experiments are the following: if T is a finite stopping time, X_T is $\mathcal{F}_T$-measurable. The **entrance time** $T_\Gamma = \inf\{s;\ X_s \in \Gamma\}$ into a set $\Gamma \in \mathcal{E}$ is a stopping time. Unfortunately, and this is one of the major technical differences between $\mathbb{R}_+$ and $\mathbb{N}$, this is in general false for continuous time processes.

Examples. (a) Let us assume that the process has trajectories continuous on the right taking values in a metric space E. Let G be an open set

$$\{T_G < t\} = \bigcup_{q<t} \{X_q \in G\} \in \mathcal{F}_t,$$
$$\{T_G \leq t\} = \bigcap_n \left\{T_G < t + \frac{1}{n}\right\} \in \mathcal{F}_t^+.$$

Thus T_G is a stopping time for the filtration $\mathbb{F}^+$, but not in general for $\mathbb{F}$ except when $\mathbb{F} = \mathbb{F}^+$, i.e. for $\mathbb{F}$ continuous on the right.

(b) Let us assume that the above process is continuous. Let F be a closed set. It is the intersection of the open sets $G_n = \{x;\ d(x,F) < (1/n)\}$, if $d(x,F)$ is the distance of x from F. Here

$$\{T_F \leq t\} = \bigcap_n \{T_{G_n} < t\} \in \mathcal{F}_t.$$

The entrance time of a continuous process into a closed set is a stopping time.

(c) Let us now assume the measurability of X_T for T a finite stopping time. If T takes its values in a countable set D,

$$X_T \cap \{T = d\} = X_d \cap \{T = d\} \in \mathcal{F}_d.$$

It follows that X_T is $\mathcal{F}_T$-measurable. A finite stopping time T is always the decreasing limit of a sequence $T^{(n)}$ of stopping times taking their values in a countable set. In fact, let us set

$$T^{(n)} = \sum_{k \geq 0} \frac{k+1}{2^n} 1_{[k/2^n \leq T < (k+1)/2^n]} .$$

It is easily shown that $T^{(n)}$ is a stopping time and that the sequence $T^{(n)}$ decreases to T. Let X be a process taking values in a topological space, with trajectories continuous on the right. $X_T = \lim \downarrow X_{T^{(n)}}$ is thus an r.v. $\cap_n \mathcal{F}_{T(n)}$-measurable.

We have shown $\mathcal{F}_T = \cap_n \mathcal{F}_{T(n)}$ when $\mathbb{F}$ is continuous on the right.

Theorem 7.1.4. *If $\mathbb{F}$ is continuous on the right and if X is a process with trajectories continuous on the right, taking values in a metric space E, then:*

(1) *For every finite stopping time T, X_T is $\mathcal{F}_T$-measurable.*
(2) *The entrance time into an open set of E is a stopping time.*
(3) *The set of subsets A of E such that the entrance time T_A of X into A is a stopping time is closed under countable unions.*

The last part of the theorem follows from the relation:

$$T_{\cup A_n} = \inf T_{A_n} .$$

In fact, if $A = \cup A_n$, we have $T_A \leq T_{A_n}$ for all n. $t > T_A$

implies that there exists an s lying between t and T_A such that $X_s \in A$, hence such that $X_s \in A_n$ for an n; $t > \inf T_{A_n}$.

Therefore, for $E = \mathbb{R}$, the entrance time of X into an interval $[a,b]$ or an interval $[a,\infty[$ is a stopping time.

These results will satisfy us. At the cost of a much greater effort we can show the following more beautiful result, and other more general results (cf. Dellacherie-Meyer [1]). Let $\mathbb{F}$ be a filtration continuous on the right such that $\mathcal{F}_0$ is complete. Consider an $\mathbb{F}$-adapted process $(\Omega,\mathcal{A},P,(X_t)_{t\geq 0})$ with trajectories continuous on the right taking values in the metric space E provided with its Borel σ-algebra $\mathcal{E}$. Then for every Borel set $\Gamma \in \mathcal{E}$,

$$T_\Gamma = \inf\{s;\ s \geqslant 0,\ X_s \in \Gamma\}$$

is an $\mathbb{F}$-stopping time.

7.2. Martingales in Continuous Time

In this section we try to extend some of the results of Chapter 2 to (sub-) martingales $X = \{\Omega, \mathcal{A}, P, (X_t)_{t\geqslant 0}\}$ adapted to a filtration $\mathbb{F} = (\mathcal{F}_t)_{t\geqslant 0}$ defined in [6.2.2].

7.2.1. Operations

As for discrete time martingales [2.2.4], we obtain the following results.

(a) If X and Y are two martingales adapted to $\mathbb{F}$, $\alpha X + \beta Y$ is also a martingale adapted to $\mathbb{F}$ (α and β real numbers).

(b) If X and Y are two submartingales adapted to $\mathbb{F}$, $(\sup(X_t, Y_t))_{t\geqslant 0} = \sup(X,Y)$ is also a submartingale.

(c) Let X be a martingale and let ϕ be a convex function from $\mathbb{R}$ into $\mathbb{R}$, such that $\phi(X_t)_+$ is integrable for all t. Then $\phi(X) = (\phi(X_t))_{t\geqslant 0}$ is a submartingale. The result remains true if X is a submartingale and if ϕ is increasing and convex.

7.2.2. Stopping Theorem and Regularity of Trajectories

Let us consider $X = (X_t)_{t\in\mathbb{R}_+}$, a submartingale adapted to $\mathbb{F} = (\mathcal{F}_t)_{t\in\mathbb{R}_+}$. Do the theorems of Chapter 2 remain true?

Recall the proof of stopping Theorem 2.3.8: only the fact that the stopping times take a countable number of values is important. Therefore, if S and T are two stopping times taking values in $\mathbb{Q}_+$ and if $S \leqslant T \leqslant H$, for a constant $H > 0$, we have

$$E[X_T \mid \mathcal{F}_S] \geqslant X_S$$

(with equality in the case of martingales).

In order to use only stopping times with a countable number of values, we observe first of all the submartingale on the set Δ of dyadics of $\mathbb{R}_+$, or on $\Delta_n = \{k/2^n;\ k \in \mathbb{N}\}$ ($\Delta = \cup\Delta_n$). Here is the principal result which leads back to submartingales, the trajectories of which are continuous on the right, and have a limit on the left.

Doob's Theorem 7.2.5. *Let* $X = (\Omega, A, P, (X_t)_{t\in\mathbb{R}+})$ *be a submartingale adapted to* $\mathbb{F}$. *For almost all* $\omega \in \Omega$, *the following limits exist for any* t:

$$X_t^+(\omega) = \lim_{\substack{s\in\Delta\cap]t,\infty[\\ s\to t}} X_s,$$

$$X_t^-(\omega) = \lim_{\substack{s\in\Delta\cap[0,t[\\ s\to t}} X_s(\omega).$$

The process $X^+ = (X_t^+)_{t\in\mathbb{R}+}$ *satisfies the following properties:*

(1) *Almost all trajectories of* X^+ *are continuous on the right, and have a limit on the left.*
(2) X^+ *is a submartingale adapted to* $\mathbb{F}^+$.
(3) *If* $\mathbb{F}$ *is continuous on the right and if* $t \longmapsto E(X_t)$ *is a continuous function from* $\mathbb{R}^+$ *into* $\mathbb{R}$, X^+ *is a modification of* X.

Proof. Consider n and H in $\mathbb{N}$ and $a < b$. We look for the oscillations of $(X_s)_{0\leqslant s\leqslant H}$ from the value b to the value a. For this, we define

$$T_0^n = \inf\{s;\ s \in \Delta_n,\ X_s \geqslant b\} \wedge H$$

$$T_1^n = \inf\{s;\ s \in \Delta_n,\ s > T_0^n,\ X_s \leqslant a\} \wedge H$$

$$T_{2p}^n = \inf\{s;\ s \in \Delta_n,\ s > T_{2p-1}^n,\ X_s \geqslant b\} \wedge H$$

$$T_{2p+1}^n = \inf\{s;\ s \in \Delta_n,\ s > T_{2p}^n,\ X_s \leqslant a\} \wedge H.$$

We have

$$(b - a)P[T_{2p+1}^n < H]$$

$$\leqslant E[(X_{T_{2p}^n} - X_{T_{2p+1}^n})1_{(T_{2p+1}^n<H)}]$$

$$= E[(X_{T^n_{2p}} - X_{T^n_{2p+1}})1_{(T^n_{2p}<H)}]$$

$$- E[(X_{T^n_{2p}} - X_H)1_{(T^n_{2p}<H\leqslant T^n_{2p+1})}]$$

Now

$$E[X_{T^n_{2p+1}} 1_{(T^n_{2p}<H)}]$$

$$= E[E(X_{T^n_{2p+1}} \mid \mathcal{F}_{T^n_{2p}})1_{(T^n_{2p}<H)}]$$

$$\geqslant E[X_{T^n_{2p}} 1_{(T^n_{2p}<H)}].$$

Thus

$$(b-a)P[T^n_{2p+1} < H] \leqslant E[(X_H - b)1_{(T^n_{2p}<H\leqslant T^n_{2p+1})}].$$

Summing over p and denoting

$$D^{n,H}_{a,b} = \sum_{p=0}^{\infty} 1_{(T^n_{2p+1}<H)}$$

the number of descents from b to a of the trajectories $(X_s)_{s\in\Delta_n\cap[0,H]}$, we have

$$(b-a)E(D^{n,H}_{a,b}) \leqslant E[X_H - b]_+ < \infty.$$

Let $D^H_{a,b}$ be the number of descents from b to a of the trajectories $s \longmapsto X_s$ when s increases from 0 to H staying in Δ; $D^H_{a,b} = \lim \uparrow D^{n,H}_{a,b}$ and $(b-a)E(D^H_{a,b}) \leqslant E(X_H - b)_+$.

Now let N be the negligible set, the union of negligible sets $\{D^H_{a,b} = \infty\}$ for all $H \in \mathbb{N}$, and $(a,b) \in \mathbb{Q}^2$; $a < b$. Let $\omega \in N^c$ and let $t \in \mathbb{R}_+$. The limits $X^+_t(\omega)$ and $X^-_t(\omega)$ defined above exist, if not two numbers a and b of Δ can be found, $a < b$, such that there is an infinity of oscillations of $s \longmapsto X_s(\omega)$ between a and b for $s < H$, with $H > t$. From which the first part of the theorem follows.

Let us study X^+. For $\omega \in N$, $t \longmapsto X_t^+(\omega)$ is a right continuous function with a limit on the left, from which property (1) follows. The process X^+ is adapted to $\mathbb{F}^+$. Let $t \in \mathbb{R}_+$, and let (t_n) be a sequence of Δ which decreases strictly to t. (X_{t_n}) is an (F_{t_n})-submartingale which converges a.s. to X_t^+. From Theorem 2.4.17 it also converges in L^1. Now let $u < t$ and let $A \in F_u^+$. We take (u_n), a sequence of Δ which decreases to u, with $u < u_n < t$. Then

$$E[1_A X_{u_n}] \leqslant E[1_A X_t] \leqslant E[1_A X_{t_n}],$$

and the convergence in L^1 allows passage to the limit,

$$E[1_A X_u^+] \leqslant E[1_A X_t] \leqslant E[1_A X_t^+].$$

From which

$$X_u^+ \leqslant E[X_t^+ | F_u^+],$$

and X^+ is a submartingale. Moreover $X_t \leqslant E[X_t^+ | F_t]$, and when $E(X_t^+) = \lim E(X_{t_n})$ equals $E(X_t)$, we have $X_t = E(X_t^+ | F_t)$ a.s.; if $F_t = F_t^+$, $X_t = X_t^+$ a.s. From which property 3 follows.

Consequences. With respect to a filtration $\mathbb{F}$, let X be a martingale or a submartingale continuous in probability. The following changes can be made:

– replace X by a modification with trajectories continuous on the right and with limits on the left (from now on this is abbreviated to **CAD-LAG**, from the French "continu à droite - limité à gauche"); it is sufficient to take X^+ on $\Omega \backslash N$ and 0 on N.

– replace $\mathbb{F}$ by $\mathbb{F}^+$.

Therefore, in what follows we shall study cad-lag martingales adapted to a filtration continuous on the right. With these two hypotheses, we therefore have no problem in speaking about the entrance time into an open set.

7.2.3. Properties of Right Continuous Submartingales

For a right continuous submartingale, the principal results of Chapter 2, inequalities, a.s. convergence criteria, and stopping theorem remain true.

Inequalities identical to those of [2.3.3] are obtained by noting that, for a cad-lag process $X = (X_t)_{t \geq 0}$, $\sup(X_s; s \leq t) = \sup(X_s; s \text{ rational}, s \leq t)$. Let us state the analogue of Corollary 2.3.11 and of Theorem 2.4.13 which will be useful in what follows.

Theorem 7.2.6. *Let M be a cad-lag martingale, $\lambda > 0$ and $p \geq 1$ denote*

$$X_t^* = \sup_{s \leq t} |X_s|$$

and $\|\cdot\|_p$ the norm in L^p. Then we have the following inequalities,

$$P(X_t^* \geq \lambda) \leq \frac{1}{\lambda^p} E(|X_t|^p),$$

$$\|X_t^*\|_p \leq \frac{p}{p-1} \|X_t\|_p .$$

Theorem 7.2.7 (*Almost sure convergence*). *Let $(X_t)_{t \in \mathbb{R}_+}$ be a right continuous submartingale, for which* $\sup E(X_t)_+ < \infty$. *It converges* a.s. *as $t \to \infty$ to an* r.v. *X_∞ taking values in $[-\infty, \infty[$.*

Proof. Let $a < b$, and let $D_{a,b}(\omega)$ be the number of descents of the trajectory $t \longmapsto X_t(\omega)$ from b to a. From the continuity on the right of the trajectories $D_{a,b} = \sup_{H \in \mathbb{N}} D^H_{a,b}$ with the notations in the proof of Theorem 7.2.5 and

$$(b - a)E(D_{a,b}) \leq \sup_t E(X_t - b)_+ < \infty.$$

Therefore taking N to be the union of negligible sets $\{D_{a,b} = \infty\}$ for a and b rational numbers and $a < b$, for every $\omega \in N^c$, the trajectory $t \longmapsto X_t(\omega)$ cannot have an infinity of oscillations between two values, and it converges a.s. in $\overline{\mathbb{R}}$ to an r.v. X_∞. By Fatou's theorem

$$E[\sup(X_\infty, 0)] \leq \underline{\lim}_{t \to \infty} E(X_t)_+ < \infty,$$

and X_∞ takes its values in $[-\infty,\infty[$.

Note. This proof also applies to discrete martingales, but we have given another for this in Chapter 2.

Finally the generalization of [2.4.4] is immediate. A process $\{\Omega,A,P,(X_t)_{t\geqslant 0}\}$ taking real values is said to be **equiintegrable** if

$$\lim_{a\to\infty} \overline{\lim_{t\to\infty}}\ E(|X_t|1_{(|X_t|>a)}) = 0.$$

For this, it is necessary for $\sup E(|X_t|)$ to be finite; and it is sufficient that, for a $p > 1$, $\sup E(|X_t|^p)$ is finite.

Theorem 7.2.8. *Let* $(X_t)_{t\in\mathbb{R}_+}$ *be a right continuous, equiintegrable submartingale. It converges* a.s. *and in* L^1 *to an* r.v. X_∞.

If T *is a stopping time, we then denote by* X_T *the* r.v. *such that* $X_{T(\omega)}(\omega) = X_T(\omega)$, *for* $T(\omega) \leqslant \infty$ *if* S *and* T *are two stopping times and* $S \leqslant T$, *we have*

$$X_S \leqslant E(X_T|\ F_S)$$

with equality when X *is a martingale.*

Proof. It can always be stated that T is the decreasing limit of

$$T^{(n)} = \sum_{k\geqslant 1} \frac{k+1}{2^n} 1_{(k/2^n\leqslant T<(k+1)/2^n)} + \infty 1_{T=\infty)}\ ;$$

$T^{(n)}$ takes its values in $\mathbb{N}/2^n$ and it is a stopping time for the filtration $\mathbb{F}^n = (F_{k/2^n})$. Hence we apply the stopping theorem 2.4.19 to the discrete submartingale $(X_{k/2^n})$ adapted to $\mathbb{F}^n$. For $A \in F_T \subset F_{T(n)}$:

$$E[1_A X_{T(n)}] \leqslant E[1_A X_{S(n)}].$$

Set $G_{-n} = F_{T(n)}$ and $y_{-n} = X_{T(n)}$; $(y_n)_{n\leqslant 0}$ is a submartingale adapted to the filtration $(G_n)_{n\leqslant 0}$. From Theorem 2.4.17, it converges a.s. and in L^1 to X_T. From which

$$E[1_A X_T] \leqslant E[1_A X_S].$$

Here is, as a corollary, the analogue of part (b) of Theorem 2.4.19.

Corollary 7.2.9. *If $X = (X_t)$ is a right continuous, $\mathbb{F}$-martingale, and T an $\mathbb{F}$-stopping time, $X^T = (X_{t\wedge T})$ is also an $\mathbb{F}$-martingale. We say that X^T is the stopped martingale X at stopping time T.*

Proof. Let us assume that the martingale is equiintegrable. Then

$$
\begin{aligned}
E(X_\infty - X_T | \mathcal{F}_{t\vee T}) &= X_{t\vee T} - X_T \\
&= 1_{(T\leqslant t)}(X_t - X_{t\wedge T}) \\
&= X_t - X_{t\wedge T}.
\end{aligned}
$$

This r.v. is $\mathcal{F}_t$-measurable; thus $E(X_\infty - X_T| \mathcal{F}_t) = X_t - X_{t\wedge T}$ and

$$E(X_T| \mathcal{F}_t) = X_{t\wedge T}\ .$$

If X is an arbitrary martingale, the stopped martingale X at a fixed time s is equiintegrable and adapted to $(\mathcal{F}_t)_{t\leqslant s}$; $s \wedge T$ is a stopping time for this filtration. By applying the above, we thus obtain, for $t \leqslant s$,

$$E(X_{s\wedge T}| \mathcal{F}_t) = X_{t\wedge T}\ .$$

Finally, here is the analogue in continuous time of Proposition 2.6.29. The hypotheses are less artificial than they appear. This will be better understood after reading [8.1], but we already have examples of applications to point processes following [6.2].

Proposition 7.2.10. *Let $\mathbb{F}$ be a filtration continuous on the right, $M = (M_t)_{t\geqslant 0}$ an $\mathbb{F}$-martingale and $A = (A_t)_{t\geqslant 0}$ an increasing $\mathbb{F}$-process. The trajectories of M and A are assumed to be* cad-lag *and $A_0 = M_0 = 0$. Finally for all t, M_t^2 and A_t are integrable and $(M_t^2 - A_t)$ is an $\mathbb{F}$-martingale. Then for any $\varepsilon > 0$, $\eta > 0$ and stopping time T,*

(a) $P\left[\sup_{s\leqslant T} |M_s| \geqslant \varepsilon| \mathcal{F}_0\right] \leqslant \dfrac{1}{\varepsilon^2} E(A_T| \mathcal{F}_0)$;

(b) *If the jumps* $A_t - A_{t-}$ *are, for* $t \leqslant T$, *majorized by a constant* c,

$$P\left[\sup_{s \leqslant t} |M_s| \geqslant \varepsilon | \mathcal{F}_0\right] \leqslant \frac{1}{\varepsilon^2} E(A_T \wedge (\eta + c) | \mathcal{F}_0) + P(A_t > \eta | \mathcal{F}_0).$$

Proof. (a) It is the same proof as in discrete time. The result is close to the first inequality of Theorem 2.7.6. Let $S = \inf\{u; |M_u| \geqslant \varepsilon\}$:

$$\varepsilon^2 P(S \leqslant T | \mathcal{F}_0) \leqslant E(M_S^2 1_{(S \leqslant T)} | \mathcal{F}_0) \leqslant E(M_{S\wedge T}^2 | \mathcal{F}_0) \leqslant E(A_T | \mathcal{F}_0).$$

For T bounded, the last inequality follows from $E[M_{S\wedge T}^2 | \mathcal{F}_0] = E[A_{S\wedge T} | \mathcal{F}_0]$. For arbitrary T, the inequality is applied to $T \wedge n$ and we let n tend to ∞.

(b) Let $S' = \inf\{u; A_u \geqslant \eta\}$:

$$\left\{\sup_{u \leqslant T} |M_u| \geqslant \varepsilon\right\} \subset \{A_T \geqslant \eta\} \cup \{S \leqslant T \cup S'\};$$

$$P\left[\sup_{u \leqslant T} |M_u| \geqslant \varepsilon | \mathcal{F}_0\right] \leqslant P[A_T \geqslant \eta | \mathcal{F}_0] + P[S \leqslant T \wedge S' | \mathcal{F}_0] \leqslant P[A_T \geqslant \eta | \mathcal{F}_0] + \frac{1}{\varepsilon^2} E[A_{T\wedge S'} | \mathcal{F}_0].$$

The proof is completed by noting that $A_{T\wedge S'} \leqslant A_T \wedge (\eta + c)$.

7.2.4. Processes with Independent Increments

Definition 7.2.11. Let $(\Omega, \mathcal{A}, P)$ be a probability space equipped with a filtration $\mathbb{F} = (\mathcal{F}_t)_{t \geqslant 0}$. A process $X = (X_t)_{t \geqslant 0}$ taking values in $\mathbb{R}^k$ is a **process with independent increments adapted to** $\mathbb{F}$ (abbreviated $\mathbb{F}$-PII) if

- it is a process adapted to $\mathbb{F}$;
- for every $t > 0$ and every $h > 0$, $X_{t+h} - X_t$ is independent of $\mathcal{F}_t$. This process is **homogeneous** if the distribution of $X_{t+h} - X_t$ does not depend on t.

Let X be a homogeneous $\mathbb{F}$-PII. Assume $X_0 = 0$ and

$$X_t \xrightarrow[t\to 0]{P} 0.$$

To every $u \in \mathbb{R}^k$ and $t \geqslant 0$, we associate,

$$\phi_u(t) = E[\exp i\langle u, X_t\rangle].$$

We have

$$\begin{aligned}\phi_u(t+s) &= E[(\exp i\langle u, X_{t+s} - X_t\rangle)(\exp i\langle u, X_t\rangle)]\\ &= \phi_u(s)\phi_u(t).\end{aligned}$$

Moreover

$$E[\exp i\langle u, X_s\rangle] \xrightarrow[s\to 0]{} 1;$$

thus ϕ_u is continuous. We are dealing with an exponential $\phi_u(t) = \exp(\psi(u)t)$, for $\psi(u) \in \mathbb{R}$. Let us set

$$M_t^u = \frac{\exp i\langle u, X_t\rangle}{\exp(\psi(u)t)}\ ;$$

$(M_t^u)_{t\geqslant 0}$ is an $\mathbb{F}$-martingale (or, rather, its real and imaginary parts are martingales). In fact,

$$\begin{aligned}E(M^u_{t+s}|\;_t) &= \frac{\exp i\langle u, X_t\rangle}{\exp(\psi(u)t)}\, E\left[\frac{\exp i\langle u, X_{t+s} - X_t\rangle}{\exp(\psi(u)s)}\right]\\ &= M_t^u .\end{aligned}$$

Taking $u_1, \ldots, u_k$, k independent vectors in $\mathbb{R}^k$, we can apply Theorem 7.2.6. For almost all ω, the following limits exist for any t and j

$$(M_t^{u_j})^+ = \lim_{\substack{s\to t\\ s\in]t,\infty[\cap\Delta}} M_s^{u_j}(\omega)$$

and

$$(M_t^{u_j})^- = \lim_{\substack{s\to t\\ s\in [0,t[\cap\Delta}} M_s^{u_j}(\omega) = \lim_{\substack{s\to t\\ s<t}} (M_s^{u_j})^+.$$

Thus the following limits exist a.s.:

$$X_t^+(\omega) = \lim_{\substack{s \to t \\ s \in]t,\infty[\cap\Delta}} X_s(\omega) \quad \text{and} \quad X_t^-(\omega) = \lim_{\substack{s \to t \\ s \in [0,t[\cap\Delta}} X_s(\omega);$$

$X_t^-(\omega)$ is the left limit in t of $s \longmapsto X_s^+(\omega)$. However, since the process is homogeneous, $X_{t+h} - X_t$ has the distribution of X_h and tends to 0 in probability if h decreases to 0. Thus X_t and X_t^+ are a.s. equal. As a result, $X^+ = (X_t^+)$ is a modification of X.

Moreover, for all u, $(M_t^u)^+$ is a martingale adapted to $\mathbb{F}^+$ and we have, for any positive u and h,

$$E[\exp i\langle u, X_{t+h}^+ - X_t^+\rangle | \mathcal{F}_t^+) = \exp(h\psi(u)).$$

As a result, $X_{t+h}^+ - X_t^+$ is independent of $\mathcal{F}_t^+$. From which the following theorem results.

Theorem 7.2.12. *Let X be a homogeneous $\mathbb{F}$-process with independent increments such that $X_0 = 0$ and*

$$X_t \xrightarrow[t\to 0]{P} 0.$$

This process has a modification X^+, having the following properties:

(a) *X^+ is an $\mathbb{F}^+$-process with independent increments;*
(b) *the trajectories of X^+ are* cad-lag *and zero at* 0.

Note. Let $(\mu_t)_{t\geq 0}$ be a **convolution semigroup** on $\mathbb{R}^k$, i.e. a family of distributions on $\mathbb{R}^k$ satisfying $\mu_{s+t} = \mu_s * \mu_t$ for every $(s,t) \in \mathbb{R}_+^2$. We assume

$$\mu_t \xrightarrow[t\to 0]{\mathcal{D}} \delta_0.$$

According to [0.2.3], we can associate with it a canonical homogeneous PII to which Theorem 7.2.12 applies. We can moreover deduce from the above a strong Markov property analogous to that of [4.1.2]. Let T be a bounded stopping time. If X is an $\mathbb{F}$-PII with cad-lag trajectories, we have from the stopping theorem

$$E[M_{T+t}^u | \mathcal{F}_T] = M_T^u.$$

From which it follows that

$$E[\exp(i<u, X_{T+t} - X_T>) | \mathcal{F}_T] = \phi_u(t).$$

Let T be a finite $\mathbb{F}$-stopping time and let $A \in \mathcal{F}_T$. For every integer n, $A \cap \{T \leq n\}$ is in $\mathcal{F}_{T\wedge n}$ and

$$E[\exp(i<u, X_{t+T\wedge n} - X_{T\wedge n}>) 1_{A\cap\{T\leq n\}}]$$
$$= P(A \cap \{T \leq n\})\phi_u(t).$$

We can apply Lebesgue's theorem and let n tend to ∞. We have obtained the following result.

Proposition 7.2.13. *Let X be an $\mathbb{F}$-PII with* cad-lag *trajectories. If T is a finite $\mathbb{F}$-stopping time, then $\{X_{t+T} - X_T\}_{t\geq 0}$ is a process independent of $\mathcal{F}_T$ and with the same distribution as X.*

7.3. Processes with Continuous Trajectories

7.3.1. The Space C_k of Continuous Functions From $\mathbb{R}_+$ into $\mathbb{R}^k$

To consider all the possible functions from $\mathbb{R}_+$ into $\mathbb{R}^k$ is often unnecessary, we can restrict ourselves to subspaces of more regular functions. The simplest is the space C_k of continuous functions from $\mathbb{R}_+$ into $\mathbb{R}^k$. Let us denote it by C if there is no doubt about k. It is a subset of $(\mathbb{R}^k)^{\mathbb{R}_+}$ and a function $\omega \in C$ is again denoted $t \longmapsto X_t(\omega)$. C is equipped with the **topology of uniform convergence on every compact set of** $\mathbb{R}_+$, which may be defined by the distance

$$\rho_c(\omega_1,\omega_2) = \sum_{n\geq 1} \frac{1}{2^n} \frac{\sup_{t\leq n} |X_t(\omega_1) - X_t(\omega_2)|}{1 + \sup_{t\leq n}|X_t(\omega_1) - X_t(\omega_2)|}.$$

Proposition 7.3.14. *The space C equipped with the topology of convergence on every compact set of $\mathbb{R}_+$ is Polish,* i.e. *metric, complete and separable.*

Proof. It is easily seen to be complete. It is separable. In fact, every function $\omega \in C$ is the limit in C of piecewise linear functions $\omega^{(n)}$ defined by setting, for every $j \in \mathbb{N}$,

$$X_t(\omega^{(n)}) = (1 - tn + h)X_{j/n}(\omega) + (tn - j)X_{(j+1)/n}(\omega)$$

for $j/n \leq t \leq (j+1)/n$. Such a piecewise linear function on the intervals $[j/n, (j+1)/n]$ is the uniform limit of functions, the values of which at the endpoints of the intervals are in $\mathbb{Q}^k$: C is separable.

Proposition 7.3.15. *The Borel σ-algebra $\mathcal{C}$ of C is the trace on C of $(\mathcal{B}_{\mathbb{R}^k})^{\mathbb{R}_+}$: thus a probability P on $(C,\mathcal{C})$ is characterized by its finite distribution functions.*

Proof. It is required to prove that $\mathcal{C}$ is the σ-algebra generated by the coordinate functions X_t, for $t \in \mathbb{R}_+$. Since these functions are continuous, the inclusion $\mathcal{C} \supset \sigma(X_t; t \geq 0)$ is clear. However $\mathcal{C}$ is generated by the open sets

$$B(\alpha,n,\varepsilon) = \left\{\omega;\ \sup_{t \leq n} |X_t(\omega) - X_t(\alpha)| < \varepsilon\right\}.$$

In fact, if (α_p) is a sequence dense in C, every open set is a union of balls $B(\alpha_p,n,q)$ which it contains (with q rational). However a ball $B(\alpha,n,\varepsilon)$ is an element of $\sigma(X_t; t \geq 0)$ by virtue of the following relations:

$$B(\alpha,n,\varepsilon) = \left\{\omega;\ \sup_{q \in \mathbb{Q} \cap [0,n]} |X_q(\omega) - X_q(\alpha)| < \varepsilon\right\}$$

$$= \bigcap_{q \in \mathbb{Q} \cap [0,n]} \left\{\omega;\ |X_q(\omega) - X_q(\alpha)| < \varepsilon\right\}.$$

Examples of Continuous Functions Defined on C. (a) Let $(t_1, \ldots, t_p) \in \mathbb{R}_+^p$. The function $\omega \longmapsto (X_{t_1}(\omega), \ldots, X_{t_p}(\omega))$ is continuous from C_k into $\mathbb{R}^{kp}$.

(b) To every function $\omega \in C_1$, we associate $\bar{\omega}$ defined for all $t \in \mathbb{R}_+$ by $\bar{\omega}(t) = \sup_{s \leq t} \omega(s)$. The function $\omega \longmapsto \bar{\omega}$ is continuous. In fact, for $N \in \mathbb{N}$ and $(\omega_1,\omega_2) \in C^2$:

$$\sup_{t \leq N} |\bar{\omega}_1(t) - \bar{\omega}_2(t)| = \sup_{t \leq N} \left|\sup_{u \leq t} \omega_1(u) - \sup_{u \leq t} \omega_2(u)\right|$$

$$\leq \sup_{t \leq N} \sup_{u \leq t} |\omega_1(u) - \omega_2(u)|$$

$$\leq \sup_{u \leq N} |\omega_1(u) - \omega_2(u)|.$$

The mapping $\omega \longmapsto \bar{\omega}$ is thus a contraction of C_1.

For $\omega \in C_k$, let us denote by $|\omega|$ the function $t \longmapsto |\omega(t)|$. The mappings $\omega \longmapsto |\omega|$ and $\omega \longmapsto \omega^* = \overline{|\omega|}$ are continuous from C_k into C_1.

(c) Let $0 \leqslant a < b < \infty$. The mapping

$$\omega \longmapsto \int_a^b \omega(s)ds$$

is continuous from C into $\mathbb{R}^k$. The mapping $\omega \longmapsto \int\omega$ is continuous from C into C by setting

$$\left[\int\omega\right](t) = \int_0^t \omega(s)ds.$$

Compact Sets of C. In order to describe these, we resort to the following **Ascoli's Theorem.** We consider $C(K)$, the set of continuous functions from a compact metric set K into $\mathbb{R}^k$, equipped with the topology of uniform convergence. A subset H of $C(K)$ is relatively compact if and only if:

(a) for every $t \in K$, $\{\omega(t); \omega \in H\}$ is bounded in $\mathbb{R}^k$;

(b) H is **equicontinuous**, i.e., for every ε, there exists a δ such that when t_1 and t_2 are two points of K at a distance of less than δ from one another, we have

$$\sup_{\omega \in H} |\omega(t_1) - \omega(t_2)| \leqslant \varepsilon.$$

This theorem from classical analysis is assumed and we apply it to $C[0,N]$. Combined with (b) condition (a) is written: $\{\omega(0); \omega \in H\}$ is bounded in $\mathbb{R}^k$. We set, for $\omega \in C$ and $\delta > 0$,

$$V^N(\omega,\delta) = \sup\{|\omega(t) - \omega(t')|; |t - t'| \leqslant \delta, t \leqslant N, t' \leqslant N\}.$$

Part (b) may be written

$$\lim_{\delta \downarrow 0} \sup_{\omega \in H} |V^N(\omega,\delta)| = 0.$$

Finally, a subset H of C is compact in C if, and only if, its restriction to $C[0,N]$ is compact on $C[0,N]$ for all N. In fact if we can, for all N, extract from a sequence a uniformly convergent sequence of $[0,N]$, we can extract from it a convergent sequence in C by the diagonal process.

Proposition 7.3.16. *A relatively compact subset H of C is a subset such that,*

(a) $\{X_0(\omega);\ \omega \in H\}$ *is bounded in* $\mathbb{R}^k$;

(b) *for all* N, $\lim\limits_{\delta \downarrow 0} \sup\limits_{\omega \in H} |V^N(\omega,\delta)| = 0.$

7.3.2. Narrow Convergence of Probabilities

A certain number of topological properties of $\mathbb{R}^k$ apply to a Polish space E. Denote by $\mathcal{E}$ its Borel σ-algebra and by $\mathcal{P}(E)$ the set of probabilities on $(E,\mathcal{E})$.

Definition 7.3.17. A subset A of $\mathcal{P}(E)$ is **tight** if, for every $\varepsilon > 0$, there exists a compact set K_ε of E such that, for every $P \in A$, we have

$$P(K_\varepsilon^c) \leqslant \varepsilon .$$

Proposition 7.3.18. *If* E *is Polish, every probability of* $\mathcal{P}(E)$ *is tight.*

Proof. For $x \in E$ and $r > 0$, we denote $B(x,r)$ the open ball with center x and radius r, and $\bar{B}(x,r)$ its closure. Since E is separable, there exists a sequence (x_n) dense in E. For every $p \in \mathbb{N}$ and every $\varepsilon > 0$, we have

$$\bigcup_{n=1}^{\infty} \bar{B}(x_n, (1/p)) = E,$$

thus there exists a finite set I_p of $\mathbb{N}$ such that

$$P\left[E\setminus \bigcup_{n\in I_p} \bar{B}(x_n,(1/p))\right] \leqslant \frac{\varepsilon}{2^p}.$$

Let

$$K_\varepsilon = \bigcap_p \bigcup_{n\in I_p} \bar{B}(x_n,(1/p));$$

we have $P(K_\varepsilon^c) \leqslant \varepsilon$ The set K_ε is a compact set. To prove it let s be a sequence contained in K_ε. For every p, it has an infinity of terms in

$$\bigcup_{n\in I_p} \{\bar{B}(x_n,(1/p))\},$$

thus a certain infinite subsequence is contained in $\overline{B}(x_{n_p},(1/p))$ for an $n_p \in I_p$. Therefore, by the diagonal procedure, we construct a subsequence of s contained in $\cap_p \overline{B}(x_{n_p},(1/p))$: it is a Cauchy convergent sequence.

Criterion of Relative Compactness of a Sequence of Probabilities on a Polish Space. Here is a theorem which generalizes Theorem 3.4.26 of Volume I.

Prokhorov's Theorem 7.3.19. *If E is a Polish space, we can extract from every tight sequence (P_n) of $\mathcal{P}(E)$ a subsequence which converges narrowly. If the sequence (P_n) does not converge narrowly, we can extract two subsequences which converge narrowly to distinct probabilities.*

Proof. Recall, first of all, two results from analysis, which hold in the particular case where E is a compact metric space, denoted K:

(a) *The Stone-Weierstrass theorem*: if $\mathcal{A}$ is an algebra of continuous functions from K into $\mathbb{R}$, which contains the constants and separates the points, is dense in $\mathcal{C}(K)$, the set of continuous functions from K into $\mathbb{R}$.

(b) *Riesz's theorem*: every positive linear form $f \longmapsto \lambda(f)$ defined on $\mathcal{C}(K)$ has a unique representation of the form $\lambda(f) = \int f\, d\mu$ for μ a bounded measure on K.

Let E be Polish.

(a) Recall the diagonal procedure of Theorem 3.4.26 of Volume 1. Let (x_n) be a sequence dense in E. A countable family of functions of $\mathcal{C}_b(E)$ is constructed which separates points by taking, for every integer n and p, a continuous $f_{n,p}$ satisfying

$$1_{\overline{B}(x_n,1/[p+1])} \leq f_{n,p} \leq 1_{B(x_n,\, 1/p)}$$

(for example

$$f_{n,p} = [1 - p(p+1)d(\cdot,\overline{B}(x_n,\, 1/[p+1]))]_+ ,$$

denoting by d the distance from E).

Let K be a compact set of E; the algebra $\mathcal{A}'(K)$ of linear combinations with rational coefficients of functions $1_k f_{n,p}$ is dense in $\mathcal{C}(K)$ and countable. By the diagonal procedure, we can thus extract from (P_n) a sub-sequence (P'_n) such that $(P'_n(f))$ converges for all $f \in \mathcal{A}'(K)$ to a number $\lambda_K(f)$. Then, for all $f \in \mathcal{C}(K)$, $P'_n(f)$ converges to $\lambda_K(f)$, and $f \longmapsto \lambda_K(f)$ is a positive linear form defined on $\mathcal{C}(K)$. Thus $\lambda_K(f) = \int f \, d\mu_K$, μ_K is a bounded measure on K. Its total mass is

$$\lim P'_n(1_K) \leqslant 1.$$

(b) The sequence (P_n) being tight, for every p we can find a compact set K_p, such that, for every n, $P_n(K_p) \geqslant 1 - 1/p$. The sequence (K_p) may be taken to be increasing. Using the diagonal procedure again, we can extract from (P_n) a subsequence (P''_n) such that, for every p and every $f \in \mathcal{C}(K_p)$, the sequence $P''_n(f)$ converges to $\mu_{K_p}(f)$. Let us denote by μ_p the measure on E, concentrated on K_p, and equal to μ_{K_p} on K_p. Let f be a positive continuous function from E into $\mathbb{R}$,

$$\begin{aligned} \mu_p(f) = \mu_p(f 1_{K_p}) &= \lim_n P''_n(f 1_{K_p}) \\ &\leqslant \lim_n P''_n(f 1_{K_{p+1}}) = \mu_{p+1}(f). \end{aligned}$$

Thus the sequence of measures (μ_p) is increasing with mass $\leqslant 1$. Let $\mu = \lim \uparrow \mu_p$. If (A_k) is a sequence of disjoint Borel sets of E of union A, we have $\mu_p(A) = \Sigma \mu_p(A_k)$. If p tends to ∞, we obtain from the Beppo-Levi theorem relative to series, i.e. to the measure $\Sigma \delta_k$,

$$\mu(A) = \Sigma \mu(A_k).$$

Thus μ is σ-additive. It is a measure on E of mass at most equal to 1,

$$\begin{aligned} |P''_n(f) - \mu(f)| &\leqslant |P''_n(f 1_{K_p}) - \mu_p(f)| \\ &\quad + \|f\| \{P''_n(K_p^c) + (\mu - \mu_p)(E)\}, \end{aligned}$$

$$\overline{\lim_n} \, |P''_n(f) - \mu(f)| \leqslant \|f\| \left\{ \frac{1}{p} + (\mu - \mu_p)(E) \right\}.$$

This being true for any p: $(P''_n) \xrightarrow{n} \mu$.

7.3.3. Convergence in Distribution of a Sequence of Continuous Processes

Let C be the space defined in [7.3.1]. In order to apply Prokhorov's theorem to a sequence of probabilities defined on C we shall make use of the following tightness criterion (with the notations of [7.3.1]).

Theorem 7.3.20. *A sequence (P_n) of probabilities on C is tight if:*

(a) $\lim_{a\to\infty} \overline{\lim_{n}}\ P_n(\omega;\ |X_0(\omega)| \geqslant a) = 0.$

(b) *For every $\varepsilon > 0$ and $N \in \mathbb{N}$*

$$\lim_{\delta\downarrow 0} \overline{\lim_{n}}\ P_n(\omega;\ V^N(\omega,\delta) \geqslant \varepsilon) = 0.$$

Proof. From (a), for every $\varepsilon > 0$ we can find $M_1 > 0$ and n_0 such that

$$\sup_{n\geqslant n_0} P_n(\omega;\ |X_0(\omega)| \geqslant M_1) \leqslant \frac{\varepsilon}{2}.$$

For all integers p and $N \geqslant 1$, we can find n_1 and $\delta^{(1)}_{p,N} > 0$ such that

$$\sup_{n\geqslant n_1} P_n\left[\omega;\ V^N(\omega,\ \delta^{(1)}_{p,N}) \geqslant \frac{1}{p}\right] \leqslant \frac{\varepsilon}{2^{p+N+1}}.$$

However we can find M_2 and $\delta^{(2)}_{p,N}$ such that,

$$\sup_{n<n_0} P_n(\omega;\ |X_0(\omega)| \geqslant M_2) + \sup_{n<n_1} P_n\left[\omega;\ V^N(\omega,\delta^{(2)}_{p,N}) \geqslant \frac{1}{p}\right]$$

$$\leqslant \frac{\varepsilon}{2^{p+N+1}}.$$

Let $\delta_{p,N} = \inf(\delta^{(1)}_{p,N},\delta^{(2)}_{p,N})$ and let $M = \sup(M_1,M_2)$. Taking K_ε $= \{\omega;\ |X_0(\omega)| \leqslant M$ and $V^N(\omega,\delta_{p,N}) \leqslant 1/p$ for $p \geqslant 1$ and $N \geqslant 1\}$ we obtain

$$\sup_{n} P_n(K^c_\varepsilon) \leqslant \varepsilon .$$

and K_ε is relatively compact from Proposition 7.3.16. Hence (P_n) is tight in C.

As a result. In order to study narrow convergence of a sequence (P_n) of probabilities on C, we shall thus proceed in two stages:

(1) *Study of Finite Distribution Functions.* Narrow convergence in $(\mathbb{R}^k)^p$ of the distributions $(X_{t_1}, ..., X_{t_p})P_n$ for every $p \geq 1$ and $(t_1, ..., t_p) \in \mathbb{R}_+^p$ is studied using known criteria ([Vol. I, 3.4]). If for each $(t_1, ..., t_p)$ there is convergence to a probability $\mu_{t_1,...,t_p}$, we continue.

(2) *Study of Tightness.* If (P_n) is tight, the finite distributions of every closure point coincide with $\{\mu_{t_1,...,t_p}\}$. The sequence (P_n) then converges narrowly in C to the unique probability P having these finite distribution functions (Proposition 7.3.15).

Translation Relative to Processes with Continuous Trajectories

Definition 7.3.21. Let $(\Omega, \mathcal{A}, P, (X_t)_{t \in \mathbb{R}_+})$ be a process taking values in $\mathbb{R}^k$, nearly all the trajectories of which are continuous. We say that we are dealing with a **continuous process.** Its distribution is then a probability concentrated on $(C, \mathcal{C})$. It is this trace which we call the **distribution of the continuous process.** Let $(X^{(n)})$ be a sequence of continuous processes taking values in $\mathbb{R}^k$. This sequence is said to be **tight** if the sequence of its distributions is tight. If Q is a probability on C or X a continuous process taking values in $\mathbb{R}^k$, the sequence $(X^{(n)})$ **converges in distribution** to Q or to X when the sequence of its distributions tends narrowly to Q. We denote

$$(X^{(n)}) \xrightarrow{\mathcal{D}} X \quad \text{or} \quad (X^{(n)}) \xrightarrow{\mathcal{D}} Q.$$

The above statements are then easily translated relative to sequences of probabilities on C.

Notes. Two notes make the study of the tightness of $(X^{(n)})$ easier.

(a) The sequence $(X^{(n)})$ is tight if and only if its components are also tight.

(b) The convergence of finite distribution functions implies condition (a) of Theorem 7.3.20, and there is then no need to check it.

7.3.4. Modulus of Continuity of a Process

Let $(X_t)_{t\in[0,1]}$ be a process taking values in $\mathbb{R}^k$. We study the modulus of continuity of X,

$$V(X,\delta) = \sup\{|X_t - X_s|,\ 0 < t < s < 1,\ s - t \leqslant \delta\}.$$

In general this need not be an r.v. For $n \in \mathbb{N}$, denote $\Delta_n = \{q/2^n;\ 0 \leqslant q < 2^n\}$ and

$$U_n = \sup\{|X_{q/2^n} - X_{(q+1)/2^n}|;\ 0 \leqslant q < 2^n\}.$$

Let $t \in [0,1]$, and let $[2^n t]$ be the integer part of $2^n t$. Set $t_n = [2^n t]2^{-n}$. Let $t \in \Delta_m$, $m > n$;

$$|X_t - X_{t_n}| \leqslant \sum_{n\leqslant p<m} |X_{t_{p+1}} - X_{t_p}| \leqslant \sum_{n<p\leqslant m} U_p.$$

Let $\Delta = \cup\Delta_n$ be the set of dyadics of $[0,1]$. For $t \in \Delta$, we have

$$|X_t - X_{t_n}| \leqslant \sum_{p>n} U_p;\ \text{for}\ \ \delta \leqslant 2^{-n}\ \text{and}\ |s - t| \leqslant \delta$$

$$|X_t - X_s| \leqslant |X_t - X_{t_n}| + |X_{t_n} - X_{s_n}| + |X_s - X_{s_n}|$$
$$\leqslant 3 \sum_{p>n} U_p.$$

Let

$$V'(X,\delta) = \sup\{|X_t - X_s|;\ |s - t| \leqslant \delta\ \text{and}\ (s,t) \in \Delta^2\}.$$

This is an r.v. and

$$V'(X,2^{-n}) \leqslant 3 \sum_{p>n} U_p.$$

Moreover if the process X is continuous $V'(X,\delta) = V(X,\delta)$. This inequality is vital in order to obtain various criteria of continuity of X or of tightness of a sequence $X^{(n)}$ of continuous processes. Here is such an example.

Theorem 7.3.22. *Let* $X^{(n)} = \{\Omega, \mathcal{A}, P, (X_t^{(n)})_{t \geq 0}\}$ *be a sequence of continuous processes taking values in* $\mathbb{R}^k$. *Assume that there exist three strictly positive constants* α, β, γ *such that, for every* $(s,t) \in \mathbb{R}_+^2$ *and every* n,

$$E[|X_s^{(n)} - X_t^{(n)}|^\alpha] \leq \beta|s - t|^{\gamma+1}.$$

Then, when $(X_0^{(n)})$ *is a tight sequence in* $\mathbb{R}^k$, *the sequence* $(X^{(n)})$ *is tight in* C.

Proof. Denoting $V^N(X^{(n)},\delta) = \sup\{|X_t^{(n)} - X_s^{(n)}|;\ 0 < t < s < N,\ s-t \leq \delta\}$, from Theorem 7.3.20, we shall show that, for every $\varepsilon > 0$, $P[V^N(X^{(n)},\delta) \geq \varepsilon]$ tends to 0 uniformly in n if δ tends to 0. By a change of variable $t \longmapsto t/N$, we are led to study the case $N = 1$. Let

$$U_p^{(n)} = \sup\{|X_{q/2^p}^{(n)} - X_{(q+1)/2^p}^{(n)}|;\ 0 \leq q < 2^p\}.$$

We have

$$\begin{aligned} E[(U_p^{(n)})^\alpha] &\leq 2^p \sup\{E(|X_{q/2^p} - X_{(q+1)/2^p}|^\alpha);\ 0 \leq q < 2^p\} \\ &\leq \beta 2^{-p\gamma}. \end{aligned}$$

By considering the norms $\|\ \|_\alpha$ in $L^\alpha(\Omega, \mathcal{A}, P)$:

$$\|U_p^{(n)}\|_\alpha \leq 2^{-p\gamma/\alpha}\, \beta^{1/\alpha},$$

$$\begin{aligned} \|V^N(X^{(n)},2^{-p})\|_\alpha &\leq 3 \sum_{q>p} \|U^{(n)}\|_\alpha \leq 3\beta^{1/\alpha} \sum_{q>p} 2^{-q\gamma/\alpha} \\ &\leq 3\beta^{1/\alpha} \frac{2^{-p\gamma/\alpha}}{1 - 2^{-\gamma/\alpha}}, \end{aligned}$$

which leads to the stated result by Tchebychev's inequality.

Theorem 7.3.23. *Let* $X = \{\Omega, \mathcal{A}, P, (X_t)_{t \in \mathbb{R}_+}\}$ *be a process taking values in* $\mathbb{R}^k$. *Let us assume that there exist constants* α, β, $\gamma > 0$ *such that, for every* $(s,t) \in \mathbb{R}_+^2$:

$$E(|X_s - X_t|^\alpha) \leq \beta|s - t|^{\gamma+1}.$$

Then the process X *has a continuous modification*

Proof. Let Δ be the set of dyadics of $\mathbb{R}_+$ and, for $N \in \mathbb{N}$, set

$$V'_N(X,\delta) = \sup\{|X_s - X_t|; |s-t| \leq \delta,$$
$$(s,t) \in \Delta^2, s \leq N, t \leq N\}$$

and prove

$$V'_N(X,\delta) \xrightarrow[\delta\to 0]{\text{a.s.}} 0.$$

By the change of variable $t \longmapsto t/N$ it is sufficient to prove $V'_1(X,\delta) \xrightarrow[\delta\to 0]{\text{a.s.}} 0$. With the previous notations:

$$V'_1(X,2^{-n}) \leq 3 \sum_{p>n} U_p.$$

For $\lambda > 0$,

$$P[U_p \geq \lambda] \leq \beta\lambda^{-\alpha}2^{-p\gamma}.$$

Take $0 < \eta < \gamma/\alpha$,

$$P[U_p \geq 2^{-p\eta}] \leq \beta 2^{-p[\gamma-\alpha\eta]}.$$

Thus

$$\sum_p P[U_p \geq 2^{-p\eta}] < \infty$$

and Borel-Cantelli's theorem implies that for $\omega \notin H_1$, H_1 negligible, there exists an $N(\omega)$ such that $p \geq N(\omega)$ and $\delta \leq 2^{-p}$ implies,

$$U_p < 2^{-p^\eta}, \quad V'_1(X(\omega),\delta) \leq 3 \sum_{q>p} 2^{-p^\eta} \leq 3 \frac{2^{-p^\eta}}{1-2^{-\eta}}.$$

Thus, for any $N \in \mathbb{N}$, there exists a negligible set H_N such that, for $\omega \notin H_N$,

$$V'_N(X(\omega),\delta) \xrightarrow[\delta\to 0]{} 0.$$

Set $H = \cup H_N$. For $\omega \notin H$, the function $t \longmapsto X_t(\omega)$ defined on Δ has a unique extension as a continuous function $t \longmapsto \tilde{X}_t(\omega)$. Set for instance $\tilde{X}_t(\omega) = 0$ for $\omega \in H$. With the theorem's hypothesis X is continuous in probability; so if (t_n)

is a sequence of Δ tending to t,

$$X_{t_n} \xrightarrow{\text{a.s.}} \tilde{X}_t \quad \text{and} \quad X_{t_n} \xrightarrow{P} X_t.$$

Thus $P[X_t = \tilde{X}_t] = 1$ and $\tilde{X}$ is a continuous modification of X.

Corollary 7.3.24. *A Brownian motion has a continuous modification.*

Proof. If $(X_t)_{t \geq 0}$ is a Brownian motion, for $0 \leq t < s$, $X_s - X_t$ has distribution $N(0, s - t)$. From which it follows that $E[X_s - X_t]^4 = 3(s - t)^2$. Corollary 7.3.23 then applies.

Note. From now on, when we speak of Brownian motion, we will be dealing with a continuous version of this process. Its distribution is thus a measure on C, called **Wiener measure** and denoted W.

7.3.5. Convergence in Distribution of a Sequence of Processes with cad-lag Trajectories to a Continuous Process

When we work with martingales or PII's, we can, from [7.2], assume that the trajectories are **cad-lag**. This leads to considering the space D of cad-lag functions from $\mathbb{R}_+$ into $\mathbb{R}^k$. this space contains C and we can also equip it with the topology of uniform convergence on every compact set of $\mathbb{R}_+$. As in [7.3.1] this topology may be defined by the distance ρ_C. However it does not have very good properties on D; it is not separable. It is possible to obtain on D properties very similar to those of C by introducing a weaker topology, the **Skorokhod topology**. However in many cases the following definition is sufficient.

Definition 7.3.25. *For each integer n, let $X^{(n)} = \{\Omega_n, A_n, P_n, (X_t^{(n)})_{t \geq 0}\}$ be a process with trajectories in D.*

(a) *For each n, if $Y^{(n)} = \{\Omega_n, A_n, P_n, (Y_t^{(n)})_{t \geq 0}\}$ is another process with trajectories in D, we say that the sequences $(X^{(n)})$ and $(Y^{(n)})$ are* **contiguous** *when*

$$\rho_C(X^{(n)}, Y^{(n)}) \xrightarrow[n \to \infty]{\mathcal{D}} 0.$$

(b) *If X is a process with trajectories in C (or if Q is a distribution on C), we say that* $(X^{(n)})$ **converges strongly in distribution** *to X (or to Q) when, for a contiguous sequence $(Y^{(n)})$ with continuous trajectories, we have*

$$(Y^{(n)}) \xrightarrow[n\to\infty]{\mathcal{D}} X \quad \text{or} \quad (Y^{(n)}) \xrightarrow[n\to\infty]{\mathcal{D}} Q$$

We then denote

$$X^{(n)} \xrightarrow[n\to\infty]{\mathcal{D}(C)} X \quad \text{or} \quad X^{(n)} \xrightarrow[n\to\infty]{\mathcal{D}(C)} Q.$$

Notes. (a) Let $(\Omega,\mathcal{A},P,(X_t)_{t\geqslant 0})$ be a cad-lag process taking values in $\mathbb{R}^k$. For $k = 1$, we have, for any t,

$$\begin{aligned}\bar{X}_t &= \{\sup X_s;\ s \leqslant t\}\\ &= \sup\{X_s;\ s \text{ rational} < t \ \text{ or } \ s = t\}.\end{aligned}$$

$\bar{X}_t$ is measurable. For every k, $|X_t|$ and $X_t^* = \sup\{|X_s|;\ s \leqslant t\}$ are measurable. This is what allows us to talk about $\rho_C(X^{(n)},Y^{(n)})$ in the above statement.

(b) It is not difficult to see, by using [7.3.1], that if $(X^{(n)})$ converges strongly to a continuous X, then for any t,

– when $k = 1$, $\bar{X}_t^{(n)} \xrightarrow{\mathcal{D}} \bar{X}_t$

– for every k, $(X_t^{(n)})^* \xrightarrow{\mathcal{D}} X_t^*$.

(c) Let $x\colon\ t \longmapsto x_t$ be a function in D and $y\colon\ t \to y_t$ a function in C. Let us denote by Δx_t the jump $x_t - x_{t-}$ of x at t and

$$\Delta^* x_t = \sup\{|\Delta x_s|;\ s \leqslant t\}.$$

We have

$$\Delta^* x_N \leqslant 2 \sup\{|x_t - y_t|;\ t \leqslant N\} \leqslant 2^{N+1}\rho_C(x,y).$$

Thus, if a sequence of processes $(X^{(n)})$ is contiguous to a sequence of continuous processes $(Y^{(n)})$, and if we denote $\Delta^* X^{(n)} = (\Delta^* X_t^{(n)})_{t\geqslant 0}$ then,

$$\rho_C(\Delta^* X^{(n)},0) \xrightarrow{\mathcal{D}} 0.$$

(d) If $X = (\Omega,\mathcal{A},P,(X_t)_{t\geqslant 0})$ is a process with trajectories in D, we again set

$$V^N(X,\delta) = \sup\{|X_t - X_s|;\ 0 \leqslant t < s \leqslant N,\ s - t \leqslant \delta\}.$$

If Y is another process with trajectories in D defined on $(\Omega,\mathcal{A},P)$, we have

$$|V^N(X,\delta) - V^N(Y,\delta)| \leqslant 2^{N+1}\rho_C(X,Y).$$

Thus, if $(X^{(n)})$ is a sequence of processes taking values in D contiguous to a sequence $(Y^{(n)})$ taking vaues in C, then the sequence $(Y^{(n)})$ is tight if, and only if, we have

$$\lim_{a\to\infty} \overline{\lim_{n}}\ P_n(|X_0^{(n)}| \geqslant a) = 0$$

and for every $\varepsilon > 0$ and $N \in \mathbb{N}$,

$$\lim_{\delta\downarrow 0} \overline{\lim_{n}}\ P_n(V(X^{(n)},\delta) \geqslant \varepsilon) = 0.$$

Moreover if the finite distribution functions of $X^{(n)}$ converge to those of X (or of Q), the sequence $(X^{(n)})$ converges strongly in distribution to X (or Q).

7.3.6. Modulus of Continuity of a Cad-Lag Process

First of all we deal with a technical tool analogous to [7.3.4]. Consider a probability space $(\Omega,\mathcal{A},P)$ equipped with a filtration $\mathbb{F} = (\mathcal{F}_t)_{t\geqslant 0}$ continuous on the right. Let $X = (X_t)_{t\geqslant 0}$ be a process taking values in $\mathbb{R}^k$, continuous on the right, adapted to $\mathbb{F}$. Let us set $T_0 = 0$ and, for $p \geqslant 1$,

$$T_{p+1} = \begin{cases} \inf\{s;\ s \geqslant T_p,\ |X_s - X_{T_p}| \geqslant \varepsilon/3\} & \text{if } T_p < \infty, \\ \infty & \text{if } T_p = \infty. \end{cases}$$

Then

$$\sup_{T_p \leqslant s < s' < T_{p+1}} |X_s - X_{s'}| < \frac{2\varepsilon}{3}.$$

Let

$$V^N(X,\delta) = \sup\{|X_s - X_t|; |s - t| < \delta,$$
$$0 \leqslant s \leqslant N,\ 0 \leqslant t \leqslant N\}$$

and let $\Delta X_N^* = \sup\{|X_t - X_{t^-}|; t \leqslant N\}$;

$$\{V^N(X,\delta) \geqslant \varepsilon\} \subset \{\Delta X_N^* \geqslant \varepsilon/3\} \cup \bigcup_p \{T_{p+1} - T_p < \delta; T_p < N\}.$$

The times T_p are stopping times (see this by recurrence by noticing that

$$1_{(T_p<\infty)}(X_{T_p+t} - X_{T_p})$$

is adapted to the right continuous filtration $(F_{T_p+t})_{t\geqslant 0}$). Denote

$$K(\varepsilon,\delta,N,X) = \sup P(T_{p+1} - T_p < \delta, T_p < N).$$

We can write

$$NP(T_p < N) \geqslant E(T_p 1_{(T_p<N)})$$
$$= E\left[\sum_{i=1}^{p} (T_i - T_{i-1}) 1_{(T_p<N)}\right]$$
$$\geqslant \gamma \sum_{i=1}^{p} [P(T_p < N) - P(T_{i+1} - T_i < \delta, T_i < N)]$$
$$\geqslant [P(T_p < N) - K(\varepsilon,\gamma,N,X)]p\gamma.$$

Let us take $p\gamma > N$. We obtain

$$P(V^N(X,\delta) \geqslant \varepsilon) \leqslant P(\Delta X_N^* \geqslant \varepsilon/3)$$
$$+ P(T_p < N) + pK(\varepsilon,\delta,N,X)$$
$$\leqslant P(\Delta X_N^* \geqslant \varepsilon/3) + K(\varepsilon,\gamma,N,X)\left[\frac{p\gamma}{p\gamma - N}\right]$$
$$+ pK(\varepsilon,\delta,N,X).$$

Consequence. Let $(X^{(n)})$ be a process with trajectories in D contiguous to a sequence $(Y^{(n)})$ of continuous processes. In order that $(X^{(n)})$ converges strongly in distribution to a continuous process X, it is sufficient to verify the following two points:

– the finite distribution functions of $X^{(n)}$ converge to those of X;
– for every $\varepsilon > 0$ and $N \in \mathbb{N}$,

$$\lim_{\delta \downarrow 0} \overline{\lim_{n\to\infty}} K(\varepsilon,\delta,N,X^{(n)}) = 0.$$

This criterion will allow us to establish a functional central limit theorem in the following paragraph. We shall again have occasion to use it in [8.2.3]. Let us state here the technical proposition which will be the key to all these theorems. It can be left till later when it is required ([7.4.1]).

Proposition 7.3.26. *Consider, for every integer n,*

– *A probability space (Ω_n, A_n, P_n) given a natural filtration* $\mathbb{F}^{(n)} = (F_t^{(n)})_{t\geq 0}$.
– *A martingale $M^{(n)}$ and an increasing process $A^{(n)}$, both adapted to $\mathbb{F}^{(n)}$, zero for $t = 0$ and with* cad-lag *trajectories. The jumps of the process $A^{(n)}$ are assumed to be majorized by a constant c. Denoting $(M^{(n)})^2 = ([M_t^{(n)}]^2)$, assume finally that $(M^{(n)})^2$ is integrable and that $(M^{(n)})^2 - A^{(n)}$ is an $\mathbb{F}^{(n)}$-martingale.*
– *A process $Y^{(n)}$ with trajectories in C_1.*

Finally we are given a continuous function $t \longmapsto a_t$ from $\mathbb{R}_+$ into $\mathbb{R}_+$. Assume that $(M^{(n)})$ and $(Y^{(n)})$ are contiguous, and that, for every t,

$$A_t^{(n)} \xrightarrow[n\to\infty]{\mathcal{D}} a_t.$$

Then the sequence $(Y^{(n)})$ is tight.

Proof. Let T be an $\mathbb{F}^{(n)}$ stopping time. For every $\varepsilon > 0$, $\delta > 0$, $\alpha > 0$ apply Proposition 7.2.10 to $(M_{T+u}^{(n)} - M_T^{(n)})_{u\geq 0}$ which is a martingale adapted to $(F_{T+u}^{(n)})_{u\geq 0}$,

$$P_n\left[\sup_{0\leq u\leq \delta} |M_{T+u}^{(n)} - M_T^{(n)}| \geq \varepsilon | \; F_T^{(n)}\right]$$
$$\leq P_n[A_{T+\delta}^{(n)} - A_T^{(n)} \geq \alpha | \; F_T^{(n)}]$$

$$+ \frac{1}{\varepsilon^2} E_n[(A^{(n)}_{T+\delta} - A^{(n)}_T) \wedge (c + \alpha) | F^{(n)}_T].$$

Hence;

$$K(\varepsilon,\partial,N,M^{(n)}) \leqslant P_n\left[\sup_{u \leqslant N} (A^{(n)}_{u+\delta} - A^{(n)}_u) \geqslant \alpha\right]$$

$$+ \frac{1}{\varepsilon^2} E_n\left[\sup_{u \leqslant N} (A^{(n)}_{u+\delta} - A^{(n)}_u) \wedge (c + \alpha)\right].$$

Now let us assume that

$$\sup_{u \leqslant N} |A^{(n)}_u - a_u| \xrightarrow[n\to\infty]{\mathcal{D}} 0.$$

For $\eta > 0$, we can find a δ such that $\sup|a(u + \delta) - a(u)| \leqslant \eta$. Take $\alpha > \eta$. Then

$$\overline{\lim_{n\to\infty}} K(\varepsilon,\delta,N,M^{(n)}) \leqslant \frac{\eta}{\varepsilon^2}.$$

From which it follows that

$$\lim_{\delta\downarrow 0} \overline{\lim_{n\to\infty}} K(\varepsilon,\delta,N,M^{(n)}) = 0.$$

The proof is now completed with the following lemma, by using note (d) of [7.3.5].

Lemma 7.3.27. *Let* $A^{(n)} = \{\Omega_n, A_n, P_n, (A_n(t))_{t\geqslant 0}\}$ *be a sequence of increasing processes (Definition 6.2.12) and let* $a\colon t \longmapsto a(t)$ *be a continuous function from* $\mathbb{R}_+$ *into* $\mathbb{R}$. *Assume that, for every* t,

$$A_n(t) \xrightarrow[n\to\infty]{\mathcal{D}} a(t);$$

then

$$\rho_C(A_n,a) \xrightarrow[n\to\infty]{\mathcal{D}} 0.$$

Proof. Let $\varepsilon > 0$ and let $0 = t_0 < t_1 \dots < t_m = N$ be a partition of $[0,N]$ such that, for $0 \leqslant p < m$, $|a(t_{p+1}) - a(t_p)| \leqslant \varepsilon/4$. Then, for $t_p \leqslant t \leqslant t_{p+1}$,

$$|A_n(t) - a(t)| \leqslant \varepsilon/2 + |A_n(t_p) - a(t_p)|$$

$$+ |A_n(t_{p+1}) - a(t_{p+1})|.$$

Thus

$$\sup_{t\leqslant N}|A_n(t) - a(t)| \leqslant \frac{\varepsilon}{2} + \sum_{p=0}^{n} |A_n(t_p) - a(t_p)|$$

Since $\Sigma^m_{p=0}|A_n(t_p) - a(t_p)| \xrightarrow{\mathcal{D}} 0$, we obtain the result.

7.4. Functional Central Limit Theorems

7.4.1. Convergence of Triangular Sequences to Brownian Motion

We are concerned with a theorem analogous to Theorem 2.8.42, the vocabulary and notations of which we recall.

Theorem 7.4.28. *Let $(\Omega, \mathcal{A}, P)$ be a probability space on which we are given for each $n \geq 0$:*

- *A filtration* $\mathbb{F}^n = (\mathcal{F}_k^n)_{k \geq 0}$
- *A sequence of* $\mathbb{F}^n$*-adapted* r.v.'s $(\xi_{n,k})_{k \geq 0}$.

We are given a function a from $\mathbb{R}_+$ into $\mathbb{R}_+$ which is increasing, continuous and zero at 0. *The following hypotheses are made (denoting by* [ ] *the integer part):*

H1) *The sequence is asymptotically negligible*, i.e. *for every t and every $\varepsilon > 0$*

$$\sum_{k=1}^{[nt]} P^{n,k-1}(|\xi_{n,k}| \geq \varepsilon) \xrightarrow[n \to \infty]{P} 0.$$

H2) *For an $\varepsilon > 0$*

$$\sum_{k=1}^{[nt]} E^{n,k-1}(\xi_{n,k}^{\varepsilon}) \xrightarrow[n \to \infty]{P} 0.$$

H3)
$$\sum_{k=1}^{[nt]} V^{n,k-1}(\xi_{n,k}^{\varepsilon}) \xrightarrow[n \to \infty]{P} a(t).$$

Now, we set

$$X_n(t) = \sum_{k=1}^{[nt]-1} \xi_{n,k} + (nt - [nt])\xi_{n,[nt]}$$

and $X^{(n)} = (X_n(t))_{t \geq 0}$. *When n tends to ∞, $(X^{(n)})$ converges in distribution to W_a, where W_a is the continuous process with independent Gaussian increments such that $W_a(t)$ has distribution* $N(0, a(t))$.

If the r.v.'s $\xi_{n,k}$ *are square integrable, we can replace* H2 *and* H3 *by*

H'2)
$$\sum_{k=1}^{[nt]} E^{n,k-1}(\xi_{n,k}) \xrightarrow[n \to \infty]{P} 0$$

H'3) $\sum_{k=1}^{[nt]} V^{n,k-1}(\xi_{n,k}) \xrightarrow[n\to\infty]{P} a(t)$.

Proof. First of all let us study the finite distribution functions. From [2.8.1], under hypothesis H1),

$$\sup_{1 \leqslant k \leqslant [nt]} |\xi_{n,k}| \xrightarrow[n\to\infty]{P} 0.$$

Hence, if we set $S_n(t) = \Sigma_{k=1}^{[nt]} \xi_{n,k}$ and $S^{(n)} = (S_n(t))_{t \geqslant 0}$, $(X^{(n)})$ is a sequence of continuous processes contiguous to the sequence $S^{(n)}$. Let us therefore study the finite distribution functions of $S^{(n)}$. Let $t_0 = 0 < t_1 < t_2 < \dots < t_d$,

$$\xi_{n,k}^i = \xi_{n,k} \mathbf{1}_{([nt_i] < k \leqslant [nt_{i+1}])}$$

and

$$\xi_{n,k}^{i,\varepsilon} = \xi_{n,k}^{\varepsilon} \mathbf{1}_{([nt_i] < k \leqslant [nt_{i+1}])}\,.$$

The random vector $(S_n(t_1), S_n(t_2) - S_n(t_1), \dots, S_n(t_d) - S_n(t_{d-1}))$ is equal to

$$\left[\sum_{[nt_i] < k \leqslant [nt_{i+1}]} \xi_{n,k}^i \right]_{1 \leqslant i < d}.$$

Theorem 2.8.42 can be applied to it. The hypotheses H1 and H2 of this are clearly consequences of hypotheses H1 and H2 of Theorem 7.4.28. As for hypothesis H3, we calculate

$$\sum_{k=1}^{[nt_d]} (E^{n,k-1}(\xi_{n,k}^{i,\varepsilon} \xi_{n,k}^{j,\varepsilon}) - E^{n,k-1}(\xi_{n,k}^{i,\varepsilon}) E^{n,k-1}(\xi_{n,k}^{j,\varepsilon}))$$

$$= \begin{cases} \sum_{[nt_{i-1}]+1}^{[nt_i]} V^{n,k-1}(\xi_n^{\varepsilon}) & \text{for } i = j \\ 0 & \text{for } i \neq j. \end{cases}$$

Thus

$$(S_n(t_1), S_n(t_2) - S_n(t_1), \dots, S_n(t_d) - S_n(t_{d-1}))$$

converges in distribution when $n \to \infty$ to a Gaussian vector the components of which are independent, centered and of variance $a(t_1), a(t_2) - a(t_1), \dots, a(t_d) - a(t_{d-1})$. It remains to show the tightness of the sequence $(X^{(n)})$. Let $\varepsilon > 0$. We denote $\xi_{n,k}^{\varepsilon} = \xi_{n,k}^{\varepsilon} - E^{n,k-1}(\xi_{n,k}^{\varepsilon})$ and

$$M_n^\varepsilon(t) = \sum_{k=1}^{[nt]} \tilde{\xi}_{n,k}^\varepsilon .$$

For each n, $M_n^\varepsilon = (M_n^\varepsilon(t))$ is a martingale adapted to $(F^n_{[nt]})_{t\geq 0}$. Let us denote

$$A_n^\varepsilon(t) = \sum_{k=1}^{[nt]} V^{n,k-1}(\xi_{n,k}^\varepsilon)$$

and

$$A_n^\varepsilon = (A_n^\varepsilon(t))_{t\geq 0} .$$

The sequence $(X^{(n)})$ of continuous processes is contiguous to (M_n^ε). This follows from H1 and H2 (see [2.8.1] and [2.8.2]). We then verify that all the hypotheses of Theorem 7.3.26 are satisfied and we obtain the result.

7.4.2. Donsker's Theorem

By setting $\xi_{n,k} = \xi_k/\sqrt{n}$ we deduce from Theorem 7.4.28 the following important theorem.

Donsker's Theorem 7.4.29. *On a probability space* (Ω,A,P), *we are given a sequence* (ξ_k) *of independent* r.v.'s, *with the same distribution* F; *F is assumed to be square integrable, centered and with variance* σ^2.

Then, setting

$$X_n(t) = \frac{1}{\sigma\sqrt{n}} \left[\sum_{k=1}^{[nt]} \xi_k + (nt - [nt])\xi_{[nt]+1}\right],$$

(X_n) *converges in distribution to a Brownian motion if* $n \to \infty$.

This theorem is going to allow us to prove Theorem 4.4.25 which has been assumed in Chapter 4 in order to deal with the statistics of branching processes.

Proposition 7.4.30. *The hypotheses are those of Theorem* 7.4.29.

(a) *For any Borel set* Γ *of* C *the frontier of which is not charged by Wiener measure and any* r.v. Y *measurable with respect to* $F_\infty = \sigma(\xi_k, k \geq 1)$ *and P-integrable, we have*

$$E[1_{(X_n\in\Gamma)}Y] \xrightarrow[n\to\infty]{} W(\Gamma)E(Y).$$

(b) *Let* (a_n) *be a sequence of positive real numbers increasing to* ∞, *and let* (ν_n) *be a sequence of* $\mathcal{F}_\infty$*-measurable integer* r.v.'s *such that* ν_n/a_n *tends in probability to a finite* r.v. Z *with distribution* F_Z. *Let*

$$S_n = \frac{1}{\sigma\sqrt{n}} \sum_{k=1}^{n} \xi_k.$$

conditional on $\{Z > 0\}$, $(S_{\nu_n}, (\nu_n/a_n))$ *tends in distribution to*

$N(0,1) \otimes (F_Z|_{Z\neq 0})$, *where* $F_Z|_{Z\neq 0}$ *is the distribution of* Z *conditional on* $Z \neq 0$.

Proof. Consider X'_n defined by

$$X'_n(t) = \frac{1}{\sigma\sqrt{n}} \left[\sum_{k=[n^{1/4}]}^{[nt]} \xi_k + (nt - [nt])\xi_{[nt]} \right].$$

We have

$$\sup_t |X_n(t) - X'_n(t)| \leqslant \frac{1}{\sigma\sqrt{n}} \sum_{k \leqslant n^{1/4}} |\xi_k|.$$

The norm in $L^2(\Omega, \mathcal{A}, P)$ of the majorant is less than $(1/\sigma\sqrt{n})n^{1/4}\sigma$,

$$\sup_t |X_n(t) - X'_n(t)|$$

tends to 0 in L^2. However property (a) means that the sequence $(X_n(Y \cdot P))$ converges narrowly to $E(Y)W$. Since, for $\varepsilon > 0$,

$$E\left[Y 1_{\{\sup_t |X_n(t) - X'_n(t)| \geqslant \varepsilon\}} \right]$$

tends to 0, $\rho_C(X'_n, X_n)$ tends in measure to 0 on $(\Omega, \mathcal{A}, Y \cdot P)$. Thus it is sufficient to show (a) by replacing X_n by X'_n.

However X'_n is independent of the r.v.'s ξ_k for $k < [n^{1/4}]$. For any integer p and any integrable r.v. Y measurable with respect to $\mathcal{F}_p = \sigma(\xi_1, ..., \xi_p)$, we thus have, from a certain point onwards,

$$E[1_{(X'_n \in \Gamma)} Y] = P(X'_n \in \Gamma)E(Y).$$

From which the result follows, if Y is $\mathcal{F}_p$-measurable. If Y is measurable with respect to $\mathcal{F}_\infty = \vee \mathcal{F}_p$, the r.v.'s $E(Y | \mathcal{F}_p) = Y_p$

tend to Y in $L^1(\Omega,\mathcal{A},P)$ from Corollary 2.4.18. For every (n,p) we have

$$|E[1_{(X'_n \in \Gamma)} Y_p] - E[1_{(X'_n \in \Gamma)} Y]| \leq E(|Y_p - Y|).$$

Let n tend to ∞: the closure points of the sequence $E[1_{(X'_n \in \Gamma)} Y]$ all lie between $W(\Gamma)E(Y_p) - E(|Y_p - Y|)$ and $W(\Gamma)E(Y_p) + E(|Y_p - Y|)$. We then let p tend to ∞ and obtain the result.

(b) Let ϕ be bounded, uniformly continuous on $\mathbb{R}$. $\phi(\nu_n/a_n)$ tends to $\phi(Z)$ in $L^1(\Omega,\mathcal{A},P)$. Thus

$$|E(1_\Gamma(X_{a_n})\phi(\nu_n/a_n)) - E(1_\Gamma(X_{a_n})\phi(Z))|$$

tends to 0 and

$$\lim_n E(1_\Gamma(X_{a_n})\phi(\nu_n/a_n)) = W(\Gamma)E(\phi(Z)).$$

The sequence of distributions $(X_{a_n},(\nu_n/a_n))$ is tight and all its closure points are probabilities on $C \times \mathbb{R}$ the integral of which on $1_\Gamma \times \phi$ coincides with $W \otimes F_Z$. By the usual extension techniques, we see that only one probability has this property. Thus:

$$(X_{a_n},(\nu_n/a_n)) \xrightarrow{\mathcal{D}} W \otimes F_Z.$$

Now let us consider $\varepsilon > 0$ and $(u,v) \in \mathbb{R}^2$; $(x,(y(t))_{t\geq 0}) \longmapsto 1_{(x \geq \varepsilon)}\exp[i(uy(x) + vx)]$ is continuous from $\mathbb{R}_+ \times C$ into C for $x \neq \varepsilon$.

Let $(\Omega',\mathcal{A}',P',(B_t)_{t\geq 0})$ be a Brownian motion; its distribution $F_Z \otimes W$ is that of $(\omega,\omega') \longmapsto (Z(\omega),(B_t(\omega'))_{t\geq 0})$ on $(\Omega,\mathcal{A},P) \times (\Omega',\mathcal{A}',P')$. Thus, if $F_Z(\varepsilon)$ is zero,

$$E\left[1_{(\nu_n \geq \varepsilon a_n)}\left(\exp\left[i\left[u\sqrt{(a_n/\nu_n)}\; X_{a_n}\left(\frac{\nu_n}{a_n}\right) + v\frac{\nu_n}{a_n}\right]\right]\right)\right]$$

tends to

$$E_{P\otimes P'}\left[1_{(Z \geq \varepsilon)}\exp\left(i\left[u\frac{1}{\sqrt{Z}}B_Z + vZ\right]\right)\right].$$

By Fubini's theorem this expression equals

$$E[\mathbf{1}_{(Z\geqslant\varepsilon)}\exp(ivZ)]\exp\left[-\frac{u^2}{2}\right].$$

However, for $0 < h < \varepsilon$,

$$\mathbf{1}_{((\nu_n/a_n)\geqslant\varepsilon+h)} - \mathbf{1}_{(|(\nu_n/a_n)-Z|\geqslant h)} \leqslant \mathbf{1}_{(Z\geqslant\varepsilon)}$$
$$\leqslant \mathbf{1}_{((\nu_n/a_n)\geqslant\varepsilon-h)} + \mathbf{1}_{(|(\nu_n/a_n)-Z|\geqslant h)}.$$

From which it follows, by setting

$$U_n = \exp i\left[uS_{\nu_n} + v\frac{\nu_n}{a_n}\right],$$

if $F_Z(\varepsilon \pm h) = 0$

$$E[\mathbf{1}_{(Z\geqslant\varepsilon+h)}\exp(ivZ)]\exp\left[-\frac{u^2}{2}\right]$$
$$\leqslant \varliminf_{n\to\infty} E[\mathbf{1}_{(Z\geqslant\varepsilon)}U_n]$$
$$\leqslant \varlimsup_{n\to\infty} E[\mathbf{1}_{(Z\geqslant\varepsilon)}U_n]$$
$$\leqslant E[\mathbf{1}_{(Z\geqslant\varepsilon-h)}\exp(ivZ)]\exp\left[-\frac{u^2}{2}\right].$$

Thus

$$\lim_n E\left[\mathbf{1}_{(Z\geqslant\varepsilon)}\exp\left[iv\frac{\nu_n}{a_n}\right]\exp iu(S_{\nu_n})\right]$$
$$= E[\mathbf{1}_{(Z\geqslant\varepsilon)}\exp(ivZ)]\exp\left[-\frac{u^2}{2}\right].$$

Let $\eta > 0$; there exists an ε not charged by F_Z and such that: $P(Z > 0) - P(Z \geqslant \varepsilon) \leqslant \eta$. Then, for all n,

$$|E[\mathbf{1}_{(Z>0)}\exp(ivZ)\exp(iuS_{\nu_n})]$$
$$- E[\mathbf{1}_{(Z\geqslant\varepsilon)}\exp(ivZ)\exp(iuS_{\nu_n})]| \leqslant \eta.$$

All the closure points of the sequence

$$E[\mathbf{1}_{(Z>0)}\exp(ivZ)\exp(iuS_{\nu_n})]$$

differ from

$$E[1_{(Z>0)}\exp\ ivZ]\exp\left(-\frac{u^2}{2}\right)$$

by less than 2η. This being true for all η, we obtain,

$$E\left[\exp i\left(v\frac{\nu_n}{a_n}+uS_{\nu_n}\right)\Big|Z>0\right]$$

$$\xrightarrow[n\to\infty]{} E[\exp(ivZ)|Z>0]\exp\left(-\frac{u^2}{2}\right).$$

This implies part (b) of the proposition.

7.4.3. Kolmogorov's Test

Let (X_n) be a sample from a continuous distribution F on $\mathbb{R}$ and let F_n be its empirical distribution function defined by

$$F_n(t) = \frac{1}{n}\sum_{p=1}^{n} 1_{(X_p\leqslant t)}\ .$$

The Glivenko-Cantelli theorem ([Vol. I, 4.4.3]) states that,

$$\sup_t|F_n(t) - F(t)| \xrightarrow[n\to\infty]{\text{a.s.}} 0.$$

Here we study the r.v.'s

$$U_n = \sqrt{n}\sup_t\,(F_n(t) - F(t)),\quad V_n = \sqrt{n}\sup_t(F(t) - F_n(t)),$$

$$W_n = \sqrt{n}\sup_t|F_n(t) - F(t)|.$$

The distributions of these r.v.'s are unaltered, F being continuous, if the r.v.'s X_p are replaced by the r.v.'s $F(X_p)$, the distribution of which is uniform on [0,1] ([Vol. I, 4.5.2] and [Vol. I, E3.3.5]). Thus let us assume that F is the uniform distribution on [0,1], $\mathcal{U}_{[0,1]}$.

For $n \geqslant 1$, let us denote by $(X^n_{(k)})_{1\leqslant k\leqslant n}$ the order statistic of $(X_1, \ldots, X_n)$, and $X^n_{(0)} = 0$, $X^n_{(n+1)} = 1$. For $X^n_{(k)} \leqslant t < X^n_{(k+1)}$, we have,

$$\frac{k+1}{n} - X^n_{(k+1)} - \frac{1}{n} = \frac{k}{n} - X^n_{(k+1)} \leqslant F_n(t) - t \leqslant \frac{k}{n} - X^n_{(k)}$$

Thus, up to $2/\sqrt{n}$, we can replace U_n, V_n and W_n by

$$U_n' = \sqrt{n}\sup\left[\frac{k}{n} - X_{(k)}^n\right], \quad V_n' = \sqrt{n}\sup\left[X_{(k)}^n - \frac{k}{n}\right]$$

and

$$W_n' = \sqrt{n}\sup\left|X_{(k)}^n - \frac{k}{n}\right|.$$

However, if we consider a sequence (Y_n) of independent r.v.'s, exponentially distributed with parameter 1, and $S_n = Y_1 + ... + Y_n$, then the density of $(S_1, ..., S_{n+1})$ is

$$(s_1, ..., s_{n+1}) \longmapsto e^{-s_{n+1}} \mathbf{1}_{(0<s_1<...<s_{n+1})}$$

and S_{n+1} has Erlang's distribution with density

$$t \longmapsto e^{-t}\frac{t^n}{n!}\mathbf{1}_{(0<t)}\,.$$

Thus the density of $(S_1, ..., S_n)$ conditional on $S_{n+1} = u$ is

$$(s_1, ..., s_n) \longmapsto \frac{n!}{u^n}\mathbf{1}_{(0<s_1<...<s_n<u)}\,;$$

conditional on $S_{n+1} = u$,

$$\left(\frac{S_1}{S_{n+1}}, ..., \frac{S_n}{S_{n+1}}\right)$$

has the same distribution as $(X_{(k)}^n)_{1\leq k\leq n}$. This is true for any u, so

$$\left(\frac{S_1}{S_{n+1}}, ..., \frac{S_n}{S_{n+1}}\right)$$

has the same distribution as $(X_{(k)}^n)_{1\leq k\leq n}$. Hence $X_{(k)}^n - k/n$ has the same distribution as,

$$\frac{S_k}{S_{n+1}} - \frac{k}{n} = \frac{1}{S_{n+1}}\left[(S_k - k) - \frac{k}{n+1}(S_{n+1} - (n+1))\right].$$

Let $(\Omega,\mathcal{A},P,(B_t)_{t\geq 0})$ be a Brownian motion. From Donsker's theorem,

$$\frac{1}{\sqrt{n}}\left[(S_{[nt]} - [nt]) - t(S_1 - 1)\right]_{t\geq 0} \xrightarrow[n\to\infty]{\mathcal{D}} (B_t - tB_1)_{t\geq 0}.$$

Since $S_n/n \xrightarrow{\mathcal{D}} 1$, we obtain

$$U_n' \xrightarrow{\mathcal{D}} \sup_{0\leq t\leq 1}(tB_1 - B_t), \quad V_n' \xrightarrow{\mathcal{D}} \sup_{0\leq t\leq 1}(B_t - tB_1),$$

$$W'_n \xrightarrow{\mathcal{D}} \sup_{0 \leqslant t \leqslant 1} |B_t - tB_1|.$$

Since (B_t) and $(-B_t)$ have the same distribution, (U'_n) and (V'_n) have the same limit distribution.

Definition 7.4.31. If $(\Omega, \mathcal{A}, P, (B_t)_{t \geqslant 0})$ is a Brownian motion, the process $(B_t - tB_1)_{0 \leqslant t \leqslant 1}$ is a **Brownian bridge**. Thus it is a centered Gaussian process the time space of which is $[0,1]$. The trajectories are continuous and the covariance is $(s,t) \longmapsto (s \wedge t)[1 - s \vee t]$.

Theorem 7.4.32. *Let $(\Omega, \mathcal{A}, P, (X_t)_{0 \leqslant t \leqslant 1})$ be a Brownian bridge and let*

$$\bar{D} = \sup_t X_t, \quad D^* = \sup_t |X_t|.$$

If (X_n) is a sample from a continuous distribution F on $\mathbb{R}$ and if F_n is the empirical distribution function of $(X_1, \ldots, X_n)$, we have,

$$U_n = \sqrt{n} \sup_t (F_n(t) - F(t)) \xrightarrow[n\to\infty]{\mathcal{D}} \bar{D}$$

$$V_n = \sqrt{n} \sup_t (F(t) - F_n(t)) \xrightarrow[n\to\infty]{\mathcal{D}} \bar{D}$$

$$W_n = \sqrt{n} \sup_t |F(t) - F_n(t)| \xrightarrow[n\to\infty]{\mathcal{D}} D^*.$$

Consequence. These limit distributions have densities which will be calculated in [8.4.4]. They are tabulated and called **Kolmogorov distributions.** They are used in large samples for some nonparametric tests, called Kolmogorov tests, relative to F ([Vol. I, 4.5.3]). Let $\bar{D}_\alpha$ and D^*_α be the quantiles of order α of $\bar{D}$ and D^* ($P(\bar{D} \geqslant \bar{D}_\alpha) = P(D^* \geqslant D^*_\alpha) = \alpha$). The following tests then have a level close to α when we are dealing with families of continuous distributions on $\mathbb{R}$.

(a) In order to test "$F = F_0$" against "$F \neq F_0$", we take the rejection region

$$\left\{\sqrt{n} \sup_t |F_n(t) - F_0(t)| \geqslant D^*_\alpha\right\}.$$

(b) In order to test "$F = F_0$" against "the distribution of F is larger than that of F_0", we take the rejection region

$$\left\{\sqrt{n}\ \sup_t\,(F_0(t) - F_n(t)) \geq \bar{D}_\alpha\right\}.$$

(c) In order to test "$F = F_0$" against "the distribution F is smaller than F_0", we take the rejection region

$$\left\{\sqrt{n}\ \sup_t\,(F_n(t) - F_0(t)) \geq \bar{D}_\alpha\right\}.$$

Bibliographic Notes

The tools of [7.1] and [7.2] on stopping times and regularity of martingale trajectories have been partly introduced in Doob; see Neveu [2]. Since Doob, many mathematicians have worked on constructing a good theory of continuous time processes; Dellacherie-Meyer [1] and [2] is the essential manual for this "general theory of processes", as well as the Strasbourg series of probability seminars. Proposition 7.2.10 is due to Lenglart.

For convergence in distribution of processes with continuous or cad-lag trajectories the indispensable books are Billingsley [2] and Parthasarathy. See also Skorokhod, Gikhman-Skorokhod [1]. Neveu [2] gives a criterion for the continuity of processes based on similar ideas to those in [7.3.4]. We have avoided introducing the "Skorokhod topology" which can be found in the above works.

The tightness criterion of a sequence of martingales obtained in [7.3.6] is due to Rebolledo [2]. It allows Donsker's theorem to be obtained, which can be proved without using martingales (see the books referred to above). See also Hall-Heyde for another proof based on martingales. Proposition 7.4.30 is proved in Billingsley [2]. For Theorem 7.4.32 on Kolmogorov distributions we have used a proof from Breiman.

Chapter 8
STOCHASTIC INTEGRALS

Objectives

The space of cad-lag martingales convergent in L^2 has a Hilbert space structure which allows a simple study to be made of it. We introduce a notion of stochastic integral generalizing that of Chapter 6. This integral is an isometry analogous to the spectral process of second order stationary processes introduced in Chapter 1.

Stochastic calculus related to this integral with respect to a cad-lag martingale has various applications:

· the existence of an increasing process $[M,M]$ such that $M^2 - [M,M]$ is a martingale (Doob decomposition);

· the construction of exponential martingales and supermartingales;

· asymptotic theorems (laws of large numbers and central limit theorems) for martingales.

We apply these results to point processes and to Brownian motion. Finally diffusions are looked at briefly.

8.1. Stochastic Integral with Respect to a Square Integrable Martingale

8.1.1. A Hilbert Space of Martingales

Let $(\Omega,\mathcal{A},P)$ be a probability space equipped with a filtration $\mathbb{F} = (\mathcal{F}_t)_{t\geq 0}$ continuous on the right.

Let $M = (M_t)_{t\geq 0}$ be an $\mathbf{F}$-martingale with trajectories zero at 0, continuous on the right and provided with a limit on the left (cad-lag). The function $t \longmapsto E(M_t^2)$ is increasing. Let us assume that

$$\|M\|_2 = \sup_t \sqrt{E(M_t^2)}$$

is finite. Then, from Theorem 7.2.7, the martingale (M_t) converges a.s. if $t \to \infty$ to an r.v. M_∞. Moreover, for $s < t$, $E(M_t - M_s)^2 = E(M_t^2) - E(M_s^2)$ and, if s and t tend to ∞, this expression tends to 0. Thus (M_t) tends a.s. and in L^2 to an r.v. M_∞: from which it follows that $E(M_\infty^2) = \|M\|_2^2$.

If $\|M\|_2$ is zero, M is zero up to a modification. Further, we have here, due to the right continuity of M,

$$P[M_t = 0 \text{ for all } t] = P[M_q = 0 \text{ for } q \text{ rational}] = 1.$$

Definition 8.1.1. Two processes defined on the same probability space are **indistinguishable** if their trajectories are a.s. equal.

Indistinguishability is therefore an equivalence relation on processes defined on the same probability space and having the same state and time spaces. Let $X = (X_t)_{t\in T}$ and $Y = (Y_t)_{t\in T}$ be such processes. They are "indistinguishable" if

$$P[X_t = Y_t \text{ for all } t] = 1$$

whereas Y is a "modification" of X if

$$\text{for all } t,\ P[X_t = Y_t] = 1.$$

So the first relation is stronger than the second unless T is countable.

Definition 8.1.2. The space $\mathcal{M}^2(\mathbb{F})$ is the space of equivalence classes under indistinguishability of cad-lag $\mathbb{F}$-martingales zero at 0, $M = (M_t)_{t \geq 0}$, such that

$$\|M\|_2^2 = \sup_t E(M_t^2)$$

is finite.

In what follows $\mathbb{F}$ is to be understood and we denote $\mathcal{M}^2$ instead of $\mathcal{M}^2(\mathbb{F})$. The hypotheses $M_0 = 0$ is intended to avoid constants in the calculations that follow. This can always be achieved by replacing a martingale $(M_t)_{t \geq 0}$ by $(M_t - M_0)_{t \geq 0}$. We denote $\Delta M_t = M_t - M_{t-}$.

Theorem 8.1.3. (a) *Every martingale $(M_t)_{t \geq 0}$ in $\mathcal{M}^2$ converges in L^2 and* a.s. *to an* r.v. *M_∞ such that $\|M\|_2^2 = E(M_\infty^2)$.*

(b) *If $(M^{(p)})$ is a sequence in $\mathcal{M}^2$ which converges under the norm $\|\cdot\|_2$ to $M \in \mathcal{M}^2$, then, when $p \to \infty$:*

$$\sup_t |M_t^{(p)} - M_t| \quad \textit{and} \quad \sup_t |\Delta M_t^{(p)} - \Delta M_t|$$

tend in L^2 to 0. Hence $\mathcal{M}_c^2$, the set of martingales in $\mathcal{M}^2$ with continuous trajectories, is a closed subspace of $\mathcal{M}^2$.

(c) *The space $\mathcal{M}^2$ is a Hilbert space if it is equipped with the scalar product*

$$(M,N) \longmapsto E(M_\infty N_\infty) = \langle\langle M,N \rangle\rangle_2.$$

From every Cauchy sequence $(M^{(p)})$ of $\mathcal{M}^2$ we can extract a subsequence which converges uniformly a.s.

Proof. Part (a) has been proved at the start of 8.1.1. The fact that $\langle\langle \cdot,\cdot \rangle\rangle$ is a scalar product associated with the norm $\|\cdot\|_2$ is simple. From Theorem 7.2.6, if $M^{(p)}$ and M are in $\mathcal{M}^2$:

$$\left[E \sup_t (M_t^{(p)} - M_t)^2\right]^{1/2} \leq 2\|M^{(p)} - M\|_2,$$

and

$$\sup_t |\Delta M_t^{(p)} - \Delta M_t| \leq 2 \sup_t |M_t^{(p)} - M_t|.$$

From which part (b) of the theorem follows. It leads on to part (c). Let $M^{(p)}$ be a Cauchy sequence. We can extract a subsequence $N^{(q)}$ such that $\Sigma\|N^{(q+1)} - N^{(q)}\|_2$ converges and $N^{(0)} = 0$. Then

$$\Sigma\left[E\left(\sup_t |N_t^{(q+1)} - N_t^{(q)}|\right)^2\right]^{1/2} \leqslant 2\Sigma\|N^{(q+1)} - N^{(q)}\|_2 < \infty.$$

For almost all ω, the series $\Sigma_q(N_t^{(q+1)} - N_t^{(q)})(\omega)$ converges uniformly at t. Denoting its limit by $M_t^{(\infty)}(\omega)$, the trajectories $t \longmapsto M_t^{(\infty)}(\omega)$ are cad-lag, and continuous if those of $N^{(a)}$ were.

Moreover, for $t \leqslant \infty$, $M_t^{(p)}$ is a Cauchy sequence in $L^2(\Omega,A,P)$ which thus converges in $L^1(\Omega,A,P)$ to $M_t^{(\infty)}$. For $A \in F_s$ and $s < t$, we can pass to the limit in the equality $E[1_A M_t^{(p)}] = E[1_A M_s^{(p)}]$: $M^{(\infty)}$ is a cad-lag martingale, $(M^{(p)})$ tends to $M^{(\infty)}$ in M^2, and the subsequence $N^{(q)}$ converges a.s. uniformly to $M^{(\infty)}$.

8.1.2. Stochastic Integrals

Let A and $\tilde{A}$ be two increasing ([6.2.5]) square integrable $\mathbb{F}$-processes such that $V = A - \tilde{A}$ is an $\mathbb{F}$-martingale and let C be a bounded $\mathbb{F}$-predictable process. Then we have seen in [6.2.5] that:

$$C \cdot V = \left[\int_0^t C_u dV_u\right]_{t\geqslant 0}$$

is an $\mathbb{F}$-martingale.

In particular, for $C = 1_{\Gamma\times]s,t]}$ with $0 \leqslant s < t$ and $\Gamma \in F_s$, we have

$$(C \cdot V)_u = 1_\Gamma(V_{t\wedge u} - V_{s\wedge u}).$$

In this section we are concerned with generalizing these formulae and defining a "stochastic integral" $\int_0^t C_s dM_s = (C \cdot M)_t$ for $M \in M^2$. We no longer assume that the trajectories of M have bounded variation, i.e. we can no longer speak of the Stieltjes integral on each trajectory. We start from the last formula and set $0 \leqslant s < t < \infty$ and $\Gamma \in F_s$,

$$(1_{\Gamma\times]s,t]} \cdot M)_u = 1_\Gamma(M_{t\wedge u} - M_{s\wedge u}).$$

It can be checked that $C \cdot M = ((C \cdot M)_u)_{u\geqslant 0}$ is a cad-lag $\mathbb{F}$-martingale. Moreover

$$(1_{\Gamma\times]s,t]} \cdot M)_u^2 = 1_\Gamma(M_{t\wedge u} - M_{s\wedge u})^2$$

$$\|1_{\Gamma\times]s,t]}\cdot M\|_2^2 = E[1_\Gamma(M_t - M_s)^2]$$
$$= E[1_\Gamma(M_t^2 - M_s^2)].$$

Now let us consider, as in [6.2], the $\mathbb{F}$-predictable σ-algebra $\mathcal{P}$ of $\Omega \times \mathbb{R}_+$, generated by $\mathcal{C} = \{\Gamma \times]s,t];\ 0 \leqslant s < t < \infty,\ \Gamma \in \mathcal{F}_s\}$. We are concerned with defining, as in [6.2], $C\cdot M$ for predictable processes. The key is the existence of a measure μ_M on $(\Omega \times \mathbb{R}_+,\mathcal{P})$ such that

$$\int 1_{\Gamma\times]s,t]}d\mu_M = E[1_\Gamma(M_t^2 - M_s^2)].$$

In certain cases this existence is clear (the particular case studied below) and will give a little more work in the general case. Assuming this existence, the stochastic integral will be the isometry from $L^2(\Omega \times \mathbb{R}_+, \mathcal{P}, \mu_M)$ into $\mathcal{M}^2$ which extends $1_{\Gamma\times]s,t]} \longmapsto 1_{\Gamma\times]s,t]}\cdot M$ according to the scheme of [1.2.3].

Let $\mathcal{B}$ be the set of finite unions of disjoint subsets of $\mathcal{C}$. Since $\mathcal{C}$ is closed under finite intersection, we see that $\mathcal{B}$ is closed under finite union and intersection and contains the empty set.

On the other hand, for $\Gamma_1 \times]s_1,t_1]$ and $\Gamma_2 \times]s_2,t_2]$ in $\mathcal{C}$, either $]s_1,t_1]$ and $]s_2,t_2]$ are disjoint, or $s_2 < t_1$, or $s_1 < t_2$. Let us assume $s_2 < t_1$:

$$(\Gamma_1 \times]s_1,t_1]) \cup (\Gamma_2 \times]s_2,t_2])$$
$$= (\Gamma_1 \times]s_1,s_2]) \cup (\Gamma_1 \cup \Gamma_2 \times]s_2,t_1])$$
$$\cup (\Gamma_2 \times]t_1,t_2]).$$

Therefore every element of $\mathcal{B}$ has a decomposition of the form

$$A = \bigcup_{i=1}^{n} \Gamma_1 \times]t_i ,t_{i+1}] \quad \text{with} \quad 0 \leqslant t_1 < \dots < t_{n+1},$$

and $\Gamma_i \in \mathcal{F}_{t_i}$, possibly empty. Let $\mathcal{B}'$ be the set, containing $\mathcal{B}$, of subsets of the form

$$B = \bigcup_{i=1}^{\infty} \Gamma_i \times]t_i,t_{i+1}],$$

where (t_i) is a sequence which increases strictly from 0 to ∞, and $\Gamma_i \in F_{t_i}$. By linearity, we are led to set for such a B:

$$(1_B \cdot M)_u = \sum_{i=1}^{\infty} 1_{\Gamma_i}(M_{t_{i+1}\wedge u} - M_{t_i\wedge u})$$

and we have

$$\|1_B \cdot M\|_2^2 = \sum_{i=1}^{\infty} E[1_{\Gamma_i}(M^2_{t_{i+1}} - M^2_{t_i})] \leq \|M\|_2^2.$$

First of all we are going to define the stochastic integral in a particular case and shall prove in [8.1.3] that there is nothing special about it.

Particular Case (Sufficient on First Reading). Let us assume that M is such that there exists an increasing integrable $\mathbb{F}$-process A such that $M^2 - A$ is an $\mathbb{F}$-martingale. This is the case for example if M is a centered $\mathbb{F}$-process with independent increments since $M_t^2 - tE(M_1)$ is then an $\mathbb{F}$-martingale (check this). In [6.2.5], $M = A - \tilde{A}$ has this property if A_∞ is square integrable.

For $\Gamma \times]s,t] \in C$, we have

$$E[1_\Gamma(M_t - M_s)^2] = E[1_\Gamma(M_t^2 + M_s^2 - 2M_sE(M_t|F_s))]$$

$$= E[1_\Gamma(M_t^2 - M_s^2)] = E[1_\Gamma(A_t - A_s)].$$

Consider, on $(\Omega \times \mathbb{R}_+, P)$, the measure μ_M defined by

$$\mu_M(\Delta) = E\left[\int_0^\infty 1_\Delta(s,\omega)dA_s(\omega)\right].$$

This is the unique measure such that $\mu_M(\Omega \times \{0\}) = 0$, and, for $\Gamma \times]s,t] \in C$,

$$\mu_M(\Gamma \times]s,t]) = E[1_\Gamma(M_t - M_s)^2].$$

Let us denote $L^2(\mu_M) = L^2(\Omega \times \mathbb{R}_+, P, \mu_M)$. Then let E be the vector subspace of $L^2(\mu_M)$ formed by the functions

$$Y = \sum_{i=1}^{n} \alpha_i 1_{\Gamma_i \times]t_i, t_{i+1}]}$$

for $0 \leq t_1 < \dots < t_{n+1} < \infty$ and $\Gamma_i \in F_{t_i}$. Set

$$(Y \cdot M)_u = \sum_{i=1}^{n} \alpha_i 1_{\Gamma_i}(M_{t_{i+1}\wedge u} - M_{t_i\wedge u})$$

and $(Y \cdot M) = ((Y \cdot M)_u)_{u \geq 0}$ is an element of $\mathcal{M}^2$,

$$\|Y \cdot M\|_2^2 = \sum_{i=1}^{n} \alpha_i^2 E[1_{\Gamma_i}(M_{t_{i+1}} - M_{t_i})^2] = \int Y^2 d\mu_M.$$

This means that the function $Y \longmapsto Y \cdot M$ is an isometry from $\mathcal{E}$ into $\mathcal{M}^2$. Moreover $(Y \cdot M)^2 - Y^2 \cdot A$ is an $\mathbb{F}$-martingale.

Let H be the closure of E in $L^2(\mu_M)$, and let Z be in the orthogonal to H. For $\Delta \in \mathcal{P}$, we have $\int 1_\Delta Z \, d\mu_M = 0$. This is true for $\Delta \in \mathcal{C}$, and is extended to $\mathcal{P}$ by the usual method [Vol. I, 3.1.11]. Hence Z is zero. As a result $\mathcal{E}$ is dense in $L^2(\mu_M)$, and we obtain a unique extension to the preceding isometry.

Let $Y \in L^2(\mu_M)$ and let $(Y^{(n)})$ be a sequence of $\mathcal{E}$ which converges to Y in $L^2(\mu_M)$: $(Y^{(n)} \cdot M)$ tends to $(Y \cdot M)$ in $\mathcal{M}^2$. For $Y \in \mathcal{E}$, we have $\Delta(Y \cdot M)_t = Y_t \cdot \Delta M_t$, and

$$E\left[\sup_t \Delta(Y \cdot M)_t\right]^2 = E\left[\sup_t |Y_t|^2 \Delta M_t^2\right] \leq 4\int Y^2 d\mu_M.$$

Thus, the inequality

$$E\left[\sup_t |Y_t|^2 \Delta M_t^2\right] \leq 4\int Y^2 d\mu_M$$

is true for all $Y \in L^2(\mu_M)$, and if $(Y^{(n)}) \subset \mathcal{E}$ converges in $L^2(\mu_M)$ to Y, then $\sup_t (Y_t^{(n)} - Y_t)^2 \Delta M_t^2$ converges in L^1 to 0. Moreover,

$$E\left[\sup_t \Delta(Y^{(n)} \cdot M - Y \cdot M)_t\right]^2 \leq 4\|Y^{(n)} \cdot M - Y \cdot M\|_2^2 .$$

We thus have,

$$P[\Delta(Y \cdot M)_t = Y_t \cdot \Delta M_t \text{ for any } t] = 1.$$

Also $(Y^{(n)} \cdot M)^2 - (Y^{(n)})^2 \cdot A$ is an $\mathbb{F}$-martingale, and, for all t, $(Y^{(n)} \cdot M)_t^2 - ((Y^{(n)})^2 \cdot A)_t$ tends in L^1 to $(Y \cdot M)_t^2 - (Y^2 \cdot A)_t$. Hence $(Y \cdot M)^2 - (Y^2 \cdot A)$ is an -martingale.

Theorem 8.1.4. *Let $M \in \mathcal{M}^2$ and let A be an increasing $\mathbb{F}$-process such that $M^2 - A$ is an $\mathbb{F}$-martingale. Let $\mathcal{P}$ be the σ-algebra of $\mathbb{F}$-predictable sets and let μ_M be the measure defined on $(\Omega \times \mathbb{R}_+, \mathcal{P})$ by*

$$\mu_M(\Delta) = E\left[\int_0^\infty 1_\Delta(\omega,s) dA_s(\omega)\right].$$

(a) *To M is associated a unique isometry from $L^2(\mathbb{R}_+ \times \Omega, \mathcal{P}, \mu_M)$ into $\mathcal{M}^2$, called the stochastic integral with respect to M*

and denoted

$$Y \longmapsto Y \cdot M = \left[\int_0^u Y_s dM_s \right]_{u \geq 0},$$

such that, for $s < t$, $\Gamma \in F_s$, *and* $Y = \Gamma_{]s,t] \times \Gamma}$, *we have*

$$(Y \cdot M)_u = 1_\Gamma (M_{t \wedge u} - M_{s \wedge u}).$$

The processes $(\Delta(Y \cdot M)_t)$ *and* $(Y_t \cdot \Delta M_t)$ *are indistinguishable. Thus, if* M *is continuous,* $Y \cdot M$ *is continuous.*

(b) *The processes* $(Y \cdot M)$ *and* $(Y \cdot M)^2 - (Y^2 \cdot A)$ *are* $\mathbb{F}$ *-martingales.*

General Case. The above increasing process has allowed us to obtain the existence on $(\Omega \times \mathbb{R}_+, P)$ of a measure μ_M which does not charge $\Omega \times \{0\}$ such that, for $\Gamma \times]s,t] \in C$, we have

$$\mu_M(\Gamma \times]s,t]) = E(1_\Gamma (M_t - M_s)^2).$$

From the following theorem such a measure always exists.

Theorem 8.1.5. *To every martingale* $M \in M^2$ *we can associate a unique measure* μ_M *on* $(\Omega \times \mathbb{R}_+, P)$ *such that* $\mu_M(\Omega \times \{0\}) = 0$ *and that, for* $0 < s < t < \infty$ *and* $\Gamma \in F_s$,

$$\mu_M(\Gamma \times]s,t]) = E(1_\Gamma (M_t - M_s)^2).$$

The **total mass** *of* μ_M *is* $\|M\|_2^2$. *Part* (a) *of Theorem* 8.1.4. *is satisfied.*

Proof. If the measure μ_M exists, the proof of part (a) of Theorem 8.1.4 still holds. It remains to show that we can extend μ_M to a measure on P.

Let $B = \cup_{i=1}^\infty \Gamma_i \times]t,t_{i+1}]$ be an element of B' with $\Gamma_i \in F_{t_i}$, $0 \leq t_1 < \ldots < t_i < \ldots$. Set

$$\mu_M(B) = \sum_{i=1}^\infty \mu_M(\Gamma_i \times]t_i,t_{i+1}]) = \sum_{i=1}^\infty E(1_\Gamma (M_{t_{i+1}}^2 - M_{t_i}^2)).$$

It is easily seen that μ_M is therefore well defined (two representations of the same form of B give $\mu_M(B)$ the same value) and that μ_M is additive on B'. For $\Gamma \in F_s$, $\mu_M(\Gamma \times]s,\infty[) = E(1_\Gamma (M_\infty^2 - M_s^2))$.

Let $\tau_B(\omega) = \inf\{t; (\omega,t) \in B\}$; $\{\tau_B < \infty\}$ is the set

$$\tilde{B} = \{\omega; \{\omega\} \times \mathbb{R}_+ \cap B \neq \phi\}.$$

τ_B takes, besides ∞, values contained in $\{t_i; 1 \leq i < \infty\}$;

$$\{\tau_B = t_i\} = \Gamma_i \cap \left[\bigcup_{j<i} \Gamma_j^c\right]$$

is in F_{t_i}. Hence τ_B is an $\mathbb{F}$-stopping time and

$$B \subset \sum_{i=1}^{\infty} (\tau_B = t_i) \times]t_i,\infty[;$$

$$\begin{aligned}\mu_M(B) &\leq \sum_{i=1}^{\infty} E[1_{(\tau_B=t_i)}(M_\infty^2 - M_{t_i}^2)]\\ &= E[1_{(\tau_B<\infty)}(M_\infty^2 - M_{\tau_B}^2)] \leq E[1_{\tilde{B}}\, M_\infty^2].\end{aligned}$$

Finally let us note that to every $\varepsilon > 0$ we can associate a $B^\varepsilon \in \mathcal{B}$, defined, for an $h > 0$ and an $n \in \mathbb{N}$, by

$$B^\varepsilon = \bigcup_{i=1}^{n} \Gamma_i \times]t_i + h, t_{i+1}]$$

such that $\mu_M(B\backslash B^\varepsilon) \leq \varepsilon$. The existence of h results from the right continuity of M.

The family of sets $\mathcal{B}'$ is closed under finite union or intersection, contains ϕ and $\Omega \times]0,\infty[$. Thus μ_M will have a unique σ-additive extension to the σ-algebra generated by $\mathcal{B}'$ on $\Omega \times]0,\infty[$, i.e. to the trace of $\mathcal{P}$ on $\Omega \times]0,\infty[$, if the following property can be shown: "when (B_n) is a sequence of $\mathcal{B}'$ decreasing to ϕ, $\mu_M(B_n)$ decreases to 0" (Neveu [2]).

To prove this, let us denote $\hat{B}_n = B_n^{\varepsilon/2n}$ and

$$C_n(\omega) = \overline{\{t; (\omega,t) \in \hat{B}_n\}}:$$

$C_n(\omega)$ is a compact set of $\mathbb{R}$. Now, for all ω, $\cap_n C_n(\omega) = \phi$ and there exists an integer k such that

$$\bigcap_{1 \leq n \leq k} C_n(\omega) = \phi.$$

Now

$$\widetilde{\bigcap_{n \leq k} \hat{B}_n} = \left\{\omega; \{\omega\} \times \mathbb{R}_+ \cap \left[\bigcap_{n \leq k} \hat{B}_n\right] \neq \phi\right\}$$

is contained in

$$D_k = \left\{\omega;\ \bigcap_{n\leqslant k} C_n(\omega) \neq \phi\right\}.$$

From which it follows that

$$\mu_M\left[\bigcap_{n\leqslant k} \hat{B}_n\right] \leqslant E(1_{D_k} M_\infty^2).$$

The sequence (D_k) decreases to ϕ, thus $\mu_M(\cap_{n\leqslant k}\hat{B}_n)$ tends to 0. Also

$$\mu_M(B_k) \leqslant \mu_M\left[\bigcap_{1\leqslant n\leqslant k} \hat{B}_n\right] + \sum_{1\leqslant n\leqslant k} \frac{\varepsilon}{2^n}$$

$$\leqslant \mu_M\left[\bigcap_{1\leqslant n\leqslant k} \hat{B}_n\right] + \varepsilon.$$

Therefore

$$\overline{\lim_{k\to\infty}}\ \mu_M(B_k) \leqslant \varepsilon \quad \text{for all} \quad \varepsilon > 0.$$

From which the stated result follows: μ_M is σ-finite on the trace of $\mathcal{P}$ on $\Omega \times]0,\infty[$. It can be extended to $\mathcal{P}$ by setting $\mu_M(\Omega \times \{0\}) = 0$.

8.1.3. Approximations of Stochastic Integrals and Associated Increasing Processes

For two stopping times $T \leqslant S$, we use the stochastic intervals $]T,S] = \{(\omega,u);\ T(\omega) < u \leqslant S(\omega)\}$, and similarly $[T,S]$, $]T,S[$, $[T,S[$.

Let T be a finite stopping time and let Y_T be an $\mathcal{F}_T$-measurable and bounded r.v. Then

$$(Y_T 1_{]T,\infty[} \cdot M)_t = Y_T(M_t - M_{t\wedge T}).$$

This is clear if T takes its values in a countable D, since

$$Y_T 1_{]T,\infty]} = \sum_d Y_d 1_{(T=d)} 1_{]d,\infty[}$$

and $Y_d 1_{(T=d)}$ is $\mathcal{F}_d$-measurable. If not, T is the limit of a decreasing sequence (T_n) of stopping times of this form, and $1_{]T,\infty[} = \lim 1_{]Tn,\infty[}$. Let S be another stopping time, with $T < S < \infty$,

$$Y_T \mathbf{1}_{]T,S]} = Y_T \mathbf{1}_{]T,\infty[} - Y_T \mathbf{1}_{]S,\infty[} \cdot$$

We obtain

Formula 8.1.6. $(Y_T \mathbf{1}_{]T,S[} \cdot M)_t = Y_T(M_{S\wedge t} - M_{T\wedge t})$.

Proposition 8.1.7. Approximation of $Y \cdot M$ by Random Cuts. *Consider, for every $p \in \mathbb{N}$, a sequence of finite stopping times (T^p_n) which increases to ∞ and such that $T^p_0 = 0$. Assume that $\sup_n(T^p_{n+1} - T^p_n)$ tends to 0 if $p \to \infty$.*

Then, for every real process Y, $\mathbb{F}$-adapted bounded and provided with limits on the left,

$$\left\{\sum_{n=0}^{\infty} Y_{T^p_n}(M_{T^p_{n+1}\wedge t} - M_{T^p_n \wedge t})\right\}_{t\geq 0}$$

is in $\mathcal{M}^2$ and converges in $\mathcal{M}^2$ to $Y^- \cdot M$ with $Y^- = (Y_{t-})_{t\geq 0}$.

Proof. Let

$$Z^p = \sum_{n=0}^{\infty} Y_{T^p_n} \mathbf{1}_{]T^p_n, T^p_{n+1}]} .$$

We have

$$(Z^p \cdot M)_t = \sum_{n=0}^{\infty} Y_{T^p_n}(M_{T^p_{n+1}\wedge t} - M_{T^p_n \wedge t}).$$

However, if $Z^p = (Z^p_t)_{t\geq 0}$, Z^p_t tends to Y_{t-} if $p \to \infty$. Y being bounded, the sequence Z^p is bounded by $\sup_s|Y_s|$. Hence (Z^p) tends to Y in $L^2(\mu_M)$, and $(Z^p \cdot M)$ tends to $(Y^- \cdot M)$ in $\mathcal{M}^2$.

Stopping. Let $X = (X_t)_{t\geq 0}$ be a process adapted to $\mathbb{F}$, and let T be an $\mathbb{F}$-stopping time. Denote by X^T the process X stopped at T, $X^T = (X_{t\wedge T})_{t\geq 0}$. From Corollary 7.2.9, if M is in $\mathcal{M}^2$, then the same is true for M^T. The left continuous process $\mathbf{1}_{]T,\infty[}$ is predictable. If Y is a process in $L^2(\mu_M)$, $Y\mathbf{1}_{[0,T]} = Y - Y\mathbf{1}_{]T,\infty]}$ is also predictable. We then have

The Localization Formula 8.1.8.

$$(Y\mathbf{1}_{[0,T]}) \cdot M = Y \cdot M^T = (Y \cdot M)^T = Y^T \cdot M^T.$$

It is sufficient to check this for $Y = \mathbf{1}_{\Gamma \times]s,t]}$ and $\Gamma \in \mathcal{F}_s$, $s < t$.

Then

$$Y1_{[0,T]} = 1_{\Gamma\cap(s<T)}1_{]s\wedge T,t\wedge T]}$$

and

$$\Gamma \cap (s < T) \in F_s \cap F_T = F_{s\wedge T}.$$

Formula 8.1.6 is applied

$$Y1_{[0,T]}\cdot M = 1_\Gamma(M_{t\wedge T} - M_{s\wedge T}) = Y\cdot M^T = (Y\cdot M)^T.$$

We have defined the stochastic integral first of all for martingales M such that there exists an increasing process A for which $M^2 - A$ is a martingale. Then we have defined the stochastic integral with respect to M without imposing this condition. We are now going to deduce from the existence of this integral that of an increasing process $[M,M]$ such that $M^2 - [M,M]$ is a martingale. This result generalizes Doob's decomposition of the submartingale M^2 associated with a martingale M in discrete time ([2.2.4]).

Theorem 8.1.9. *Let $M \in \mathcal{M}^2$. There exists an increasing $\mathbb{F}$-process, denoted $[M,M]$, satisfying the following properties:*

(a) *$M^2 - [M,M]$ is an $\mathbb{F}$-martingale.*

(b) *For every $p \in \mathbb{N}$, let (T_n^p) be a sequence of $\mathbb{F}$-stopping times which increase to ∞ and such that $T_0^p = 0$. If*

$$\lim_{p\to\infty} \sup_n |T_{n+1}^p - T_n^p| = 0,$$

then, for any t, the **quadratic variation**

$$\sum_{n=0}^{\infty} (M_{T_{n+1}^p\wedge t} - M_{T_n^p\wedge t})^2$$

converges in probability to $[M,M]_t$ if $p \to \infty$.

(c) *$([M,M]_t - \Sigma_{s\leqslant t}\Delta M_s^2)_{t\geqslant 0}$ is a continuous increasing $\mathbb{F}$-process. Hence if M is continuous, $[M,M]$ is continuous.*

Proof. Let $A < \infty$ and let $T_A = \inf\{s;\ |M_s| \geqslant A\}$; the process $(M_{s-}^{T_A})$ is left continuous thus predictable and bounded in modulus by A. Let us denote for ease of writing, N instead of M^{T_A}:

$$N_t^2 = \sum_{n=0}^{\infty} (N^2_{T^p_{n+1}\wedge t} - N^2_{T^p_n\wedge t})$$

$$= \sum_{n=0}^{\infty} (N_{T^p_{n+1}\wedge t} - N_{T^p_n\wedge t})^2$$

$$+ 2 \sum_{n=0}^{\infty} N_{T^p_n\wedge t}(N_{T^p_{n+1}\wedge t} - N_{T^p_n\wedge t}).$$

From Proposition 8.1.7, we have

$$\sum_{n=0}^{\infty} (N_{T^p_{n+1}\wedge t} - N_{T^p_n\wedge t})^2 \xrightarrow{L^2} N_t^2 - 2\int_0^t N_{s-}\, dN_s.$$

From formula 8.1.8, this limit equals

$$[M,M]_t^A = M^2_{t\wedge T_A} - 2\int_0^{t\wedge T_A} M_{s-}\, dM_s\,.$$

Let $s < t$; we can take sequences (T^p_n) two of the terms of which equal s and t respectively. Then,

$$\sum_{s\leq T^p_n<t} (N_{T^p_{n+1}} - N_{T^p_n})^2 \xrightarrow{P} [M,M]_t^A - [M,M]_s^A.$$

This proves that $[M,M]^A$ is increasing.

The trajectories $t \longmapsto [M,M]^A_{t\wedge T_A}$ are cad-lag; and we have a.s. for all t:

$$\Delta[M,M]_t^A = \Delta M^2_{t\wedge T_A} - 2M_{t\wedge T_A-}\,\Delta M_{t\wedge T_A} = (\Delta M_{t\wedge T_A})^2\,.$$

However from Proposition 7.2.6, we have

$$P[T_A \leq t] \leq \frac{1}{A^2} E(M_t^2).$$

Thus, for all $\varepsilon > 0$, we can find an A such that $P(T_A \leq t)$ is less than ε. On $\{T_A > t\}$, the processes N and M coincide a.s. Thus a process $[M,M]$ may be defined with cad-lag trajectories such that, for every $A \in \mathbb{N}$,

$$[M,M]_t = [M,M]_t^A \quad \text{on} \quad \{T_A > t\}.$$

We have

$$P[\Delta[M,M]_t = (\Delta M_t)^2 \ \text{ for all } \ t] = 1.$$

Thus

$$[M,M]_t - \sum_{s\leqslant t} (\Delta M_s)^2 = [M,M]_t - \sum_{s\leqslant t} \Delta[M,M]_s$$

is a continuous increasing process.

Consequence. The particular case studied in [8.1.2] is in fact the general case. We can use Theorem 8.1.4 by taking $A = [M,M]$.

Let M and N be two martingales in $\mathcal{M}^2$. By Minkowski's inequality

$$\Sigma\left(M_{T^p_{n+1}\wedge t} - N_{T^p_{n+1}\wedge t} - M_{T^p_n\wedge t} + N_{T^p_n\wedge t}\right)^2$$

$$= \Sigma\left(M_{T^p_{n+1}\wedge t} - M_{T^p_n\wedge t}\right)^2 + \Sigma\left(N_{T^p_{n+1}\wedge t} - N_{T^p_n\wedge t}\right)^2$$

$$- 2\Sigma\left(M_{T^p_{n+1}\wedge t} - M_{T^p_n\wedge t}\right)\left(N_{T^p_{n+1}\wedge t} - N_{T^p_n\wedge t}\right)$$

$$\geqslant \Sigma\left(M_{T^p_{n+1}\wedge t} - M_{T^p_n\wedge t}\right)^2 + \Sigma\left(N_{T^p_{n+1}\wedge t} - N_{T^p_n\wedge t}\right)^2$$

$$- 2\left[\Sigma\left(M_{T^p_{n+1}\wedge t} - M_{T^p_n\wedge t}\right)^2 \Sigma\left(N_{T^p_{n+1}\wedge t} - N_{T^p_n\wedge t}\right)^2\right]^{1/2}$$

Passing to the limit for $p \to \infty$,

$$[M-N,\ M-N]_t \geqslant \left(\sqrt{[M,M]_t} - \sqrt{[N,N]_t}\right)^2.$$

Thus, if a sequence (M^p) of $\mathcal{M}^2$ converges in $\mathcal{M}^2$ to M, then

$$\sqrt{[M^p,M^p]_t} \xrightarrow[p\to\infty]{L^2} \sqrt{[M,M]_t}\ .$$

for any t.

Proposition 8.1.10. *Let* $M \in \mathcal{M}^2$.

(a) *For every predictable process* Y *of* $L^2(\mu_M)$, *we have*

$$[Y\cdot M,\ Y\cdot M] = Y^2\cdot[M,M].$$

(b) *Assume* M *is bounded. Let* Y *be a real process adapted to* $\mathbb{F}$, *bounded, and having left limits on each trajectory. For every family* (T_n^p) *of stopping times such as that of Theorem* 8.1.9 (b), *we have,*

$$\sum_{n=0}^{\infty} Y_{T_n^p}\left[M_{T_n^{p+1}\wedge t} - M_{T_n^p\wedge t}\right]^2 \xrightarrow[p\to\infty]{P} \int_0^t Y_{s^-}d[M,M]_s.$$

Proof. From the note which precedes Proposition 8.1.10 the set of Y which satisfy property (a) is a closed set of $L^2(\mu_M)$. Hence it is sufficient to prove it for stepped Y of the form

$$\sum_{i=1}^{n} \alpha_i 1_{\Gamma_i\times]t_i,t_{i+1}]} \quad \text{with} \quad 0 = t_1 < \dots < t_{n+1} < \infty.$$

Then however, assuming $t_j < t \leq t_{j+1}$,

$$\sum_{1\leq i<j} ((Y\cdot M)_{t_{i+1}\wedge t} - (Y\cdot M)_{t_i\wedge t})^2$$
$$= \sum_{1\leq i<j} \alpha_i^2 1_{\Gamma_i}(M_{t_{i+1}} - M_{t_i})^2 + \alpha_j^2 1_{\Gamma_j}(M_t - M_{t_j})^2.$$

Let us take a sequence of partitions of $[0,t]$, finer than $t_1, \dots, t_j, t$ and the width of which tends to 0. The sum of squares of the increments of $Y\cdot M$ along this partition tends in probability to

$$\sum_{1\leq i<j} \alpha_i^2 1_{\Gamma_i}([M,M]_{t_{i+1}} - [M,M]_{t_i})$$
$$+ \alpha_j^2 1_{\Gamma_j}([M,M]_t - [M,M]_{t_j})$$
$$= \int_0^t Y_s^2 d[M,M]_s.$$

Let us prove (b). By separating the positive and negative parts we can assume that Y is positive, thus we prove the result for Y^2. Let

$$Z^p = \Sigma Y_{T_n^p} 1_{]T_n^p,T_{n+1}^p]};$$

from Proposition 8.1.7, $(Z^p \cdot M)$ tends in $\mathcal{M}^2$ to

$$Y^- \cdot M = \left[\int_0^t Y_{s-} dM_s\right]_{t \geq 0} .$$

However,

$$(Z^p \cdot M)_t^2 = \sum_{n=0}^{\infty} Y^2_{T_n^p} \left[M_{T_{n+1}^p \wedge t} - M_{T_n^p \wedge t}\right]^2$$

$$+ 2 \sum_{n=0}^{\infty} Y_{T_n^p} [Z^p \cdot M]_{T_n^p} \left[M_{T_{n+1}^p \wedge t} - M_{T_n^p \wedge t}\right].$$

From which, letting p tend to ∞,

$$\sum_{n=0}^{\infty} Y^2_{T_n^p} \left[M_{T_{n+1}^p \wedge t} - M_{T_n^p \wedge t}\right]^2$$

$$\xrightarrow{L^1} \left[\int_0^t Y_{s-} dM_s\right]^2 - 2 \int_0^t Y_{s-} (Y^- \cdot M)_{s-} \, dM_s .$$

This limit equals

$$[Y^- \cdot M, Y^- \cdot M]_t = \int_0^t Y^2_{s-} \, d[M,M]_s.$$

The Case of Continuous Processes

Proposition 8.1.11. *Let $M \in \mathcal{M}^2_c$. The process $[M,M]$ is the unique continuous increasing process such that $M^2 - [M,M]$ is an $\mathbb{F}$-martingale.*

Proof. Let A be an increasing continuous $\mathbb{F}$-process such that $M^2 - A$ is an $\mathbb{F}$-martingale: $[M,M] - A$ is an $\mathbb{F}$-martingale. Let $N \in \mathbb{N}$ and let $T_N = \inf\{t;\ [M,M]_t + A_t \geq N\}$; $[M,M] - A$, stopped at T_N, is a bounded $\mathbb{F}$-martingale. From Proposition 6.2.15, this means that

$$[M,M]^{T_N} = A^{T_N}.$$

Letting N tend to ∞, we obtain $[M,M] = A$.

Consequences. If τ is a partition $0 = t_1 < \ldots < t_{n+1} = t$ of $[0,t]$, let us denote its width by $|\tau|$, and

$$Q_T^M = \sum_{i=1}^{n} (M_{t_{i+1}} - M_{t_i})^2 .$$

When $|T| \to 0$, we have

$$Q_T^M \xrightarrow{P} [M,M]_t;$$

there exists a sequence $T_k = (t_i^k)$ of partitions of $[0,t]$ such that $|T_k| \to 0$ and

$$Q_{T_k}^M \xrightarrow{a.s.} [M,M]_t.$$

Moreover, if M is continuous,

$$\sup \left| M_{t_{i+1}} - M_{t_i} \right| \xrightarrow[|T|\to 0]{a.s.} 0.$$

Since

$$Q_T^M \leqslant \left(\sum_{i=1}^{n} |M_{t_{i+1}} - M_{t_i}| \right) \sup |M_{t_{i+1}} - M_{t_i}|$$

we obtain, a.s. on $\{[M,M]_t > 0\}$:

$$\lim_{k\to\infty} \sum_{i=1}^{n} |M_{t_{i+1}^k} - M_{t_i^k}| = \infty .$$

A continuous square integrable martingale has almost no trajectories with bounded variation on $\{[M,M]_\infty > 0\}$.

8.1.4. Localization

Definition 8.1.12. We denote by $\mathcal{M}_{loc}$ the set of **local $\mathbb{F}$-martingales** i.e. real processes M, with cad-lag trajectories, such that there exists a sequence (T_n) of $\mathbb{F}$-stopping times increasing to ∞ for which M^{T_n} is an $\mathbb{F}$-martingale for any n. $\mathcal{M}^2_{loc}$ is the set of $M \in \mathcal{M}_{loc}$ such that there exists a sequence (T_n) of $\mathbb{F}$-stopping times increasing to ∞ for which M^{T_n} is in $\mathcal{M}^2$, for any n.

A real process Y adapted to $\mathbb{F}$ is **locally bounded** if there exists a sequence (T_n) of stopping times increasing to ∞ such that Y^{T_n} is bounded, for any n.

Notes. (a) If M is a local martingale and if $E[\sup_{s\leq t}|M_s|]$ is finite for all t (in particular if $E(M_t^2)$ is finite for all t), then it is an $\mathbb{F}$-martingale. In fact the sequence $(M_{T_n \wedge t})$ then tends a.s. and in L^1 to M_t. For $\Gamma \in \mathcal{F}_s$ and $s < t$,

$$E[1_\Gamma 1_{(T_n > s)} M_{T_n \wedge t}] = E[1_\Gamma 1_{(T_n > s)} M_s].$$

From which it follows that $E[M_t | \mathcal{F}_s] = M_s$.

(b) The hypothesis $M \in \mathcal{M}^2_{loc}$ is much weaker than $M \in \mathcal{M}^2$. For example, every martingale the jumps of which are bounded by a constant C is in $\mathcal{M}_2^{loc}$. In fact, we can consider $T_n = \inf\{t; |M_t| \geq n\}$; we then have

$$|M_{t \wedge T_n}| \leq n + C$$

for all t.

(c) If M is a local martingale and if T is a stopping time such that $\sup_{t\leq T}|M_t|$ is integrable, then $M_T \in L^1$ and $E(M_T|\mathcal{F}_t) = M_{t\wedge T}$, for any $t \geq 0$. To see this apply (a) to the local martingale $M^T = (M_{T\wedge t})_{t\geq 0}$.

Now let us consider $M \in \mathcal{M}^2_{loc}$ and Y predictable and locally bounded. There exists an increasing sequence of stopping times (T_n) such that M^{T_n} is in $\mathcal{M}^2$ and Y^{T_n} is bounded. The stochastic integral

$$Y^{T_n} \cdot M^{T_n}$$

exists. However the localization formula 8.1.8 allows us to write

$$(Y^{T_{n+1}} \cdot M^{T_{n+1}})^{T_n} = Y^{T_n} \cdot M^{T_n}.$$

Set

$$(Y \cdot M)_t = (Y^{T_n} \cdot M^{T_n})_t, \quad \text{for} \quad T_n > t.$$

$Y \cdot M$ is therefore defined as an element of $\mathcal{M}^2_{loc}$. This definition does not depend on the sequence chosen, because of the localization formula. In fact, if T is a stopping time such that Y^T is bounded and $M^T \in \mathcal{M}^2$ then

$$(Y\cdot M)^{T_n\wedge T} = Y^{T_n\wedge T} \cdot M^{T_n\wedge T};$$

and by letting n tend to ∞, $(Y\cdot M)_t^{T\wedge T_n}$ and

$$(Y^{T_n\wedge T} \cdot M^{T_n\wedge T})_t$$

converge a.s. and in L^2 to $(Y\cdot M)_{t\wedge T}$ and $(Y^T\cdot M^T)_t$ respectively.

Starting from Theorems 8.1.10, 8.1.4 and Proposition 8.1.11, we easily obtain the following results, by localization:

Theorem 8.1.13. *Let* $M \in \mathcal{M}^2_{loc}$. *There exists an increasing* $\mathbb{F}$*-process* $[M,M]$ *satisfying the following properties:*

(a) $M^2 - [M,M]$ *is a local* $\mathbb{F}$*-martingale,*

(b) *Property* (b) *of Theorem* 8.1.9.

(c) $\{[M,M]_t - \Sigma_{s\leqslant t}(\Delta M_s)^2\}$ *is a continuous increasing* $\mathbb{F}$*-process.*

(d) *If* M *is continuous,* $[M,M]$ *is continuous, and it is the unique continuous increasing process satisfying property* (a).

There exists a unique linear mapping

$$Y\longmapsto Y\cdot M = \left[\int_0^t Y_s dM_s\right]_{t\geqslant 0}$$

from the set of locally bounded predictable processes into $\mathcal{M}^2_{loc}$ *satisfying the following two properties:*

(1) $(Y\cdot M)^2 - Y^2\cdot[M,M]$ *is in* $\mathcal{M}_{loc}$.

(2) *If* T *and* S *are two finite* $\mathbb{F}$*-stopping times and* $S < T$ *and if* Y_S *is a bounded* $\mathcal{F}_S$*-measurable* r.v., *then*

$$Y_S 1_{]S,T]} \cdot M = \{Y_S(M_{T\wedge t} - M_{S\wedge t})\}_{t\geqslant 0}.$$

We have the relation

$$M_t^2 = 2\int_0^t M_{s-}dM_s + [M,M]_t.$$

If, for all t,

$$E\left[\int_0^t Y_s^2\, d[M,M]_s\right]$$

is finite, then $Y \cdot M$ is a martingale.

The last part of the theorem follows from note (a) which precedes it.

8.2. Ito's Formula and Stochastic Calculus

8.2.1. Ito's Formula

We are given a stochastic integral relative to either a cad-lag process V of bounded variation, or an $M \in \mathcal{M}^2$. In this section we are concerned with establishing a formula analogous to Taylor's formula, for these integrals. Let us look at a few particular cases first of all.

Let f be a function from $\mathbb{R}$ into $\mathbb{R}$ of class C^1. For every partition τ: $t_0 = 0 < t_1 < \ldots < t_{n+1} = t$ of $[0,t]$ of width $|\tau| = \sup|t_{i+1} - t_i|$, we write

$$f(V_t) - f(V_0) = \sum_{i=0}^{n} f(V_{t_{i+1}}) - f(V_{t_i}),$$

$$\left|f(V_t) - f(V_0) - \int_0^t f'(V_s)dV_s\right|$$
$$\leqslant T_t^V \sup_{1 \leqslant i \leqslant n} \sup_{t_i \leqslant u,v \leqslant t_{i+1}} \left|f'(V_u) - f'(V_v)\right|$$

where

$$T_t^V = \sup\left\{\sum_{i=0}^{n} |V_{t_{i+1}} - V_{t_i}|;\ 0 = t_0 < t_1 < \ldots < t_n = t\right\}$$

is the total variation on $[0,t]$. For $|\tau| \to 0$, if V is continuous, we obtain the expected formula

$$f(V_t) - f(V_0) = \int_0^t f'(V_s)dV_s.$$

However if V is only cad-lag the above majorant does not in general tend to 0 when $|\tau| \to 0$.

Moreover the formula for integration by parts 6.2.4 gives

$$V_t^2 - V_0^2 = 2\int_0^t V_{s-}dV_s + \sum_{0<s\leqslant t} (\Delta V_s)^2.$$

Jumps are thus going to complicate the formulae even for processes with bounded variation. In this case we shall obtain the following formula:

$$f(V_t) - f(V_0) = \int_0^t f'(V_{s-})dV_s + \sum_{0<s\leqslant t}[f(V_s) - f(V_{s-}) - f'(V_{s-})\Delta V_s].$$

Do we obtain the preceding formulae again replacing V by M? The answer is negative even for continuous M. To see this it is sufficient to recall the formula,

$$M_t^2 = 2\int_0^t M_{s-}dM_s + [M,M]_t.$$

Here, even for continuous M, a supplementary term $[M,M]_t$ appears. This depends on the fact that M, instead of being of bounded variation, is of bounded quadratic variation. Assume f is of class C^2, f, f' and f'' being bounded. We write

$$\begin{aligned} f(M_t) - f(M_0) &= \sum_{i=0}^n [f(M_{t_{i+1}}) - f(M_{t_i})] \\ &= \sum_{i=0}^n f'(M_{t_i})(M_{t_{i+1}} - M_{t_i}) \\ &\quad + \frac{1}{2}\sum_{i=0}^n f''(M_{\theta_i})(M_{t_{i+1}} - M_{t_i})^2 \end{aligned}$$

(θ_i is a random point on $[t_i,t_{i+1}]$). Thus,

$$\Big| f(M_t) - f(M_0) - \sum_{i=0}^n f'(M_{t_i})(M_{t_{i+1}} - M_{t_i}) - \frac{1}{2}\sum_{i=0}^n f''(M_{t_i})(M_{t_{i+1}} - M_{t_i})^2 \Big|$$

$$\leqslant \frac{1}{2}\sum_{i=1}^n (M_{t_{i+1}} - M_{t_i})^2 \sup_{1\leqslant i\leqslant n}\ \sup_{t_i\leqslant u,v\leqslant t_{i+1}} |f''(M_u) - f''(M_v)|.$$

We have

$$\sum_{i=1}^n (M_{t_{i+1}} - M_{t_i})^2 \xrightarrow{P} [M,M]_t.$$

If M is continuous, the right hand term tends in probability to 0. From Propositions 8.1.7 and 8.1.10, we obtain *if M is continuous*

$$f(M_t) - f(M_0) = \int_0^t f'(M_s)dM_s + \frac{1}{2}\int_0^t f''(M_s)d[M,M]_s.$$

If M has jumps things are a little more complicated... . These preliminaries should have been of use in accepting the following formula, where D_i signifies differentiation with respect to the ith variable.

Theorem 8.2.14. Ito's or the "Change of Variable" Formula. *Let M be a martingale in $\mathbb{M}^2_{loc}$ adapted to $\mathbb{F}$, V a* cad-lag *process of bounded variation adapted to $\mathbb{F}$, and f a function of class C^2 from $\mathbb{R}^2$ into $\mathbb{R}$. We have* a.s., *for all t,*

$$\begin{aligned} f(V_t,M_t) - f(V_0,M_0) &= \int_0^t D_1 f(V_s,M_s)dV_s \\ &+ \int_0^t D_2 f(V_s,M_{s-})dM_s \\ &+ \frac{1}{2}\int_0^t D_2^2 f(V_{s-},M_{s-})d[M,M]_s \\ &+ \sum_{s\leqslant t}[f(V_s,M_s) - f(V_{s-},M_s) - D_1 f(V_s,M_s)\Delta V_s] \\ &+ \sum_{s\leqslant t}[f(V_s,M_s) - f(V_s,M_{s-}) - D_2 f(V_{s-},M_{s-})\Delta M_{s-} \\ &\qquad - \frac{1}{2}D_2^2 f(V_{s-},M_{s-})(\Delta M_s)^2]. \end{aligned}$$

Proof. First of all let us show the theorem for bounded M and V with bounded total variation $T^V = (T^V_t)_{t\geqslant 0}$ ($\sup_t|M_t|$ and $\sup_t T^V_t$ are bounded).

If c is a constant which majorizes $\sup_s|M_s|$ and $\sup_s T^V_s$, then on

$$K = \{(x,y);\ |x| \leqslant c \ \text{ and } \ |y| \leqslant c\}$$

the first and second derivatives of f are uniformly continuous. Let $\eta > 0$; we associate with it an $\varepsilon > 0$ such that, for every (x,y) and (x',y') in K and $|x - x'| \leqslant \varepsilon$ and $|y - y'| \leqslant \varepsilon$, these derivatives at (x,y) and (x',y') differ by less than η. Consider an $\alpha > 0$, and the stopping times T_n

defined by

$$T_0 = 0, \ldots, \quad T_{n+1} = \inf\{s;\ s > T_n,\ |V_s - V_{T_n}| + |M_s - M_{T_n}| \geqslant \varepsilon/2\} \wedge (T_n + \alpha) \wedge t.$$

We have

$$f(V_t, M_t) - f(V_0, 0) = \sum_p [f(V_{T_{p+1}}, M_{T_{p+1}}) - f(V_{T_p}, M_{T_{p+1}})] + \sum_p [f(V_{T_p}, M_{T_{p+1}}) - f(V_{T_p}, M_{T_p})].$$

Let us study the first sum. It can be compared with

$$\sum_p D_1 f(V_{T^-_{p+1}}, M_{T_{p+1}}) \cdot (V_{T_{p+1}} - V_{T_p})$$

which tends, if η tends to 0, to

$$\int_0^t D_1 f(V_{s-}, M_s) dV_s.$$

Let

$$Y_p = f(V_{T_{p+1}}, M_{T_{p+1}}) - f(V_{T_p}, M_{T_{p+1}}) - D_1 f(V_{T^-_{p+1}}, M_{T_{p+1}})(V_{T_{p+1}} - V_{T_p}),$$

and

$$\beta_t = \sum_{s \leqslant t} [f(V_s, M_s) - f(V_{s-}, M_s) - D_1 f(V_{s-}, M_s)\Delta V_s\}.$$

We have

$$\left|\beta_t - \sum_{\{s \leqslant t, |\Delta V_s| \geqslant \varepsilon/2\}} \{f(V_s, M_s) - f(V_{s-}, M_s) - D_1 f(V_{s-}, M_s)\Delta V_s\}\right| \leqslant \eta \sum_{s \leqslant t} |\Delta V_s| \leqslant \eta T_t^V;$$

$$\left|\Sigma Y_p 1_{\{|\Delta V_{T_{p+1}}| < \varepsilon/2\}}\right| \leqslant \eta \Sigma |V_{T_{p+1}} - V_{T_p}| \leqslant \eta T_t^V.$$

On each trajectory of V, there are before t at most a finite number of jumps of modulus greater than $\varepsilon/2$ which occur at certain of the times T_p;

$$\Sigma Y_p \mathbf{1}_{\{|\Delta V_{T_{p+1}}| \geq \varepsilon/2\}}$$

tends, if α tends to 0, to

$$\sum_{\{s \leq t, |\Delta V_s| \geq \varepsilon/2\}} (f(V_s, M_s) - f(V_{s-}, M_s) - D_1 f(V_{s-}, M_s)\Delta V_s).$$

We let α tend to 0, then η tend to 0:

$$\Sigma[f(V_{T_{p+1}}, M_{T_{p+1}}) - f(V_{T_p}, M_{T_p})]$$

tends to

$$\int_0^t D_1 f(V_{s-}, M_s) dV_s + \beta_t$$
$$= \int_0^t D_1 f(V_s, M_s) dV_s + \sum_{s \leq t} [f(V_s, M_s) - f(V_{s-}, M_s) - D_1 f(V_s, M_s)\Delta V_s].$$

Let us study the second sum

$$\Sigma f(V_{T_p}, M_{T_{p+1}}) - f(V_{T_p}, M_{T_p}).$$

It is going to be compared with its second order Taylor approximation

$$\Sigma D_2 f(V_{T_p}, M_{T_p})(M_{T_{p+1}} - M_{T_p})$$
$$+ \frac{1}{2} \Sigma D_2^2 f(V_{T_p}, M_{T_p})(M_{T_{p+1}} - M_{T_p})^2.$$

If η tends to 0, this approximation tends, from Propositions 8.1.7 and 8.1.10 to

$$\int_0^t D_2 f(V_{s-}, M_{s-}) dM_s + \frac{1}{2} \int_0^t D_2^2 f(V_{s-}, M_{s-}) d[M,M]_s.$$

Consider the r.v.'s,

$$Z_p = f(V_{T_p}, M_{T_{p+1}}) - f(V_{T_p}, M_{T_p})$$
$$- D_2 f(V_{T_p}, M_{T_p})(M_{T_{p+1}} - M_{T_p})$$
$$- \frac{1}{2} D_2^2 f(V_{T_p}, M_{T_p})(M_{T_{p+1}} - M_{T_p})^2,$$

$$\gamma_t = \sum_{s \leqslant t} \Big[f(V_{s-},M_s) - f(V_{s-},M_{s-}) - D_2 f(V_{s-},M_{s-}) \Delta M_s - \frac{1}{2} D_2^2 f(V_{s-},M_{s-})(\Delta M_s)^2 \Big].$$

We have

$$\sum Z_p \mathbf{1}_{\{|\Delta M_{T_{p+1}}| < \varepsilon/2\}} \leqslant \frac{1}{2} \eta \sum (M_{T_{p+1}} - M_{T_p})^2.$$

If α tends to 0, $\Sigma(M_{T_{p+1}} - M_{T_p})^2$ tends to $[M,M]_t$, and the majorant tends in probability to $(1/2)\eta[M,M]_t$. Moreover, on a given trajectory of M, there is before t a finite number of jumps $\geqslant \varepsilon/2$ which occur at certain of the times T_p. If $\alpha \to 0$,

$$\sum Z_p \mathbf{1}_{\{|\Delta M_{T_{p+1}}| \geqslant \varepsilon/2\}}$$

tends to

$$\sum_{\{s \leqslant t; |\Delta M_s| \geqslant \varepsilon/2\}} \Big[f(V_{s^-},M_s) - f(V_{s^-},M_{s^-}) - D_2 f(V_{s^-},M_{s^-}) \Delta M_s - \frac{1}{2} D_2^2 f(V_{s^-},M_{s^-})(\Delta M_s)^2 \Big].$$

Finally

$$\sum_{\{s \leqslant t; |\Delta M_s| < \varepsilon/2\}} \Big[f(V_{s^-},M_s) - f(V_{s^-},M_{s^-}) - D_2 f(V_{s^-},M_{s^-}) \Delta M_s - \frac{1}{2} D_2^2 f(V_{s^-},M_{s^-})(\Delta M_s)^2 \Big]$$

$$\leqslant \frac{\eta}{2} \sum_{s \leqslant t} (\Delta M_s)^2 \leqslant \frac{\eta}{2} [M,M]_t$$

(from Theorem 8.1.9(d)).

Hence, if η and ε tend to 0, this expression tends to 0. Letting α tend to 0, then η and ε to 0, we obtain

$$\sum Z_p - \gamma_t \xrightarrow{P} 0.$$

We have therefore shown the theorem if $\sup_t |M_t|$ and $\sup_t |T_t^V|$ are bounded r.v.'s.

Let us take $M \in \mathcal{M}_2$ and V a cad-lag process of bounded variation;

$$P\left[\sup_s |M_s| \geq n\right] \leq \frac{1}{n^2}E[M_\infty^2].$$

Set $S_n = \inf\{s;\ |M_s| \geq n \text{ or } T_s^V \geq n\}$, denoting by T_s^V the total variation (finite) of $(V_t)_{t \leq s}$.

If we go back to the preceding calculations, we obtain the formula for $t < S_n$. Since (S_n) increases a.s. to ∞, Ito's formula is true for all t.

Likewise for $M \in \mathcal{M}_{loc}^2$, we can find a sequence (R_n) of stopping times which increases to ∞, such that M^{R_n} is in $\mathcal{M}_2$ for any n. We apply the theorem to M^{R_n} and V^{R_n}, which gives the formula for $t \leq R_n$. Then we let n tend to ∞.

Note. Let f be a function of class C^2 from $\mathbb{R}^2$ into $\mathbb{R}$ and let $M \in \mathcal{M}_{loc}^2$. Let $[M,M]^c$ be the continuous part of $[M,M]$,

$$[M,M]_t^c = [M,M]_t - \sum_{s \leq t} \Delta M_s^2 .$$

We have

$$f(M_t) - f(0) = \int_0^t f'(M_{s^-})dM_s$$

$$+ \frac{1}{2}\int_0^t f''(M_{s^-})d[M,M]_s^c$$

$$+ \sum_{s \leq t}[f(M_s) - f(M_{s^-}) - f'(M_{s^-})\Delta M_s].$$

Example. Integration by Parts Formula. Taking $f\colon (x,y) \longmapsto xy$ we obtain

$$V_tM_t = V_0M_0 + \int_0^t M_s dV_s + \int_0^t V_{s-}dM_s.$$

8.2.2. Gaussian Martingales

Theorem 8.2.15. *Let $a\colon t \longmapsto a_t$ be a continuous increasing function from $\mathbb{R}_+$ into $\mathbb{R}$, zero at 0, and M a centered continuous local $\mathbb{F}$-martingale such that $M^2 - a$ is a local martingale. Then*

M is a process with independent increments adapted to $\mathbb{F}$ and, for $s < t$, $M_t - M_s$ has distribution $\mathcal{N}(0, a_t - a_s)$. Thus we are dealing with a continuous Gaussian process with covariance $(s,t) \longmapsto a_{s \wedge t}$.

We shall denote by W_a the distribution on C of this Gaussian process. If $a_t = t$ for any t, W_a is the Wiener measure W.

Proof. The result is analogous to that of [6.3] on the Poisson process and is based likewise on the introduction of an exponential martingale.

Let $u \in \mathbb{R}$. Let us apply Ito's formula to the function

$$(x,y) \longmapsto \exp\left[iux + \frac{u^2}{2}y\right].$$

We have

$$Z_t^u = \exp\left[iuM_t + \frac{u^2}{2}a_t\right] = 1 + iu\int_0^t Z_s^u\, dM_s.$$

Thus $(Z_t^u - 1)$ is in $\mathcal{M}_{loc}^2$; since $(Z_t^u)^2$ is integrable for all t, (Z_t^u) is a martingale. From which it follows, for $s < t$,

$$E[Z_t^u \mid \mathcal{F}_s] = Z_s^u E\left[\exp iu(M_t - M_s) + \frac{u^2}{2}(a_t - a_s) \mid \mathcal{F}_s\right]$$

$$= Z_s^u;$$

$$E[\exp iu(M_t - M_s) \mid \mathcal{F}_s] = \exp\left[-\frac{u^2}{2}(a_t - a_s)\right].$$

Hence $M_t - M_s$ is independent of $\mathcal{F}_s$ and has distribution $(0, a_t - a_s)$.

By a similar proof, we obtain the following proposition which is a tool used in [8.2.3] for asymptotic theorems.

Proposition 8.2.16. *Let $M \in \mathcal{M}_{loc}^2$. For $\lambda > 0$ set*

$$\phi_c(\lambda) = \frac{1}{c^2}(e^{\lambda c} - 1 - \lambda c) \qquad \textit{for } c > 0,$$

and $\phi_0(\lambda) = \lambda^2/\lambda$.

(a) *If $|\Delta M| \leqslant c$, then*

$$(\exp\{\lambda M_t - e^{\lambda c}\phi_c(\lambda)[M,M]_t\})$$

is a positive supermartingale.

(b) *If* $|\Delta M| \leq c$ *and if* $(A_t)_{t\geq 0}$ *is a continuous increasing* $\mathbb{F}$*-process such that* $M^2 - A$ *is a local martingale, then* $Z^c_\lambda = (\exp\{\lambda M_t - \phi_c(\lambda)A_t\})$ *is a positive supermartingale.*

Proof. M can be replaced by $-M$, thus assume $\lambda \geq 0$. For $\mu \in \mathbb{R}$, set $Y_t = \exp(\lambda M_t + \mu A_t)$ with A an increasing $\mathbb{F}$-process.

From Ito's formula,

$$Y_t = 1 + \mu\int_0^t Y_s dA_s + \sum_{s\leq t} Y_s(1 - \exp(-\mu\Delta A_s) - \mu\Delta A_s)$$

$$+ \lambda\int_0^t Y_{s-}dM_s + \frac{\lambda^2}{2}\int_0^t Y_s d[M,M]^c_s$$

$$+ \sum_{s\leq t} Y_{s-}(\exp(\lambda\Delta M_s) - 1 - \lambda\Delta M_s).$$

For $c > 0$ and $|y| \leq c$, we have,

$$e^{\lambda y} - 1 - \lambda y = \lambda^2 y^2 \sum_{n\geq 2} \frac{(\lambda y)^{n-2}}{n!} \leq y^2\phi_c(\lambda).$$

From which, by taking $|\Delta M_s| \leq c$,

$$Y_{s-}(\exp(\lambda\Delta M_s) - 1 - \lambda\Delta M_s) \leq Y_{s-}\phi_c(\lambda)(\Delta M_s)^2$$

$$\leq Y_s e^{\lambda c}\phi_c(\lambda)(\Delta M_s)^2.$$

If A is continuous and such that $M^2 - A$ is a martingale, let us take $\mu = -\phi_c(\lambda)$. Set

$$B_t = \frac{\lambda^2}{2}\int_0^t Y_s d[M,M]^c_s + \sum_{s\leq t} Y_{s-}(\exp(\lambda\Delta M_s) - 1 - \lambda\Delta M_s)$$

$$- \phi_c(\lambda)\int_0^t Y_{s-}d[M,M]_s.$$

The process $(B_t)_{t\geq 0}$ is decreasing and,

$$Y_t = 1 - \phi_c(\lambda)\left[\int_0^t Y_{s-}dA_s - \int_0^t Y_{s-}d[M,M]_s\right]$$

$$+ \lambda\int_0^t Y_{s-}dM_s + B_t;$$

Y is the sum of a local martingale and a decreasing process.

We deduce from this, by a proof analogous to note (a) of [8.1.4], that it is a positive supermartingale. If $A = [M,M]$, take $\mu = -e^{\lambda c}\phi_c(\lambda)$. We set

$$D_t = \frac{\lambda^2}{2}\int_0^t Y_s d[M,M]_s^c + \sum_{s \leq t} Y_{s^-}(\exp(\lambda\Delta M_s) - 1 - \lambda\Delta M_s)$$
$$- e^{\lambda c}\phi_c(\lambda)\int_0^T Y_s d[M,M]_s$$
$$- \sum_{s \leq t} Y_s(\exp(-\mu\Delta[M,M]_s) - 1 + \mu\Delta[M,M]_s).$$

The process (D_t) is decreasing and

$$Y_t = 1 + \lambda\int_0^t Y_{s^-} dM_s + D_t.$$

We again conclude that (Y_t) is a positive supermartingale. When $c = 0$ we use similar arguments with $\phi_0(\lambda) = \lambda^2/2$.

8.2.3. Asymptotic Theorems

Theorem 8.2.17. Law of Large Numbers. *Let M be a local martingale adapted to $\mathbb{F}$ and A a continuous increasing $\mathbb{F}$-process such that $M^2 - A$ is a local martingale.*

(a) *On $\{A_\infty < \infty\}$, (M_t) converges* a.s. *to a finite* r.v. *M_∞ when $t \to \infty$.*

(b) *On $\{A_\infty = \infty\}$,*

$$\frac{M_t}{f(A_t)} \xrightarrow[t\to\infty]{\text{a.s.}} 0$$

for every function f from $\mathbb{R}_+$ into $\mathbb{R}_+$ increasing, tending to ∞ at infinity and such that

$$\int_0^\infty \frac{dt}{f^2(t)} < \infty.$$

In particular,

$$\frac{M_t}{A_t} \xrightarrow[t\to\infty]{\text{a.s.}} 0 \quad \text{on} \quad \{A_\infty = \infty\}.$$

Proof. Let $T_n = \inf\{t;\ A_t \geq n\}$. From Theorem 8.1.3, M^{T_n} is in $\mathcal{M}^2$ and converges a.s. for $t \to \infty$. Thus (M_t) converges a.s. for $t \to \infty$ on $\{T_n = \infty\}$. We conclude by noting

$$\{A_\infty = \infty\} = \bigcup_n \{T_n = \infty\}.$$

(b) Set $V_t = f(A_t)$ and $Z_t = \int_0^t (dM_s/V_s)$. By the integration by parts formula of [8.2.1], we have

$$V_t Z_t = M_t + \int_0^t Z_s dV_s$$

and

$$\frac{M_t}{V_t} = \frac{1}{V_t}\int_0^t (Z_t - Z_s)dV_s.$$

For $0 < u < t < \infty$,

$$\left|\frac{M_t}{V_t}\right| \leqslant \frac{1}{V_t}\left|\int_0^u (Z_t - Z_s)dV_s\right| + \frac{1}{V_t}(V_t - V_u)\sup_{u<s\leqslant t}|Z_t - Z_s|.$$

However

$$\left(Z_t^2 - \int_0^t \frac{dA_s}{V_s}\right)$$

is a martingale. Setting $\tau(t) = \inf\{s; A_s > t\}$, we have $A(\tau(t)) = t$ since A is continuous. Moreover, by the change of time formula 6.2.10,

$$\int_0^\infty \frac{dA_s}{f^2(A_s)} = \int_0^{A_\infty} \frac{ds}{f^2(s)} \leqslant \int_0^\infty \frac{ds}{f^2(s)}.$$

Thus (Z_t) has a finite limit a.s. for $t \to \infty$.

Letting t then u tend to ∞, we obtain

$$M_t/V_t \xrightarrow{\text{a.s.}} 0 \quad \text{on } \{A_\infty = \infty\}.$$

Theorem of Large Deviations 8.2.18. *We make the hypotheses of Theorem* 8.2.16(b) *and assume that, for a constant* K, $A_t \leqslant Kt$ *for any* t. *Then for all* $a > 0$,

$$P\left[\sup_{s\leqslant t} M_s \geqslant at\right] \leqslant \exp\left[-\frac{K}{c^2} g\left(\frac{ac}{K}\right)t\right]$$

with $g(\varepsilon) = (1+\varepsilon)\text{Log}(1+\varepsilon) - \varepsilon$. *If* M *is continuous, we obtain*

$$P\left[\sup_{s\leqslant t} M_s \geqslant at\right] \leqslant \exp\left[-\frac{a^2}{2K}t\right].$$

Proof. Let $\lambda > 0$; from Proposition 8.2.16,

$$P\left[\sup_{s\leqslant t} M_s \geqslant at\right] \leqslant P\left[\sup_{s\leqslant t}(\text{Log } Z_\lambda^c(s) + \phi_c(\lambda)Kt) \leqslant a\lambda t\right]$$

$$\leqslant P\left[\sup_{s\leqslant t} Z_\lambda^c(s) \geqslant \exp(a\lambda t - \phi_c(\lambda)Kt)\right]$$

$$\leqslant \exp\{-[a\lambda - \phi_c(\lambda)K]t\}.$$

For $c > 0$: $\sup\limits_{\lambda>0} (a\lambda - \phi_c(\lambda)K) = \dfrac{K}{c^2} g\left(\dfrac{ac}{K}\right)$

For $c = 0$: $\sup\limits_{\lambda>0} \left(a\lambda - \dfrac{\lambda^2}{2} K\right) = \dfrac{a^2}{2K}$.

From which the result follows.

Central Limit Theorem 8.2.19. *Let $(M^{(n)})$ be a sequence of martingales of $\mathbb{M}^2_{loc}$. The following hypotheses are made.*

(a) *The jumps of the martingales $M^{(n)}$ are majorized in modulus by a constant.*
(b) *The sequence $(M^{(n)})$ is contiguous to a sequence $(Y^{(n)})$ of continuous processes.*
(c) *For every t, the sequence $([M^{(n)},M^{(n)}]_t)$ is majorized by a constant $\alpha(t)$.*
(d) *There exists a continuous function $t \longmapsto a_t$ from $\mathbb{R}_+$ into $\mathbb{R}_+$ such that, for all t,*

$$[M^{(n)},M^{(n)}]_t \xrightarrow{\mathcal{D}} a_t.$$

Then the sequence $M^{(n)}$ converges strongly in distribution to a centered Gaussian process with independent increments the distribution of which at time t is $N(0,a_t)$. In other words,

$$M^{(n)} \xrightarrow{\mathcal{D}(C)} W_a.$$

Proof. From Proposition 7.3.19, the sequence $(Y^{(n)})$ is tight. From the fact that $(M^{(n)})$ is contiguous to $(Y^{(n)})$, if Q is a distribution on C, a closure point of $(Y^{(n)})$, the finite distribution functions of a subsequence of $(M^{(n)})$ tend to those of Q.

We have to prove that the canonical process with distribution Q, $(C,\mathcal{C},(X_t)_{t\geqslant 0},Q)$ is a centered Gaussian martingale associated with the increasing process $t \longmapsto a_t$, i.e. $Q = W_a$. From Theorem 8.2.15, this is equivalent to showing that, on $(C,\mathcal{C},Q)$, $(X_t)_{t\geqslant 0}$ and $(X_t^2 - a_t)_{t\geqslant 0}$ are martingales. In other words, we have to prove that, for $s < t$, $0 = u_0 < u_1 < \dots < u_m = s$ and h a bounded r.v. on $\mathbb{R}^{m+1}$,

$$\int h(X_{u_0}, X_{u_1}, \dots, X_{u_m})X_t dQ$$
$$= \int h(X_{u_0}, X_{u_1}, \dots, X_{u_m})X_s dQ$$
$$\int h(X_{u_0}, X_{u_1}, \dots, X_{u_m})[X_t^2 - a_t]dQ$$
$$= \int h(X_{u_0}, X_{u_1}, \dots, X_{u_m})[X_s^2 - a_s]dQ.$$

However, from note (a) which follows definition 8.1.12, $[M^{(n)},M^{(n)}]_t \leqslant \alpha(t)$ implies that $M^{(n)}$ and $(M^{(n)})^2 - [M^{(n)},M^{(n)}]$ are martingales for all n.

From which

$$\int h(M_{u_0}^{(n)}, M_{u_1}^{(n)}, \dots, M_{u_m}^{(n)})M_t^{(n)}dP_n$$
$$= \int h(M_{u_0}^{(n)}, M_{u_1}^{(n)}, \dots, M_{u_m}^{(n)})M_s^{(n)}dP_n;$$
$$\int h(M_{u_0}^{(n)}, M_{u_1}^{(n)}, \dots, M_{u_m}^{(n)})((M_t^{(n)})^2 - [M^{(n)},M^{(n)}]_t)dP_n$$
$$= \int h(M_{u_0}^{(n)}, M_{u_1}^{(n)}, \dots, M_{u_m}^{(n)})((M_s^{(n)})^2 - [M^{(n)},M^{(n)}]_s)dP_n.$$

If a subsequence of $(M^{(n)})$ tends to Q, we can pass to the limit along this subsequence on condition that the corresponding subsequences of $(M_t^{(n)})$ and $(M_s^{(n)})$ are equi-integrable (property 5 of [0.3.1]). This is the case however because of the exponential supermartingale of Proposition 8.2.16 (applied to the martingales $M^{(n)}$ and $-M^{(n)}$), for $\lambda > 0$:

$$\exp(\lambda|M_t^{(n)}|) \leqslant \exp(\lambda M_t^{(n)}) + \exp(-\lambda M_t^{(n)});$$

$$E[\exp(\lambda|M_t^{(n)}|)] \leqslant 2\exp[e^{\lambda c}\phi_c(\lambda)\alpha(t)] = \beta.$$

Let $p \geqslant 1$ and $k \in \mathbb{N}$:

$$E[|M_t^{(n)}|^p 1_{(k-1 \leqslant |M_t^{(n)}| < k)}] \leqslant k^p P[k-1 \leqslant |M_t^{(n)}| < k]$$
$$\leqslant k^p \beta e^{-\lambda(k-1)}.$$

From which

$$E[|M_t^{(n)}|^p 1_{(|M_t^{(n)}| \geqslant a)}] \leqslant \beta \sum_{k \geqslant a} k^p e^{-\lambda(k-1)},$$

and the majorant tends to 0 if $a \to \infty$. Hence the only closure point Q of $Y^{(n)}$ is W_a, and

$$(Y^{(n)}) \xrightarrow[n\to\infty]{\mathcal{D}} W_a.$$

Corollary 8.2.20. *In the statement of Theorem* 8.2.19 *we can replace* $[M^{(n)},M^{(n)}]$ *by a continuous increasing process* $A^{(n)}$ *such that* $(M^{(n)})^2 - A^{(n)}$ *is a martingale.*

Proof. The same proof is followed as for the theorem with the majorization

$$E[\exp(\lambda|M_t^{(n)}|)] \leq 2\exp[\phi_c(\lambda)\alpha(t)].$$

Note. This theorem can be improved by contiguity arguments. This has been done for example in [7.4.1], by replacing the sequence $(\xi_{n,k})$ by $(\xi^{\varepsilon}_{n,k})$.

Corollary 8.2.21. Multidimensional Central Limit Theorem. *Consider, for* $1 \leq i \leq d$, *a sequence* $(M^{(n,i)})$ *of martingales of* $\mathcal{M}^2_{loc}$ *and a continuous function* a^i: $t \longmapsto a^i_t$ *satisfying the hypotheses of Theorem* 8.2.19, *with*

$$[M^{(n,i)}, M^{(n,i)}]_t \xrightarrow[n\to\infty]{\mathcal{D}} a^i_t.$$

Assume that, for $1 \leq i < j \leq d$:

$$[M^{(n,i)} + M^{(n,j)}, M^{(n,i)} + M^{(n,j)}]_t \xrightarrow[n\to\infty]{\mathcal{D}} a^i_t + a^j_t\,.$$

Then

$$(M^{(n,i)})_{1\leq i\leq d} = M^{(n)} \xrightarrow[n\to\infty]{\mathcal{D}(C)} W_{a_1} \otimes \cdots \otimes W_{a_d}\,.$$

Proof. Let $(u_1, \ldots, u_d) = u$ be in $\mathbb{R}^d$. Set

$$M^{(n,u)}_t = \sum_{i=1}^{d} u_i M^{(n,i)}_t.$$

Then $(M^{(n,u)}_t)_{t\geq 0} = M^{(n,u)}$ is in $\mathcal{M}^2$. For two martingales M and N of $\mathcal{M}^2$, denote

$$[M,N] = \frac{1}{2}([M+N, M+N] - [M,M] - [N,N]).$$

We have, for $s < t$,

$$(M^{(n,u)}_t - M^{(n,u)}_s)^2 = \sum_{i=1}^{d}\sum_{j=1}^{d} u_i u_j (M^{(n,i)}_t - M^{(n,i)}_s)(M^{(n,j)}_t - M^{(n,j)}_s)$$

From which, by using the approximation by quadratic variations of Theorem 8.1.9,

$$[M^{(n,u)},M^{(n,u)}]_t = \sum_{i=1}^{d}\sum_{j=1}^{d} u_i u_j [M^{(n,i)},M^{(n,j)}]_t.$$

With the hypothesis made in the corollary,

$$[M^{(n,u)},M^{(n,u)}]_t \text{ tends to } \sum_{i=1}^{d} u_i^2 a_t^i, \quad \text{if } n \to \infty.$$

For each i, the sequence $(M^{(n,i)})$ is contiguous to a sequence $(Y^{(n,i)})$ of continuous processes. Let us set $Y^{(n)} = (Y^{(n,i)})_{1 \leqslant i \leqslant d}$; $(M^{(n)})$ is contiguous to $(Y^{(n)})$ which is tight. If Q is a closure point of $(Y^{(n)})$ and if $(C,\mathcal{C},Q,(X_t)_{t \geqslant 0})$ is a canonical process with distribution Q, then $(<u,X_t>)$ and

$$\left[<u,X_t>^2 - \sum_{i=1}^{d} u_i^2 a_t^i\right]$$

are martingales. Hence $(X_t - X_s)$ is independent of $\sigma(X_v; v \leqslant s)$ and its distribution is

$$\bigotimes_{i=1}^{d} N(0, a_t^i - a_s^i).$$

The finite distribution functions of Q arc those of $W_{a_1} \otimes \ldots \otimes W_{a_d}$. From which the result follows.

8.3. Asymptotic Study of Point Processes

8.3.1. General Layout

Consider, for $1 \leqslant i \leqslant d$, point processes $N^i = (N_t^i)_{t \geqslant 0}$ which do not have jumps in common ([6.2]). From Theorem 6.2.7, we know how to construct an increasing process $\tilde{N}^i = (\tilde{N}_t^i)_{t \geqslant 0}$, for $1 \leqslant i \leqslant d$, such that $\tilde{N}_0^i$ is zero and satisfying the following condition: for each i, $(N^i - \tilde{N}^i)$ is a martingale with respect to the filtration $\sigma(N_u^j; u \leqslant t, 1 \leqslant j \leqslant d) = \mathbb{G}$.

In the principal examples, the compensators $\tilde{N}^i$ are continuous processes, and we make this hypothesis in what follows. $\mathbb{G}$ is replaced by $\mathbb{F} = \mathbb{G}^+$ continuous on the right. From Theorem 6.2.17, if N_t^i is square integrable for all t, then the processes $(N^i - \tilde{N}^i) - \tilde{N}^i$ and $(N^i - \tilde{N}^i)(N^j - \tilde{N}^j)$, for $i \neq j$, are $\mathbb{F}$-martingales. However we have,

$$N_t^i = \sum_{s \leq t} [\Delta(N_s^i - \widehat{N}_s^i)]^2.$$

Thus $[N^i - \widehat{N}^i, N^i - \widehat{N}^i] - N^i$ is an increasing process, zero for $t = 0$, and at the same time a martingale. This process is zero. We thus have $N^i = [N^i - \widetilde{N}^i, N^i - \widetilde{N}^i]$ and

$$[N^i + N^j - \widehat{N}^i - \widehat{N}^j, N^i + N^j - \widehat{N}^i - \widetilde{N}^j] = 0,$$

for $i \neq j$.

Theorem 8.3.22. *Let $(N^i)_{1 \leq i \leq d}$ be a point processes without common jumps adapted to a right continuous filtration $\mathbb{F}$. Assume N_t^i is square integrable for all t and for all i.*

(a) *Let $J = (J_t)_{t \geq 0}$ be a predictable process bounded by $\|J\|$. Assume that there exists a constant $L^i(J) > 0$ such that*

$$\frac{1}{t} \int_0^t J_s d\widehat{N}_s^i \xrightarrow[t \to \infty]{\text{a.s.}} L^i(J).$$

Then

$$\frac{1}{t} \int_0^t J_s dN_s^i \xrightarrow[t \to \infty]{\text{a.s.}} L^i(J).$$

Moreover, if $\widehat{N}_t^i \leq Kt$, for all t and for all $a > 0$ we have

$$P\left[\left| \int_0^t J_s dN_s^i - \int_0^t J_s d\widehat{N}_s^i \right| \geq at\right]$$

$$\leq 2 \exp\left[-tKg\left(\frac{a}{K\|J\|}\right)\right],$$

with the function $g\colon x \longmapsto (1 + x)\mathrm{Log}(1 + x) - x$.

(b) *Assume that, for all t and every i, $\widetilde{N}_t^i \leq Kt$, K constant > 0. For each i let $(J_t^i)_{t \geq 0}$ be a bounded predictable process. Assume*

$$\frac{1}{t} \int_0^t (J_s^i)^2 d\widehat{N}_s^i \xrightarrow[n \to \infty]{\text{a.s.}} H^i,$$

H^i constant > 0. Let

$$M_t^{(n,i)} = \frac{1}{\sqrt{nH^i}} \left(\int_0^{nt} J_s^i dN_s^i - \int_0^{nt} J_s^i d\widehat{N}_s^i \right)$$

and let $M^{(n,i)} = (M_t^{(n,i)})$. Then

$$(M^{(n,i)}) \xrightarrow{\mathcal{D}(C)} W^{\otimes d}.$$

Proof. Part (a) is a direct consequence of Theorems 8.2.17 and 8.2.18.

Let us show part (b). Let (T^i_k) be the sequence of $\mathbb{F}$-stopping times such that

$$N^i_t = \Sigma 1_{(T^i_k \leqslant t)} .$$

Consider the process $Y^{(n,i)}$ defined by setting, for $T^i_k \leqslant t < T^i_{k+1}$,

$$Y^{(n,i)}_t = M^{(n,i)}_t + \frac{1}{\sqrt{nH^i}} \frac{t - T^i_k}{T^i_{k+1} - T^i_k} J^i_{T_{k+1}} .$$

The processes $Y^{(n,i)}$ are continuous and the sequence $(Y^{(n,i)})$ is contiguous to $(M^{(n,i)})$. Let us set $A^{(n,i)} = (A^{(n,i)}_t)$ with

$$A^{(n,i)}_t = \frac{1}{n} \int_0^{nt} (J^i_s)^2 d\widetilde{N}^i_s .$$

Then $(M^{(n,i)})^2 - A^{(n,i)}$ is a martingale, and part (b) of the theorem follows easily from Corollaries 8.2.20 and 8.2.21.

8.3.2. Poisson Processes and The M/M/1 Queue

The following proposition is a corollary of Theorem 8.3.22 (see Vol. 1, E4.3.6 and E4.4.12 for some analogous results).

Proposition 8.3.23. *Let* $N = (\Omega, A, P, (N_t)_{t \geqslant 0})$ *be a stationary Poisson process with intensity* $(\lambda t)_{t \geqslant 0}$.

(a) $\dfrac{Nt}{t} \xrightarrow[t\to\infty]{\text{a.s.}} \lambda$; *and, for all* $\varepsilon > 0$,

$$P\left[\left|\frac{Nt}{t} - \lambda\right| \geqslant \varepsilon\right] \leqslant 2 \exp(-tg(\varepsilon)).$$

(b) $[N_{nt} - \lambda nt]/\sqrt{\lambda n} \xrightarrow[n\to\infty]{\mathcal{D}(C)} W.$

Now let us consider an M/M/1 queue associated with exponential interarrival and service distributions with respective parameters λ and μ. We have defined in [6.2.3] N^+_t the number of arrivals before t and N^-_t the number of departures. The compensators are $N^+_t = \lambda t$ and

$$\tilde{N}_t^- = \mu \int_0^t 1_{(X_s \neq 0)} ds.$$

If $t \to \infty$,

$$\frac{1}{t} \int_0^t 1_{(X_s \neq 0)} ds$$

tends a.s., for any initial distribution, to 1 for $\lambda \geqslant \mu$, and to $(1 - (\lambda/\mu))$ for $\lambda < \mu$. Let us assume this result which will be proved in [8.3.3]

$$\frac{N_t^+}{t} \xrightarrow[t\to\infty]{\text{a.s.}} \lambda \quad \text{and} \quad \frac{N_t^-}{t} \xrightarrow[t\to\infty]{\text{a.s.}} \begin{cases} \mu & \text{if} \quad \lambda \geqslant \mu \\ \lambda & \text{if} \quad \lambda < \mu. \end{cases}$$

For $\lambda > \mu$, $\int_0^\infty 1_{(X_s=0)} ds$ is a.s. finite and

$$\frac{\tilde{N}_{tn}^- - \mu tn}{\sqrt{\mu n}} \xrightarrow{\text{a.s.}} 0.$$

Thus

$$\left(\frac{N_{tn}^+ - \lambda tn}{\sqrt{\lambda n}}, \frac{N_{tn}^- - \mu tn}{\sqrt{\mu n}}\right) \xrightarrow[n\to\infty]{\mathcal{D}(C)} W \otimes W.$$

At time t, the length of the queue is $X_t = X_0 + N_t^+ - N_t^-$. From which the next proposition follows.

Proposition 8.3.24. *Let there be an* M/M/1 *queue of which the interarrival distributions are* $\mathcal{E}(\lambda)$ *and the length of service distributions are* $\mathcal{E}(\mu)$. *Let us call* X_t *the length of the queue at time* t.

(a) $\dfrac{X_t}{t} \xrightarrow[t\to\infty]{\text{a.s.}} (\lambda - \mu)_+.$

(b) For $\lambda > \mu$,

$$\frac{1}{\sqrt{n(\lambda + \mu)}} [X_{tn} - (\lambda - \mu)tn]_{n\geqslant 0} \xrightarrow[n\to\infty]{\mathcal{D}(C)} W.$$

8.3.3. Markovian Jump Processes

Let E be a countable space, π a transition from E into E and q a positive r.v. We use the notations of [6.2.1] and [6.2.3] and study the jump processes $\{\Omega, A, (P_x)_{x\in E}, (X_t)_{0\leqslant t<\infty}\}$ associated with q and π. Let (T_n) be the sequence of jump times. Let $\Gamma \subset E$, and let $N_t^{i,\Gamma}$ be the number of jumps going from i to Γ and τ_t^i the time passed in i up till time t. Here, $\tilde{N}_t^{i,\Gamma} = q(i)\pi(i,\Gamma)\tau_t^i$. Let us assume that E is a recurrent class of the Markov chain with transition π. Let μ be a measure invariant under π.

For every initial distribution ν, consider the sequence $(S_p)_{p\geq 0}$ of passage times to i. Then we see, by similar arguments to those of [4.2], that, for $p \geq 1$, the sequences $(T_{n+S_p} - T_{S_p}, Z_{n+S_p})_{n\geq 0}$ are independent of F_{S_p} on (Ω,A,P_ν) and are distributed as $(T_n,Z_n)_{n\geq 0}$ on (Ω,A,P_i). Thus $(T_{S_{p+1}} - T_{S_p})_{p\geq 1}$ is a sequence of independent identically distributed r.v.'s,

$$\frac{T_{S_p}}{p} \xrightarrow[p\to\infty]{P_\nu\text{-a.s.}} E_i(T_{S_1}) = E_i\left[\sum_{p=1}^{S_1}(T_p - T_{p-1})\right].$$

However,

$$E_i\left[\sum_{p=1}^{S_1}(T_p - T_{p-1})\right] = E_i\left[\sum_{p=1}^{\infty}(T_p - T_{p-1})1_{(T_{p-1}<T_{S_1})}\right]$$

$$= \sum_{p=1}^{\infty} E_i[E_i(T_p - T_{p-1}| F_{T_{p-1}})1_{(T_{p-1}<T_{S_1})}]$$

$$= \sum_{p=1}^{\infty} E_i\left[\frac{1}{q(X_{T_{p-1}})}1_{(p-1<S_1)}\right] = \frac{\mu(1/q)}{\mu(i)}.$$

Moreover, the r.v.'s, $(T_{S_{p+1}} - T_{S_p})_{p\geq 1}$ are independent and identically distributed,

$$\frac{1}{p}\tau^i_{T_{S_p}} = \frac{1}{p}\sum_{0\leq q<p}(T_{S_{q+1}} - T_{S_q}) \xrightarrow[p\to\infty]{P_\nu\text{-a.s.}} \frac{1}{q(i)}.$$

Thus

$$\frac{\tau^i_{T_{S_p}}}{T_{S_p}} \xrightarrow[p\to\infty]{P_\nu\text{-a.s.}} \frac{(1/q(i))\mu(i)}{\mu(1/q)} > 0.$$

However, for all t, we can find a p such that $T_{S_p} \leq t < T_{S_{p+1}}$. From which it follows,

$$\frac{\tau^i_{T_{S_p}}}{T_{S_{p+1}}} \leqslant \frac{\tau^i_t}{t} \leqslant \frac{\tau^i_{T_{S_{p+1}}}}{T_{S_p}}.$$

If $\mu(1/q)$ is infinite, the left hand term tends, if $t \to \infty$ (thus $p \to \infty$), to 0. If $\mu(1/q)$ is finite, set

$$\hat{\mu} = \frac{1}{q} \cdot \frac{\mu}{\mu(1/q)}.$$

We therefore define a probability on E such that

$$\frac{\tau^i_t}{t} \xrightarrow[t\to\infty]{P_\nu\text{-a.s.}} \hat{\mu}(i).$$

Moreover, $q(i)\pi(i,\Gamma)\tau^t_i$ is always majorized by $q(i)t$. Theorem 8.3.22 applies.

Theorem 8.3.25. *We are given a Markovian jump process on a countable space E associated with a transition π and a strictly positive function q. Assume the Markov chain associated with π is recurrent with invariant measure μ and that the function $1/q$ is integrable under μ. Let $N^{i,\Gamma}_t$ be the number of jumps from i into Γ before t, τ^i_t the sojourn time at t. For every initial distribution ν we have,*

(a)
$$\frac{\tau^i_t}{t} \xrightarrow[t\to\infty]{P_\nu\text{-a.s.}} \hat{\mu}(i) = \frac{1}{q(i)} \frac{\mu(i)}{\mu(1/q)}$$

$$\frac{N^{i,\Gamma}_t}{t} \xrightarrow[t\to\infty]{P_\nu\text{-a.s.}} q(i)\pi(i,\Gamma)\hat{\mu}(i).$$

and these convergences take place at an exponential rate.

(b) *If $\pi(i,\Gamma) > 0$, set*

$$M^{(n,i,\Gamma)}_t = \frac{N^{i,\Gamma}_{nt} - q(i)\pi(i,\Gamma)\tau^i_{nt}}{\sqrt{nq(i)\pi(i,\Gamma)\hat{\mu}(i)}}$$

$M^{(n,i,\Gamma)} = (M^{(n,i,\Gamma)}_t)_{t\geqslant 0}$ converges in distribution to Wiener measure W. For $\pi(i,\Gamma_1)\pi(j,\Gamma_2) > 0$ and $i \neq j$ or $\Gamma_1 \cap \Gamma_2 = \emptyset$, the pair

$$(M^{(n,i,\Gamma_1)}, M^{(n,j,\Gamma_2)})$$

converges in distribution to the measure $W \otimes W$.

Example. M/M/1 Queue. For the M/M/1 queue studied in

[8.3.2], we have

$$\pi(0,1) = 1, \quad q(0) = \lambda$$

and, for $i \neq 0$, $\pi(i,i+1) = \lambda/(\lambda + \mu) = 1 - \pi(i,i-1)$; $q(i) = \lambda + \mu$. the recurrence at 0 can be studied directly or deduced from the behavior of the G/G/1 queue. An infinity of customers will not have to wait if, and only if, the length of the queue vanishes infinitely often. From [4.1.3], 0 is recurrent for the G/G/1 queue if, and only if, $\lambda \leqslant \mu$; 0 communicates with every point. For $\lambda > \mu$ every point is transient and, for any initial distribution ν, (X_t) returns, a.s., only a finite number of times in each finite set:

$$X_t \xrightarrow{P_\nu\text{-a.s.}} \infty.$$

In the case $\lambda \leqslant \mu$, $\mathbb{N}$ is a recurrent class of π. An invariant measure ξ under π satisfies

$$\xi(0) = \frac{\mu}{\lambda + \mu}\xi(1); \quad \xi(1) = \frac{\mu}{\lambda + \mu}\xi(2) + \xi(0);$$

$$\xi(i) = \frac{\mu}{\lambda + \mu}\xi(i + 1) + \frac{\lambda}{\lambda + \mu}\xi(i - 1) \quad \text{for } i \geqslant 2.$$

Let us set $\hat{\xi}(i) = (1/q(i))\xi(i)$. We obtain,

$$\hat{\xi}(i) = \left[\frac{\lambda}{\mu}\right]^i \hat{\xi}(0).$$

The measure $\hat{\xi}$ is bounded if and only if $\lambda < \mu$. For every i and every initial distribution, if t tends to ∞, (τ_t^i/t) tends a.s. to 0 for $\lambda \geqslant \mu$ and to $(\lambda/\mu)^i(1 - (\lambda/\mu))$ for $\lambda < \mu$. This is the result that we have used in [8.3.1].

With the help of Proposition 6.4.21, it is not difficult to deal with some statistical problems of families of jump processes. For example, having accepted that a queue is of type M/M/1, we can look for the maximum likelihood estimator of (λ,μ), a parameter in $(]0,\infty[)^2$. Theorem 8.3.24 allows the asymptotic behavior of these estimators to be studied.

8.3.4. Censored Statistics of Point Processes

Let $(\Omega,A,(P_\theta)_{\theta\in\Theta})$ be a statistical model. Assume given on (Ω,A), d point processes $(N^i)_{1\leqslant i\leqslant d}$ without jumps in common. With the notation of [8.3.1], we can define for each (i,θ) a compensator $\tilde{N}^{i,\theta}$ of N^i on (Ω,A,P_θ).

Examples. The compensators which we have studied above depend on the parameters λ of the Poisson process, or (λ,μ) of the M/M/1 queue, or (q,π) of the jump process.

Let us assume that, on (Ω,A,P_θ), we have a law of large numbers for $1 \leqslant i \leqslant d$,

$$\frac{1}{t}\tilde{N}_t^{i,\theta} \xrightarrow{P_\theta\text{-a.s.}} L^i, \quad L^i \text{ constant} > 0.$$

Then Theorem 8.3.22 implies that the point process N^i is a good approximation of $\tilde{N}^{i,\theta}$. We are led to predict or "filter" $\tilde{N}^{i,\theta}$ by N^i.

It is natural, therefore, to replace the parameter θ by the compensators $\tilde{N}^{i,\theta}$ which we conveniently know how to filter. We can for example construct goodness of fit tests of "$\theta = \theta_0$" against "$\theta \neq \theta_0$" by using rejection regions of the form

$$\left\{\sum_{i=1}^{d} \alpha_i|N_t^i - \tilde{N}_t^{i,\theta_0}| \geqslant C\right\}, \quad \alpha_1, \dots, \alpha_d, C \text{ constants} \geqslant 0.$$

Aalen's idea which we have already mentioned in [6.4.1], has the following advantage. It allows us to deal with statistical problems with censoring. A statistical problem relative to a process is said to be **censored** if the complete observation of the process is not possible; either because the statistician must sleep sometimes (a cause independent of the observed process); or because the process is observable only when it takes its values in a given region. It is often very difficult to study such a situation. This question is also called the **missing data problem.**

In the above framework, it is often possible to describe the censoring by predictable processes $(J^i)_{1\leqslant i\leqslant d}$ taking values 0 or 1, 0 if there is censoring, 1 if there is none. Then the censored observation is that of the point processes

$$N_J^i = \left(\int_0^t J_s^i dN_s^i\right)$$

which, on (Ω,A,P_θ), are compensated by

$$\tilde{N}_J^{i,\theta} = \left[\int_0^t J_s^i d\tilde{N}_s^{i,\theta}\right].$$

What can be said about N^i and $\tilde{N}^{i,\theta}$ remains true for the censored processes N_J^i and $\tilde{N}_J^{i,\theta}$.

8.4. Brownian Motion

8.4.1. Characterizations

From Theorem 8.2.15 we have various equivalent definitions of Brownian motion $(\Omega, \mathcal{A}, (B_t)_{t\geq 0}, P)$.

Definition 8.4.25. *A process $B = (\Omega, \mathcal{A}, (B_t)_{t\geq 0}, P)$ is a Brownian motion adapted to the filtration $\mathbb{F}$ if it is a process adapted to $\mathbb{F}$ with continuous trajectories, zero at $t = 0$ and satisfying one of the following equivalent conditions:*

(a) *It is a homogeneous $\mathbb{F}$-process with independent increments, associated with the convolution semigroup $(N(0,t))_{t\geq 0}$.*
(b) *The processes (B_t) and $(B_t^2 - t)$ are centered $\mathbb{F}$-martingales.*
(c) *For any $\lambda \in \mathbb{R}$,*

$$\left(\exp\left[\lambda B_t - \left(\frac{\lambda^2}{2}\right)t\right]\right)$$

is an $\mathbb{F}$-martingale.

Throughout paragraph [8.4], $\{\Omega, \mathcal{A}, P, (B_t)_{t\geq 0}\}$ is a Brownian motion adapted to a right continuous filtration $\mathbb{F}$.

8.4.2. Maxima of Brownian Motion

Proposition 8.4.26. *Let $a > 0$, $T_a = \inf\{s; B_s = a\}$ and $S_a = \inf\{s; B_s + s = a\}$. For $\lambda > 0$, we have: $E[\exp(-\lambda/T_a)] = \exp(-a\sqrt{2\lambda})$.*

The r.v. T_a *has density* $t \longmapsto \dfrac{1}{\sqrt{2\pi t^3}} a \exp[-(a^2/2t)]$.

For $\lambda \geq -1/2$, we have:

$$E[\exp(-\lambda S_a)] = \exp[a(1 - \sqrt{1 + 2\lambda}\,)].$$

Proof. We apply the stopping theorem to the exponential martingale $\{\exp(\mu B_t - (\mu^2/2)t)\}$. For every $t \geqslant 0$, the following two r.v.'s have expectation equal to 1,

$$U_t = \exp\left[\mu B_{t\wedge T_a} - \left(\frac{\mu^2}{2}\right)(t \wedge T_a)\right],$$

$$V_t = \exp\left[\mu(B_{t\wedge S_a} + t \wedge S_a) - \left(\mu + \frac{\mu^2}{2}\right)(t \wedge S_a)\right].$$

For $\mu > 0$, we thus have

$$E\left\{\exp\left[\left(-\frac{\mu^2}{2}\right)t \wedge T_a\right]\right\} \geqslant \exp(-a\mu).$$

By Lebesgue's theorem, we obtain

$$E(\exp[(-\mu^2/2)T_a]) \geqslant \exp(-a\mu);$$

by letting μ tend to 0, we have: $P(T_a < \infty) = 1$. Since $S_a \leqslant T_a$, S_a is also a.s. finite. For $\mu \geqslant 0$, U_t and V_t are majorized by $\exp(\mu a)$. We apply Lebesgue's theorem,

$$(\exp \mu a)E[\exp(-\mu^2/2)T_a] = 1,$$

$$(\exp \mu a)E\left[\exp\left(-\mu - \frac{\mu^2}{2}\right) S_a\right] = 1.$$

From which the Laplace transform of T_a follows, by setting $\lambda = \mu^2/2$, and its density by identifying the Laplace transforms.

By setting $\lambda = \mu + \mu^2/2$, we obtain

$$E[\exp(-\lambda S_a)] = \exp[a(1 - \sqrt{1 + 2\lambda})].$$

The function $x \longmapsto E(\exp xS_a)$ is analytic in the interior of the interval where it is finite. The above equality thus remains true where the function which appears in the second part is analytic, thus for $\lambda > -1/2$. By continuity it is true for $\lambda = -1/2$.

Note. The stopping times $\inf\{s; B_s = -a\}$ and $\inf\{s; B_s = -a+s\}$ have the same distribution as T_a and S_a respectively. This is seen by considering the Brownian motion $(-B_t)_{t\geqslant 0}$.

From Proposition 7.2.13, if T is a finite $\mathbb{F}$-stopping time, $(B_{t+T} - B_T)_{t\geqslant 0}$ is a Brownian motion adapted to $(\mathcal{F}_{t+T})_{t\geqslant 0}$. We deduce from this the following **reflection principle**: let $a \in \mathbb{R}$

and let $T_a = \inf\{s;\ B_s = a\}$. The process B has the same distribution as $B^1 = (B^1_t)$ with

$$B^1_t = \begin{cases} B_t & \text{for } t \leqslant T_a \\ 2a - B_t & \text{for } t \geqslant T_a. \end{cases}$$

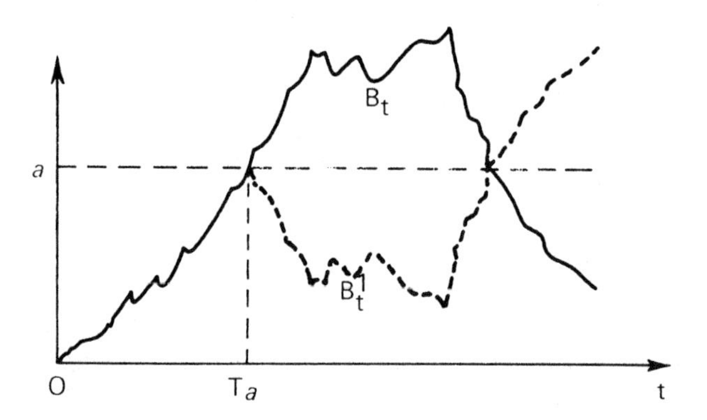

This follows from: $B_{t+T_a} - B_{T_a} = B_{t+T_a} - a$ is a Brownian motion independent of $\mathcal{F}_{T_a}$.

Proposition 8.4.27. *Let*

$$\overline{B}_t = \sup_{s \leqslant t} B_t,\quad \underline{B}_t = \inf_{s \leqslant t} B_s,\quad B^*_t = \sup_{s \leqslant t} |B_s|.$$

Let $a < c < d < b$:

$$P(\overline{B}_t \geqslant b,\ B_t \in [c,d]) = \frac{1}{\sqrt{2\pi t}} \int_{2b-d}^{2b-c} \exp\left[-\frac{x^2}{2t}\right] dx,$$

$$P(\underline{B}_t \leqslant a,\ B_t \in [c,d]) = \frac{1}{\sqrt{2\pi t}} \int_{-2a+c}^{-2a+d} \exp\left[-\frac{x^2}{2t}\right] dx,$$

$$P(\overline{B}_t \geqslant b,\ \underline{B}_t \leqslant a,\ B_t \in [c,d])$$
$$= \frac{1}{\sqrt{2\pi t}} \sum_{k=-\infty}^{\infty} \int_c^d \left[\exp\left[-\frac{1}{2t}(x + 2k(b-a))^2\right]\right.$$
$$\left. + \exp\left[-\frac{1}{2t}(x - 2b + 2k(b-a))^2\right]\right] dx.$$

For $-a < c < d < a$,

$$P(B_t^* \leqslant a, c < B_t < d) = \frac{1}{\sqrt{2\pi t}} \int_c^d \sum_{k=-\infty}^{\infty} (-1)^k \exp\left[-\frac{(u-2ka)^2}{2t}\right] du.$$

Proof. Let B^1 be the process reflected at T_b:

$$P(\bar{B}_t \geqslant b, B_t \in [c,d]) = P[T_b < t, B_t^1 \in [c,d])$$

$$= P[T_b < t, B_t \in [2b-d, 2b-c]].$$

However $d < b$, and $B_t \in [2b-d, 2b-c]$ implies $T_b < t$; from which it follows,

$$P(\bar{B}_t \geqslant b, B_t \in [c,d]) = P(B_t \in [2b-d, 2b-c]).$$

Since (B_t) and $(-B_t)$ have the same distribution:

$$P(\underline{B}_t \leqslant a, B_t \in [c,d]) = P(\bar{B}_t \geqslant -a, B_t \in [-d,-c]).$$

From which the first two relations follow (which once again give the distribution of $\bar{B}_t$ found in (b)).

Finally, for the third relation, we can iterate the reflection principle as shown in the diagram below:

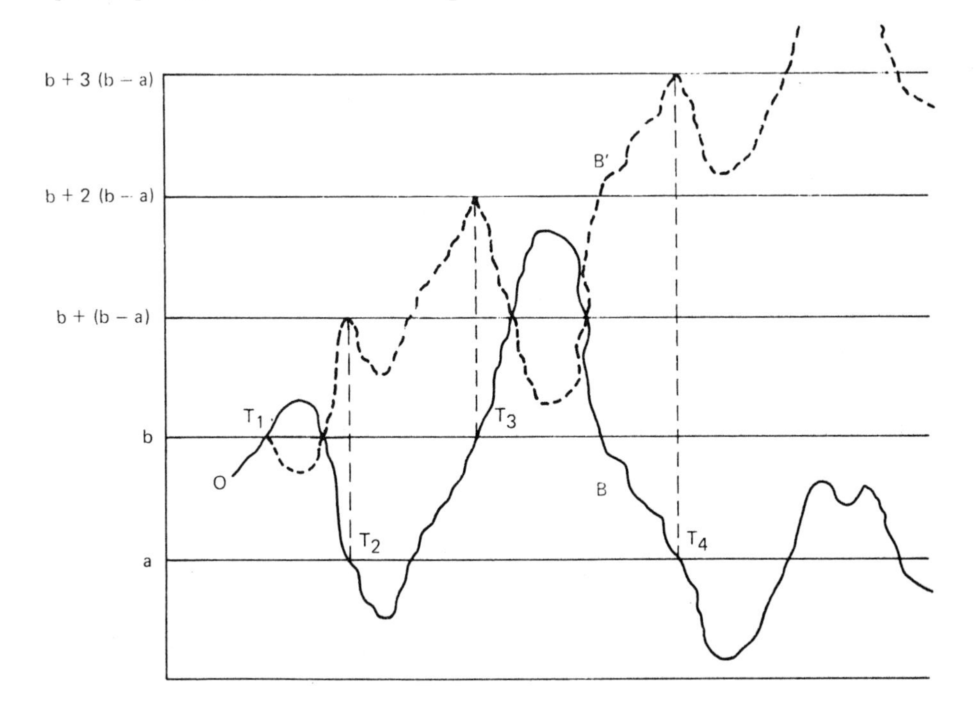

Let $T_1 = \inf\{s;\ B_s \notin [a,b]\}$;

$$\text{on } B_{T_1} = b, \text{ set} \quad T_2 = \inf\{s;\ s > T_1, B_s < a\}$$

$$\cdots\cdots$$

$$T_{2p-1} = \inf\{s;\ s > T_{2p-2}, B_s > b\}$$

$$T_{2p} = \inf\{s;\ s > T_{2p-1}, B_s < a\};$$

$$\text{on } B_{T_1} = a, \text{ set} \quad T_2 = \inf\{s;\ s > T_1, B_s > b\}$$

$$\cdots\cdots$$

$$T_{2p-1} = \inf\{s;\ s > T_{2p-2}, B_s < a\}$$

$$T_{2p} = \inf\{s;\ s > T_{2p}, B_s > b\}.$$

At each of these times, we use the reflection principle and define a Brownian motion (B_t') with

$$B_t' = \begin{cases} B_t & \text{for } t \leqslant T_1 \\ 2B_{T_1} - B_t & \text{for } T_1 \leqslant t \leqslant T_2 \\ \cdots\cdot \end{cases}$$

We have

$$P(a < \underline{B}_t < \overline{B}_t < b,\ B_t \in [c,d])$$
$$= P(B_t \in [c,d]) - \sum_{n=2}^{\infty} P(T_n \leqslant t < T_{n+1},\ c < B_t < d).$$

However on $\{B_{T_1} = b\}$, we write

$$B_t' = \begin{cases} 2b + 2(p-1)(b-a) - B_t & \text{on } T_{2p-1} \leqslant t \leqslant T_{2p} \\ 2p(b-a) + B_t & \text{on } T_{2p} \leqslant t \leqslant T_{2p+1}; \end{cases}$$

$$\{B_{T_1} = b\} \cap \{T_{2p-1} \leqslant t \leqslant T_{2p}\} \cap \{c < B_t < d\}$$
$$= \{2b + 2(p-1)(b-a) - d < B_t'$$
$$< 2b + 2(p-1)(b-a) - c\};$$

$$\{B_{T_1} = b\} \cap \{T_{2p} \leqslant t < T_{2p+1}\} \cap \{c < B_t < d\}$$
$$= \{2p(b-a) + c < B_t^1 < 2p(b-a) + d\}.$$

Similarly, we express

$$\{B_{T_1} = a\} \cap \{T_n \leqslant t \leqslant T_{n+1}\} \cap \{c < B_t < d\}$$

in the form $\{\alpha_n < B_t' < \beta_n\}$ for $\alpha_n < \beta_n < a$. By summing these probabilities we find the stated result and its consequence relative to B^*.

Consequence. (a) The following functions are continuous on the set of continuous functions from $\mathbb{R}_+$ into $\mathbb{R}$ ([7.3.1]):

$$(x_t) \longmapsto \left(\sup_{s \leqslant t}(x_s), \inf_{s \leqslant t}(x_s), x_t\right) \text{ from } C \text{ into } \mathbb{R}^3;$$
$$(x_t) \longmapsto \left(\sup_{s \leqslant t}(|x_s|), x_t\right) \text{ from } C \text{ into } \mathbb{R}^2.$$

Hence if $(X^{(n)})$ is a sequence of cad-lag processes which converge strongly in distribution to B, Proposition 8.4.27 gives the limit distributions of

$$\left(\sup_{s \leqslant t}(X_s^{(n)}), \inf_{s \leqslant t}(X_s^{(n)}), X_t^{(n)}\right)$$

and of

$$\left(\sup_{s \leqslant t}|X_s^{(n)}|, X_t^{(n)}\right).$$

This applies to triangular sequences (Theorem 7.4.28) or to point processes (Theorem 8.3.22).

(b) We are now in a position to state what are the Kolmogorov distributions obtained in Theorem 7.4.32. This will result from the following lemma.

Lemma 8.4.28. *The distribution of* $B = (B_t)_{0 \leqslant t \leqslant 1}$ *conditional on* $(0 \leqslant B_1 \leqslant h)$ *on the set* $C_{[0,1]}$ *of continuous functions from* $[0,1]$ *into* $\mathbb{R}$ *converges narrowly to that of a Brownian bridge, if* $h \to 0$.

Proof. The distribution of the Brownian bridge is that of $X = (B_t - tB_1)_{0 \leqslant t \leqslant 1}$. This Gaussian process X is independent of B_1 since each of its components is orthogonal to B_1. Let us denote by W^0 the Brownian bridge distribution. For all $h > 0$, the distribution of $(B_t - tB_1)$ conditional on $0 \leqslant B_1 \leqslant h$ is thus W^0.

However,

$$\sup_{0\leqslant t\leqslant 1} |(B_t - tB_1) - B_t| = |B_1|.$$

Thus, if f is an r.v. uniformly continuous on $C_{[0,1]}$ for uniform convergence, we have

$$\lim_{h\to 0} |E[f(X)|0 \leqslant B_1 \leqslant h] - E[f(B)|0 \leqslant B_1 \leqslant h]| = 0.$$

From which it follows that,

$$\lim_{h\to 0} E[f(B)|0 \leqslant B_1 \leqslant h] = \int f\, dW^0.$$

We then deduce from Lemma 8.4.28 and Proposition 8.4.27 (with $(c,d) = (0,h)$) the distribution functions of the Kolmogorov distributions. With the notations of [7.4.3], for $t > 0$,

$$P[\bar{D} \leqslant t] = 1 - \exp(-2t^2)$$

$$P[D^* \leqslant t] = \sum_{k=-\infty}^{\infty} (-1)^k \exp(-2k^2t^2).$$

8.4.3. Exponential Martingales Associated with Brownian Motion

Lemma 8.4.29. *Let T be a stopping time such that $E[\exp(T/2)]$ is finite. Then*

$$E(\exp(B_T - T/2)) = 1.$$

Proof. Let $Z_t = \exp(B_t - t/2)$; $(Z_t)_{t\geqslant 0}$ is a martingale and, for all $t \geqslant 0$,

$$E(Z_{t\wedge T}) = 1.$$

By Fatou's theorem, we have $E(Z_T) \leqslant 1$.

For $a > 0$, consider the time

$$S_a = \inf\{t;\, B_t \leqslant t - a\} = \inf\{t;\, B_t = t - a\}.$$

From Proposition 8.4.26 and the note following its proof we have $E[Z_{S_a}] = 1$. Let us assume $T \leqslant S_a$. Then

$$S_a = T + \inf(s; B_{s+T} - B_T \leqslant s - a - B_T) = T + T_1$$

where T_1 is a stopping time adapted to $(F_{T+t})_{t\geqslant 0}$. Applying the above to Brownian motion $(B_{t+T} - B_T)_{t\geqslant 0}$, we have

$$E[Z_{T+T_1} \mid F_T] \leqslant 1; \quad 1 = E(Z_{S_a}) \leqslant E(Z_T).$$

Hence, if $T \leqslant S_a$, we have: $E(Z_T) = 1$. In the general case, we write

$$Z_{T\wedge S_a} = Z_{S_a} 1_{(S_a \leqslant T)} + Z_T 1_{(T<S_a)}$$

with

$$Z_{S_a} 1_{(S_a \leqslant T)} = \exp[-a + S_a/2] 1_{(S_a \leqslant T)} \leqslant \exp(T/2).$$

We let a tend to ∞ and use Lebesgue and Beppo-Levi's theorems: the expectation $E(Z_{T\wedge S_a})$, equal to 1, tends to $E(Z_T)$ if $a \to \infty$.

Proposition 8.4.30. *Let C be a predictable process and T an $\mathbb{F}$-stopping time such that*

$$E\left[\exp \frac{1}{2}\int_0^T C_s^2 \, ds\right] < \infty.$$

Then

$$E\left[\exp\left(\int_0^T C_s dB_s - \frac{1}{2}\int_0^T C_s^2 \, ds\right)\right] = 1.$$

If we have

$$E\left[\exp \frac{1}{2}\int_0^t C_s^2 \, ds\right] < \infty$$

for all t, let us set

$$Z_C(t) = \exp\left\{\int_0^t C_s dB_s - \frac{1}{2}\int_0^t C_s^2 ds\right\};$$

then $Z_C = (Z_c(t))$ is an $\mathbb{F}$-martingale.

Proof. (a) First of all let us show the proposition by assuming $\int_0^\infty C_s^2 ds$ is a.s. infinite. Consider

$$\tau(t) = \inf\left\{u; \int_0^u C_s^2 ds \geq t\right\};$$

$\tau(t)$ is an $\mathbb{F}$-stopping time. If $X_t = \int_0^{\tau(t)} C_s dB_s$, $X = (X_t)_{t\geq 0}$ is a martingale adapted to $(F_{\tau(t)})$; since $\int_0^{\tau(t)} C_s^2 ds = t$, $(X_t^2 - t)$ is also an $(F_{\tau(t)})$ martingale. Hence X is a Brownian motion adapted to $(F_{\tau(t)})$. However,

$$\left\{\int_0^T C_s^2 ds \leq u\right\} = \{T \leq \tau(u)\}$$

is an event of $F_{\tau(u)\wedge T} \subset F_{\tau(u)}$: thus $\int_0^T C_s^2 ds$ is a stopping time adapted to $(F_{\tau(t)})$, and we obtain

$$E\left[\exp\left[\int_0^T C_s dB_s - \frac{1}{2}\int_0^T C_s^2\, ds\right]\right] = 1$$

by applying Lemma 8.4.29.

For $s < t$, Z_c being a supermartingale, we have $E[Z_C(t)|\ _s] \leq Z_C(s)$; since $E[Z_C(t)] = 1 = E[Z_C(s)]$, we have equality, and Z_C is a martingale.

(b) By replacing C by $\hat{C}$, with $\hat{C} = (\hat{C}_s)$ where $\hat{C}_s = C_s \mathbf{1}_{(s\leq T)} + \mathbf{1}_{(s>T)}$, we obtain $E[Z_C(T)] = 1$, and $(Z_C(T \wedge t))_{t\geq 0}$ is a martingale. From which the second part of the property follows by taking T to be an arbitrary fixed time.

8.5. Regression and Diffusions

8.5.1. Regression in Continuous Time

A general regression model in discrete time can be defined as follows: we are given a filtration $\mathbb{F} = (F_n)_{n\geq 0}$, (ε_n) a Gaussian white noise adapted to $\mathbb{F}$, and (C_n) an $\mathbb{F}$-predictable process (C_n is F_{n-1} measurable). The observation X_n is then the sum of an r.v. C_n known at time $n - 1$ and a noise ε_n: $Y_n = C_n + \varepsilon_n$. The study of (Y_n) is equivalent to the study of the cumulative effect (X_n), with $X_n = Y_1 + ... + Y_n$.

The natural analogue in continuous time is the following. We are given a filtration $\mathbb{F} = (F_t)_{t\geq 0}$, a Brownian motion $(B_t)_{t\geq 0}$, and an $\mathbb{F}$-predictable process $(C_t)_{t\geq 0}$. The sequence $(B_n - B_{n-1})_{n\geq 1}$ is a Gaussian white noise. If a signal is observed, the intensity of which is C_t at time t, its cumulative effect up to time t is $\int_0^t C_s ds$. If a Brownian noise is added to this signal, we observe

$$X_t = \int_0^t C_s ds + B_t.$$

Using $(X_s)_{s \leqslant t}$, it is necessary to decide what $(C_s)_{s \leqslant t}$ equals, i.e. to **filter** C by X. The principal theorem is the following.

Girsanov's Theorem 8.5.31. *Consider* $\{\Omega, A, P, (B_t)_{t \geqslant 0}\}$, *a Brownian motion adapted to a filtration* $\mathbb{F}$ *and an* $\mathbb{F}$*-predictable process* C. *Assume, for all* t,

$$E\left[\exp \frac{1}{2}\int_0^t C_s^2 ds\right] < \infty.$$

Let

$$X_t = \int_0^t C_s ds + B_t,$$

$$L_t = \exp\left[-\int_0^t C_s dB_s - \frac{1}{2}\int_0^t C_s^2\, ds\right]$$

$$= \exp\left[-\int_0^t C_s dX_s + \frac{1}{2}\int_0^t C_s^2\, ds\right].$$

A probability Q *is defined on* (Ω, F_∞) *by setting* $Q = L_t P$ *on* (Ω, F_t); $\{\Omega, F_\infty, Q, (X_t)_{t \geqslant 0}\}$ *is a Brownian motion adapted to* $\mathbb{F}$. *When* T *is an* $\mathbb{F}$*-stopping time such that*

$$E\left[\exp \frac{1}{2}\int_0^T C_s^2\, ds\right] < \infty,$$

the probabilities P *and* Q *are equivalent on* F_t, *and:*

$$Q = L_T P, \quad P = [1/L_T] Q.$$

Proof. Let $\lambda \in \mathbb{R}$ and $Y_\lambda(t) = \exp(\lambda X_t - (\lambda^2/2)t)$:

$$L_t Y_\lambda(t) = \exp\left[\int_0^t (\lambda - C_s) dB_s - \frac{1}{2}\int_0^t (\lambda - C_s)^2 ds\right].$$

The probability Q, defined on the Boolean algebra $\cup F_t$, has a unique extension to F_∞ (Neveu [2]). From Proposition 8.5.30, $(L_t Y_\lambda(t))$ is an $\mathbb{F}$-martingale under P. The theorem will be proved following Definition 8.4.26 if we verify that Y_λ is an $\mathbb{F}$-martingale under Q. However, denoting by E_P and E_Q expectations relative to P and Q, we have (cf. Vol. I, 7.1.5):

$$E_Q[Y_\lambda(t+h)|\mathcal{F}_t] = \frac{E_P[Y_\lambda(t+h)L_{t+h}|\mathcal{F}_t]}{E_P[L_{t+h}|\mathcal{F}_t]} = Y_\lambda(t).$$

Example. Let $\theta \in \mathbb{R}$ and $X_t = \theta t + B_t$. We obtain $L_t(\theta) = \exp(-\theta X_t + \theta^2 t/2)$. Let $a > 0$, $S_a = \inf\{t; B_t + \theta t = a\} = \inf\{t; X_t = a\}$ and $\lambda > 0$. Let us denote P_θ for Q and E_θ for E_Q,

$$E_0[\exp(-\lambda S_a)] = E_\theta[\exp(\theta a - \theta^2 S_a/2)\exp(-\lambda S_a)].$$

If $T_a = \inf\{t; B_t = a\}$,

$$E_0[\exp(-\lambda S_a)] = e^{\theta a}E_0\left[\exp - \left(\lambda + \frac{\theta^2}{2}\right)T_a\right].$$

We meet the result of Proposition 8.4.26 again:

$$E_0[\exp(-\lambda S_a)] = \exp(a[\theta - \sqrt{2\lambda + \theta^2}]).$$

If X is observed, the parameter θ being unknown, the maximum likelihood estimator of θ is X_t/t.

By the law of large numbers (Theorem 8.2.17), this estimator is consistant:

$$X_t/t \xrightarrow{P_\theta\text{-a.s.}} \theta.$$

Moreover, on $(\Omega,\mathcal{A},P_\theta)$, X_t/t has distribution $\mathcal{N}(\theta,1/t)$ and $\sqrt{t}(X_t/t - \theta)$ has distribution $\mathcal{N}(0,1)$.

8.5.2. Diffusions

A diffusion process on $\mathbb{R}$ is physically described in the following manner: the position X_t at time t being given, the infinitesimal evolution is the sum of a motion at a rate $a(X_t)$ and a centered Gaussian perturbation with variance $\sigma^2(X_t)$. In other words, we are given $\{\Omega,\mathcal{A},P,(B_t)\}$, a Brownian motion which represents a noise. The process $(X_t)_{t\geq 0}$ under study satisfies a **stochastic differential equation**

$$dX_t = a(X_t)dt + \sigma(X_t)dB_t$$

which implies

$$X_t - X_0 = \int_0^t a(X_s)ds + \int_0^t \sigma(X_s)dB_s.$$

Definition 8.5.32. We call a **diffusion** a process $X = (\Omega,\mathcal{A},P,(X_t)_{t\geq 0})$ with continuous trajectories such that there exist two real continuous functions a and σ, σ positive, and a Brownian motion $B = (B_t)$ defined on $(\Omega,\mathcal{A},P)$, for which X is the solution of the stochastic differential equation

$$dX_t = a(X_t)dt + \sigma(X_t)dB_t.$$

Example. Ornstein-Uhlenbeck Process. Let $\theta \in \mathbb{R}$ and X_0 an r.v. Set

$$X_t = e^{\theta t}X_0 + e^{\theta t}\int_0^t e^{-\theta s}dB_s.$$

By the integration by parts formula of [8.2.1], we have,

$$X_t - X_0 = \theta\int_0^t X_s ds + B_t \quad \text{or} \quad dX_t = \theta X_t dt + dB_t.$$

First of all let us study the equation $dX_t = a(X_t)dt + B_t$, assuming a is continuous. Let us show that it has a solution. Let $(C,\mathcal{C},(X_t)_{t\geq 0},W)$ be a canonical Brownian motion. Let $\mathbb{F} = (\mathcal{F}_t)$ be the filtration $(\sigma(X_u;\ u \leq t))^+$. We then deduce from Theorem 8.5.31 the following theorem, by setting $C_t = -a(X_t)$.

Existence Theorem 8.5.33. *Let a be a continuous function from $\mathbb{R}$ into $\mathbb{R}$. We assume that, for a Brownian motion $(B_t^1)_{t\geq 0}$ and every $t \geq 0$,*

$$E\left[\exp\left(\frac{1}{2}\int_0^t a^2(B_s^1)ds\right)\right]$$

is finite (in particular a may be bounded). Then we can construct a diffusion, a solution of the stochastic differential equation

$$dX_t = a(X_t)dt + dB_t$$

with the initial condition $X_0 = 0$, in the following manner. Consider the canonical Brownian motion $(C,\mathcal{C},(X_t)_{t\geq 0},W)$ set:

(a) *$P = L_tW$ on $(C,\mathcal{F}_t)$ by defining L_t on $(C,\mathcal{C},W)$ by*

$$L_t = \exp\left[\int_0^t a(X_s)dX_s - \frac{1}{2}\int_0^t a^2(X_s)ds\right];$$

(b) $B_t = X_t - \int_0^t a(X_s)ds.$

Then $(C, \mathcal{C}, (X_t)_{t\geq 0}, P)$ *is the canonical solution which we are looking for.*

In the framework of the preceding theorem, the traces of P and of W on the σ-algebras $\mathcal{F}_t$ are equivalent. On $(C, \mathcal{C}, P)$ we write,

$$L_t = \exp\left[\int_0^t a(X_s)dB_s + \frac{1}{2}\int_0^t a^2(X_s)ds\right].$$

For a process such as the Ornstein-Uhlenbeck, the existence of which is known, the following theorem gives a simple criterion for domination by Wiener measure.

Theorem 8.5.34. *Let* $(C, \mathcal{C}, (X_t)_{t\geq 0}, P)$ *be a real continuous canonical process, and let a be a continuous function from* $\mathbb{R}$ *into* $\mathbb{R}$. *We assume that* $B = (B_t)_{t\geq 0}$ *is a Brownian motion, setting*

$$B_t = X_t - \int_0^t a(X_s)ds.$$

Finally we assume that, for all t,

$$P\left[\int_0^t a^2(X_s)ds < \infty\right] = 1.$$

Then for any t, P *restricted to* $\mathcal{F}_t$ *is absolutely continuous with respect to Wiener measure* W *and we have, on* $\mathcal{F}_t$:

$$P = \left\{\exp\left[\int_0^t a(X_s)dB_s + \frac{1}{2}\int_0^t a^2(X_s)ds\right]\right\}W.$$

Proof. Let

$$T_n = \inf\left\{t;\ \int_0^t a^2(X_s)ds \geq n\right\}.$$

Let us apply Girsanov's theorem to the process $(a(X_{t\wedge T_n}))$. Set

$$L_t = \exp\left[-\int_0^t a(X_s)dB_s - \frac{1}{2}\int_0^t a^2(X_s)ds\right].$$

Then the process (X_t^n) defined by

$$X_t^n = \int_0^{t\wedge T_n} a(X_s)ds + B_t$$

is a Brownian motion on $(C,\mathcal{C},Q^n)$, with

$$Q^n = L_{t\wedge T_n} \cdot P \quad \text{on} \quad (\Omega,\mathcal{F}_t).$$

Let $0 \leqslant t_1 < \ldots < t_k \leqslant t$ and let Γ be a Borel set of $\mathbb{R}^k$:

$$Q^n[(X_{t_1}, \ldots, X_{t_k}) \in \Gamma, t \leqslant T_n]$$

$$= W[(X_{t_1}, \ldots, X_{t_k}) \in \Gamma, \quad t \leqslant T_n].$$

Hence the traces of Q^n and of W on $(\{t \leqslant T_n\},\mathcal{F}_t)$ coincide.

Example. A canonical version of the Ornstein-Uhlenbeck process zero for $t = 0$ is obtained by taking on $(C,\mathcal{C})$ the distribution P_θ such that $P_\theta = L_t(\theta)W$ on $(C,\mathcal{F}_t)$ with

$$L_t(\theta) = \exp\left[\theta \int_0^t X_s dX_s - \frac{\theta^2}{2}\int_0^t X_s^2 ds\right].$$

However, on $(C,\mathcal{C},W)$, Ito's formula gives

$$2\int_0^t X_s dX_s = X_t^2 - t.$$

From which it follows that

$$L_t(\theta) = \exp\left[\frac{\theta}{2}(X_t^2 - t) - \frac{\theta^2}{2}\int_0^t X_s^2 ds\right].$$

If the parameter θ is unknown, we have a dominated statistical model. The maximum likelihood estimator of θ at time t is

$$\hat{\theta}_t = \frac{1}{2}(X_t^2 - t)\left[\int_0^t X_s^2 ds\right]^{-1}.$$

How do we study the stochastic differential equation

$$dX_t = a(X_t)dt + \sigma(X_t)dB_t$$

where the noise is amplified in a random manner?

Let us assume that the function σ is continuous, bounded and minorized by a constant > 0. The process

$$B^\sigma = (B_t^\sigma) = \left[\int_0^t \sigma(X_s)dB_s\right]_{t\geqslant 0}$$

is in M^2, and

$$[B^\sigma, B^\sigma]_t = \int_0^t \sigma^2(X_s)ds.$$

Let τ be the "**time change**" associated with this increasing process, defined by

$$\tau_t = \inf\left\{u;\ \int_0^u \sigma^2(X_s)ds \geqslant t\right\}$$

If the processes B and X are adapted to the right continuous filtration $\mathbb{F}$, then for all t, τ_t is a finite $\mathbb{F}$-stopping time. The trajectories $t \longmapsto \tau_t$ are continuous and strictly increasing from 0 to ∞. Knowing the process X amounts to knowing the process $Y = (Y_t)$ with $Y_t = X_{\tau_t}$. However the processes

$$B^1 = (B^\sigma_{\tau_t}) \quad \text{and} \quad ((B^\sigma_{\tau_t})^2 - t)$$

are martingales adapted to the filtration $\mathbb{G} = (F_{\tau_t})$: B^1 is a Brownian motion adapted to $\mathbb{G}$. Finally the change of variables formula 6.2.10 gives

$$\int_0^{\tau_t} a(X_s)ds = \int_0^t \frac{a(Y_s)}{\sigma^2(Y_s)}ds.$$

Thus we are led to the preceding problem by studying the process Y which is a solution of the stochastic differential equation:

$$dY_t = \frac{a(Y_t)}{\sigma^2(Y_t)}\,dt + B_t^1.$$

To finish off, let us study how to estimate the amplification σ of the noise. If a trajectory $(X_s)_{0\leqslant s\leqslant t}$ of the process has been observed, we can determine its quadratic variation. Let δ be a partition $0 = t_0 < t_1 < \dots t_{n+1} = t$ of $[0,t]$ of width $|\delta|$. Set

$$X^\delta = \sum_{i=0}^n (X_{t_{i+1}} - X_{t_i})^2.$$

We have

$$\sum_{i=1}^{n}\left[\int_{t_i}^{t_{i+1}} a(X_s)ds\right]^2 + 2\sum_{i=1}^{n}\left[\int_{t_i}^{t_{i+1}} a(X_s)ds\right]\cdot$$

$$\cdot\left[\int_{t_i}^{t_{i+1}} \sigma(X_s)dB_s\right]$$

$$\leqslant \int_0^t |a(X_s)|ds + \sup\left\{\left|\int_u^v a(X_s)ds\right| + 2\left|\int_u^v \sigma(X_s)dB_s\right|;\right.$$

$$\left.|u-v| \leqslant |\delta|\right\}.$$

If $|\delta|$ tends to 0, this majorant tends a.s. to 0. Thus, from Theorem 8.1.9,

$$X^{\delta} \xrightarrow[|\delta|\to 0]{P} \int_0^t \sigma^2(X_s)ds.$$

It can even be shown in the case of stochastic integrals of Brownian motion that convergence takes place a.s. (Doob, p. 395). Thus knowledge of a trajectory of X on the time interval $[0,t]$ is sufficient to approximate as closely as we wish the integrals $\int_0^u \sigma^2(X_s)ds$, for $u \leqslant t$. The process $(\sigma(X_s))_{s \leqslant t}$ can then be considered as known.

Bibliographic Notes

A detailed study of Brownian motion, local or asymptotic behavior, stochastic integrals, diffusions... is interesting and worthwhile without a general theory on martingales. See Freedman [1], Hida, Ito-McKean, Levy, McKean, Port-Stone, and Rao. An elementary study of diffusions and some examples will be found in Karlin-Taylor [2].

In the last few years the theory of stochastic integrals with respect to martingales has made great progress. Doob already gives the elements of this. The works of Meyer and the probabilists brought together in the Strasbourg seminars have been decisive: Dellacherie-Meyer [2], Meyer [1] and [2], Strasbourg seminars. However many others have constributed to it; some of the recent books being Gikhman-Skorokhod ([1] and [2], Vol. 3), Ikeda-Watanabe, Kussmaul, Liptzer-Shiryayev, McShane, Metivier, Metivier-Pellaumail. We have limited ourselves to the simplest case of square integrable martingales. The method adopted in [8.1] was inspired by Metivier-Pellaumail, the other methods cover the Doob decomposition before the stochastic integral. We have also used Neveu [3].

Many works are based on exponential martingales or supermartingales and on the stochastic integral (Strasbourg Seminars). We have used only those which lead to asymptotic results: Lepingle for the exponential supermartingale and the law of large numbers; Portal-Touati for the theorem of large deviations, Rebolledo [2] and [3] for the central limit theorem. For extensions see Jacod-Memin.

The asymptotic study of point processes is the subject of Chapter 6 where the bibliography appears.

For stochastic differential equations Skorokhod and Gikhman-Skorokhod [1] construct solutions by successive approximations and study them analytically. The links with the theory of differential equations and numerous examples are studied in Friedman. The modern theory of stochastic differential equations with respect to a semimartingale is given in Stroock-Varadhan.

The statistics and filtering of diffusions are given in Basawa-Rao, Kallianpur, Kutoiants, Liptzer-Shiryayev, and Dacunha Castelle-Duflo (Exercises of Vol. II).

BIBLIOGRAPHY

AALEN O. O. *Nonparametric inference for family of counting processes*, Annals of Stat., Vol. 6, No. 4, 701-706 (1978).
ANDERSON T. W. *The statistical analysis of time series*, Wiley (1971).
ARAUJO A. and GINE E. *The central limit theorems for real and Banach valued random variables*, Wiley (1980).
ASTERISQUE. *Séminaires de statistique d'Orsay* (*ouvrages collectifs*).
[1] *Théorie de la robustesse et estimation d'un paramètre de translation*, Asterisque 43-44 (1977).
[2] *Grandes déviations et applications statistiques*, Asterisque 60 (1979).
BARTLETT M. S. [1] *The statistical analysis of spatial pattern*, Chapman and Hall (1975).
[2] *An introduction to stochastic processes*, 3rd ed., Cambridge Univ. Press (1978).
BASAWA I. V. and PRAKASA RAO B. L. S. *Statistical inference for stochastic processes*, Academic Press (1980).
BELIAEV Y. See GNEDENKO.
BELLMAN R. *Dynamic programming*, Princeton Univ. Press (1957).
BHARUCHA REID. *Elements of the theory of Markov Processes and their applications*, McGraw Hill (1960).
BILLINGSLEY P. [1] *Statistical inference for Markov processes*, University of Chicago Press (1961).
[2] *Convergence of probability measures*, Wiley (1968).
[3] *Ergodic theory and information*, Wiley (1975).

[4] *Probability and measure*, Wiley (1979).
BOCHNER S. *Harmonic analysis and the theory of probability*, Univ. of California Press (1955).
BOROVKOV A. A. *Stochastic processes in queuing theory*, Springer (1976).
BOX G. and JENKINS G. *Time series analysis*, Holden Day (1976).
BREMAUD P. *Point processes and queues: martingale dynamics*, Springer (1980).
BREMAUD P. and JACOD J. *Processus ponctuels et martingales*, Adv. Appl. Proba 9, 362-416 (1977).
BREIMAN L. *Probability*, Addison Wesley (1968).
BRILLINGER D. *Time series: data analysis and theory*, Holt, Rinehart, Winston (1975).
CHATFIELD C. *The analysis of time series: an introduction*, Chapman and Hall (1975).
CHERNOFF H. *Sequential design of experiments*, Ann. Math. Stat., Vol. 30, 755-770 (1959).
CHOW, Y. S., ROBBINS H. and SIEGMUND, D. *Great expectations: The theory of Optimal Stopping*, Houghton (1971).
CHUNG K. L. *Markov chains with stationary transition probabilities*, 2nd Edition, Springer (1967).
CINLAR E. *Introduction to stochastic processes*, Prentice Hall (1975).
COGBURN R. *The central limit theorem for Markov problems*, Sixth Berkeley Symposium in Prob., 485-512 (1972).
COX D. R. *Renewal theory*, Chapman & Hall (1966).
COX D. R. and LEWIS, P. A. W. *The statistical analysis of series of events*, Methuen (1966).
COX D. R. and SMITH M. L. *Queues*, Methuen (1961).
DACUNHA-CASTELLE D. *Vitesse de convergence pour certains problemes statistiques*, Lect. Notes in Math., 678, Springer (1977).
DACUNHA-CASTELLE D. and DUFLO, M. *Exercise de probabilités et statistiques*, volume II, Masson (1983).
DE GROOT M. *Optimal statistical decisions*, McGraw Hill (1970).
DELLACHERIE C. *Capacités et processus stochastiques*, Springer (1972).
DELLACHERIE C. and MEYER P. A. *Probability and potential*, North Holland (1978).
DERMAN C. *Finite state Markovian decision process*, Academic Press (1970).

DOOB J. L. *Stochastic processes*, Wiley (1953).
DUFLO M. and FLORENS-ZMIROU D. *Pas à pas*, Cours du CIMPA (1981).
DYNKIN E. B. and YUCHKEVICH A. A. [1] *Theorems and problems on Markov Processes*, in Russian, Nauka (1967).
[2] *Controlled Markov processes*, Springer (1979).
EL KAROUI N. *Les aspects probabilistes du côntrole stochastique*, Lect. Notes in Math. 876: Springer (1980).
EWENS J. *Mathematical population genetics*, Springer (1979).
FELLER W. *An introduction to probability theory and its applications*, Wiley, Volume 1, 3rd edition, (1967); Volume 2, 2nd edition, (1971).
FERGUSON T. S. *Mathematical statistics: a decision theoretic approach*, Academic Press (1967).
FLORENS-ZMIROU D. See DUFLO.
FOMINE S. See KOLMOGOROV.
FREEDMAN D. [1] *Brownian motion and diffusion*, Holden Day (1971).
[2] *Markov chains*, Holden Day (1971).
FRIEDMAN A. *Stochastic differential equations and applications*, 3 volumes, Academic Press (1975-1976).
GARSIA A. M. *Martingale inequalities*, Benjamin-Reading (1973).
GEORGIN J. P. *Côntrole des chaines de Markov dépendant d'un paramètre*. Statistique des processus stochastiques, Lect. Notes in Math. 636, Springer (1977).
GHOSH B. K. *Sequential tests of statistical hypothesis*, Addison-Wesley (1970).
GIKHMAN I. I. and SKOROKHOD A. V. [1] *Stochastic differential equations*, Springer (1972) [in Russian, Kiev (1968)].
[2] *The theory of random processes*, Springer (3 volumes, 1974, 1975, 1979) [in Russian, Nauka (1971, 1973, 1975)].
[3] *Introduction à la théorie des processus aleatoires*, Mir (1980) [in Russian, Nauka (1977)].
[4] *Controlled stochastic processes*, Springer (1979).
GILL, R. D. *Censoring and stochastic integrals*, Mathematical Centre Tracts, Amsterdam (1980).
GINE E. See ARAUJO.
GNEDENKO B., BELIAEV Y. and SOLOVIEV A. *Méthode mathématique en théorie de la fiabilité*, Mir (1972).

GNEDENKO B. W. and KOLMOGOROV A. N. *Limit distributions for sums of independent random, variables*, Addison Wesley (1954).
GOVINDARAJULU Z. *Sequential statistical procedures*, Academic Press (1975).
GRANDELL J. *Doubly stochastic Poisson processes*, Lect. Notes in Math. 529, Springer (1976).
GRENANDER U. *Abstract inference*, Wiley (1981).
GRANANDER U. and ROSENBLATT, M. *Statistical analysis of stationary time series*, Wiley (1957).
HALL P. and HEYDE C. C. *Martingale limit theory and its applications*, Academic Press (1981).
HAJEK, J. and SIDAK, Z. *Theory of rank tests*, Academic Press (1967).
HANNAN E. J. *Multiple time series*, Wiley (1970).
HARRIS T. E. *The theory of branching processes*, Springer (1963).
HAS'MINSKI R. See IBRAGIMOV.
HEYDE C. C. see HALL
HIDA T. *Brownian motion*, Springer (1980).
HOEL P., PORT S. and STONE C. *Introduction to stochastic processes*, Houghton Mifflin (1972).
HOFFMAN K. *Banach spaces analytic functions*, Prentice Hall (1962).
HOWARD R. A. *Dynamic programming and Markov processes*, MIT Press (1960).
HUNT G. A. *Martingales et processus de Markov*, Dunod (1966).
IBRAGIMOV I. A. and HAS'MINSKI R. *Statistical estimation*, Springer (1981) [in Russian, Nauka (1978)].
IBRAGIMOV I. A. and LINNIK Y. *Independent and stationnary sequences of random variables*, Wolters-Noordhorff, Groniner [in Russian, Nauka (1965)].
IBRAGIMOV I. A. and ROZANOV Y. *Processus aleatoires gaussiens*, Mir (1974).
IKEDA N. and WATANABE S. *Stochastic differential equations*, North Holland (1982).
ITO K. and MCKEAN H. P. *Diffusion processes and their sample paths*, Springer (1965).
JACOD J. *Calcul stochastique et problèmes de martingales*, Lecture Notes in Math. 714, Springer (1980).
JACOD J. See BREMAUD

JACOD J. and MEMIN J. *Sur la convergence des semimartingales vers un processus à accroissements indépendants*, Lect. Notes in Math. 784, 227-248, Springer (1980).
JENKINS G. See BOX G.
KALLIANPUR *Stochastic filtering theory*, Springer (1980).
KARLIN S. and TAYLOR H. M. [1] *A first course in stochastic processes*, Academic Press, 2nd edition (1975).
[2] *A second course in stochastic processes*, Academic Press (1981).
KEMENY J. G., SNELL J. L. and KNAPP A. W. *Denumerable Markov chains*, Springer (1976).
KENDALL M. G. and STUART, A. *The advanced theory of statistics*, Vol. 3, Griffin (1966).
KINGMAN J. F. C. *Regenerative phenomena*, Wiley (1972).
KLEINROCK L. *Queuing systems*, 2 volumes, Wiley (1976).
KNAPP A. W. See KEMENY
KOLMOGOROV A. N. See GNEDENKO
KOLMOGOROV A. and FOMINE S. *Eléments de la théorie des fonctions et de l'analyse fonctionnelle*, Mir (1974) [in Russian, Nauka (1973)].
KOOPMANS L. H. *The spectral analysis of time series*, Academic Press (1974).
KUSHNER H. J. *Introduction to stochastic control theory*, Holt, Rinehart, Winston (1971).
KUSSMAUL A. U. *Stochastic integration and generalized martingales*, Pitman (1977).
KUTOYANTS Yu. *Parameter estimation for stochastic processes*, Helderman Verlag, Berlin (1984).
LeCAM L. [1] *Théorie asymptotique de la décision statistique*, Presse Université Montréal (1969).
[2] *Notes on asymptotic methods in decision theory*, Centre de recherches mathématiques de l'université de Montreal (1974).
LEHMANN E. L. *Testing statistical hypothesis*, Wiley (1959).
LENGLART E. *Relation de domination entre deux processus*, Annales Inst. Henri Poincaré 13, 171-179 (1977).
LEPLINGLE D. *Sur le comportement asymptotique des martingales locales*. Lect. Notes 649, 148-161, Springer (1978).
LEVY P. [1] *Théorie de l'addition des v.a.*, Gauthier Villars (1937).
[2] *Processus stochastique et mouvement brownien*, Gauthier Villars (1954).
LEWIS P. A. *Stochastic point processes*, ouvrage collectif, Wiley (1972). See COX.

LINDVALL T. *A probabilistic proof of Blackwell's renewal theorem*, Annals of Prob., Vol. 5, 3, 482-485 (1977).

LINNIK Y. See IBRAGIMOV.

LIPTZER R. S. and SHIRYAYEV A. N. *Statistics of stochastic processes*, 2 volumes, Springer, New York (1977) [in Russian, Nauka (1974)].

LOEVE M. *Probability theory*, 4th edition, Springer (1977-1978).

McKEAN H. P. [1] *Stochastic integrals*, Academic Press (1969).
[2] See ITO

McSHANE E. J. *Stochastic calculus and stochastic models*, Academic Press (1974).

MAIGRET N. *Théorème de limite centrale fonctionnelle pour une chaine de Markov récurrente Harris positive*, Ann. IHP Vol. 14, No. 4, 425-440 (1978).

MANDL P. *Estimation and control in Markov chains*, Adv. Appl. Prob. 6, 40-60 (1974).

MARTIN J. J. *Bayesian decision problems and Markov chains*, Wiley (1967).

METIVIER M. *Reelle und vektor wertige quasi-martingale und die theorie der stochastichen integration*, Lect. Notes in Math. 607, Springer (1977).

MEMIN J. See JACOD J.

METIVIER, M. and PELLAUMAIL J. *Stochastic integration*, Academic Press (1979).

MEYER P. A. [1] *Martingales and stochastic integrals*, Lect. Notes in Math. 284, Springer (1972).
[2] *Un cours sur les intégrales stochastiques*, Lect. Notes in Math. 511, 245-400, Springer (1976).
[3] See DELLACHERIE.

MINE, O. *Markovian decision processes*, Elsevir (1970).

NEVEU J. [1] *Processus Gaussiens*, Presses de l'université de Montréal (1968).
[2] *Mathematical foundations of the calculus of probability*, Holden-Day (1965).
[3] *Martingales à temps discret*, Masson (1972).
[4] *Processus ponctuels*, Lect. Notes in Math., 598, Springer (1977).

OREY S. *Limit theorems for Markov chain transition probabilities*, Van Nostrand (1971).

PARTHASARATHY K. R. *Probability measures on Metric spaces*, Academic Press (1967).

PELLAUMAIL J. See METIVIER

PITMAN J. W. *Uniform rates of convergence for Markov chain transition probabilities*, Z. Wahrscheinlichkeitstheorie 29, 193-227 (1974).

PORT S. and STONE C. *Brownian motion and classical potential theory*, Academic Press (1978).

PORTAL F. and TOUATI A. *Théorèmes de grandes déviations pour des mesures aléatoires*, Z. Wahrscheinlichkeitstheorie (1983).

PRIESTLEY M. B. *Spectral analysis and time series*, 2 volumes, Academic Press (1981).

RAO C. R. *Linear statistical inference and its applications*, Wiley (1965).

RAO K. M. *Brownian motion and classical potential theory*, Aarhus Lect. Notes 47 (1977).

REBOLLEDO R. [1] *Sur les applications de la théorie des martingales à l'étude statistique d'une famille de processus ponctuels*, Lect. Notes in Math., 636, 27-70 (1978).

[2] *La méthode des martingales appliquée à l'étude de la convergence en loi des processus*, Memoire de la Société Mathématique de France 62 (1979).

[3] *Central limit theorems for local martingales*, Z. Wahrsheinlichkeitstheorie 51, 269-286 (1980).

REVUZ D. *Markov chains*, North Holland (1975).

ROSS S. M. *Applied probability models with optimization applications*, Holden Day (1970).

ROUSSAS G. *Contiguity of probability measures: some applications in statistics*, Cambridge University Press (1972).

ROZANOV Yu. A. [1] *Stationary random processes*, Holden Day (1967) [in Russian, Nauka (1963)].

[2] *Processus aléatoires*, Mir (1975).

[3] See IBRAGIMOV.

RUDIN W. [1] *Real and complex analysis*, McGraw Hill (1966).

[2] *Fourier analysis on groups*, Interscience (1970).

SEMINARES DE STRASBOURG 1-14. Lect. Notes in Math., Springer (1967-1981).

SERFLING J. R. *Approximation theorems of mathematical statistics*, Wiley (1980).

SHIRYAYEV A. N. [1] *Optimal stopping rules*, Springer (1978) [in Russian, Nauka (1969)].

[2] See LIPTZER.

SIDAK Z. See HAJEK J.

SKOROKHOD. [1] *Studies in the theory of random processes*, Addison Wesley (1965) [in Russian, Kiev (1961)].
[2] See GIKHMAN.
SMITH M. L. See COX.
SNYDER D. L. *Random point processes*, Wiley (1975).
SOLOVIEV A. See GNEDENKO.
SPITZER F. L. *Principles of random walks*, Springer (1976).
STOUT W. *Almost sure convergence*, Academic Press (1974).
STROOCK E. W. and VARADHAN S. R. S. *Multidimensional diffusion processes*, Springer (1979).
TAKACS L. *Introduction to the theory of queues*, Oxford Univ. Press, 1962.
TOUATI. See PORTAL.
VARADHAN S. R. S. See STROOCK.
WALD A. [1] *Statistical decision functions*, Wiley (1950).
[2] *Sequential analysis*, Wiley (1947); Dover (1973).
WATANABE S. See IKEDA.
WETHERILL, G. B. *Sequential methods in statistics*, Methuen (1966).
WILLIAMS D. *Diffusions, Markov processes and martingales*, Wiley (1979).
YAGLOM A. M. *Stationary random functions*, Prentice Hall (1962) [in Russian (1952)].
YOR M. *Sur la theorie de filtrage*, Lect. Notes 876, Springer (1979).

NOTATION AND CONVENTIONS

Mathematical Notations (Volumes I and II)

$\mathbb{R}$ real numbers

$\overline{\mathbb{R}}$ extended real line

$\mathbb{C}$ complex numbers

$\mathbb{Q}$ rational numbers

$\mathbb{N}$ integers $\geqslant 0$

$\mathbb{Z}$ integers

$\mathfrak{S}$ or $\mathfrak{S}_n$ set of permutations of $\{1, ..., n\}$

$n!$ n factorial

$\binom{n}{p}$ number of combinations of n objects, taken p at a time

$\geqslant 0$ means positive; > 0 means strictly positive;

$\leqslant 0$ means negative; < 0 means strictly negative

For a and b real, and f and g real-valued functions:

$\wedge$ minimum, $a \wedge b = \inf(a,b)$, $f \wedge g = \inf(f,g)$;

$\vee$ maximum, $a \vee b = \sup(a,b)$, $f \vee g = \sup(f,g)$;

$a_+ = a \vee 0$ or $f_+ = f \vee 0$, positive part;

$a_- = -a \wedge 0$ or $f_- = -f \wedge 0$, negative part;

$|a|$ or $|f|$, modulus of a or f.

For $A \subset \mathbb{R}$ or F a set of real-valued functions:

$A_+ = \{a; a \in A, a \geqslant 0\}$, $F_+ = \{f; f \in F, f \geqslant 0\}$, positive parts of A or F.

For (a_n) a sequence of real numbers of (f_n) or a sequence of real functions.

$\overline{\lim}\, a_n$ or $\overline{\lim}\, f_n$ is the upper limit

$\underline{\lim}\, a_n$ or $\underline{\lim}\, f_n$ is the lower limit

$o(x)$, of order less than x, for $x \to 0$

$O(x)$, of order x, for $x \to 0$

C_0, C_b, C_k 120

x a vector of $\mathbb{R}^n$ is also the $n \times 1$ matrix of which it is the column vector

$\|x\|$ Euclidean norm of x

$\bar{x} = \frac{1}{n}(x_1 + \ldots + x_n)$ for $x = (x_1, \ldots, x_n)$

$<x,y>$ scalar product of x and y

$\mathbf{1}_n$ or $\mathbf{1}$ is the vector $(1, \ldots, 1)$ of $\mathbb{R}^n$

I_n or I is the $n \times n$ identity matrix

${}^tM$ transposed matrix of M; t transposition

$|M|$ determinant of the square matrix M

J_ϕ Jacobian of ϕ

Mathematical Notation (Volume II)

$T_n(h)$ Toeplitz matrix 51

C_k 304

C 304

D 315

Measure Theory (Vol. I)		Measure Theory (Vol. II)	
$A \cup B$	104	$\underset{i \in I}{\otimes} \mathcal{E}_i$	4
$A \cap B$	104	$(E, \mathcal{E})^T$	4
$A \supset B$	104	$(E, \mathcal{E}, F)^N$	5
$A \subset B$	104	L^2_C	13
$A \backslash B$	104	$dQ_{\mathcal{F}} / dP_{\mathcal{F}}$	63
A^c	104	$\mathbf{F} = (\mathcal{F}_n)$	64

$\overline{\lim}\, A_n$ 107

$\underline{\lim}\, A_n$ 112

$\bar{A}$ closure of A

ess sup A_i 123

$f^{-1}(A) = (f \in A) = \{\omega;\ f(\omega) \in A\}$ 105

Probabilities (Volume I)

Probabilities (Volume II)

Statistics (Volume I)

$\overline{X}_n$ 173

$\overline{F}_n$ 177

S_n^2 179

Distributions (Volume I)

F 111

F^{-1} 136

F_X 124

ϕ_F 120

g_F 57

h_F 134

q_α^-, q_α^+ 86

F_α^-, F_α^+ 180, 181

ϕ_α, $\chi^2_{n,\alpha}$, $t_{n,\alpha}$, $F_{n_1,n_2,\alpha}$ 202

$b(p)$ 55

$b(p,n)$ 61

$\beta(a,b)$ 139

$\chi^2(n)$ 171, 202

$\chi'^2(n,\lambda)$ 172, 105

δ_a 55

$\mathcal{D}(\alpha_1, \ldots, \alpha_k)$ 326

Convergence for Sequences of r.v.'s (Volume I)

Queues (Volume II)

Abbreviations (Volumes I and II)

INDEX

DATE DUE
